CHARLES BABBAGE

Pioneer of the Computer

Copyright © R. A. Hyman 1982
All rights reserved

Reprinted 1983

Published by Princeton University Press,
41 William Street, Princeton, New Jersey

Library of Congress Card: 81–48078
ISBN 0–691–08303–7

Printed in Hong Kong

ACKNOWLEDGEMENTS

The greater part of this book was written during 1977 and 1978 while holding the Alistair Horne fellowship, and I am greatly indebted to Alistair Horne, and to the Warden and Fellows of St Antony's College, Oxford, for provison of congenial working conditions and valuable help. I am also indebted to St John's College for their hospitality while working on archives in Cambridge. During years of research on Babbage and his Engines I have incurred far more obligations than can be individually acknowledged, but to all who have drawn my attention to relevant material, I take the opportunity to offer my heartfelt thanks. Mr Alfred Van Sinderen has transcribed manuscripts from his collection and kindly permitted me to quote from them; Michael Havinden joined me in exploring Babbage's home territory in South Devon; I have enjoyed many discussions about Babbage's Engines with the late Maurice Trask; Professor Christopher Platt read chapter eight, and Professor Brian Randell all the chapters on the Engines, and made valuable suggestions; Professor Margaret Gowing provided much help, and Professor Peter Matthias lent a helping hand in tackling the intractable problem of securing access to banking records; Michael Laithwaite identified John Babbage's house; Mr Erwin Tomash provided photocopies of American doctoral theses, and Mr V. Bonham-Carter showed me the typescript of his book *Authors as a Profession*; Mr C. H. Wills kindly provided photographs of the Babbage family; Ms Jane Raimes of the Science Museum and the Director of the Science Museum Library placed invaluable facilities at my disposal. I am also greatly indebted to the librarians and staffs of The Royal Society of London, The Institute for Historical Research, The Crawford Library, the libraries of St John's and Trinity College, Cambridge, The Oxford Museum of the History of Science, the manuscript rooms of the British Library and the Bodleian Library, the Devon County Record Office, and the Southwark Room of the Southwark Public Library. Mr A. K. Corry of the Science Museum kindly investigated the gear trains of the Difference Engine.

For permission to quote from manuscripts I am also indebted to: The Royal Society of London, Miss Elizabeth Buxton, Trinity and St John's Colleges in Cambridge, The Director of the Science Museum, The British Libraries Board, Lord Knebworth and the Earl of Lytton, and the Burndy Library. This work has further been assisted by grants from the Leverhulme Foundation, The Royal Society, and The British Academy.

ANTHONY HYMAN

CONTENTS

———— ○●○ ————

List of Illustrations   ix
Abbreviations   xiii
Principal Sources   xv

Prelude: Child of Two Revolutions   1
 1. Childhood   5
 2. Cambridge   20
 3. Marriage: Early Years of a Philosopher in London   31
 4. Science in Action: Start on the First Engine   47
 5. The Death of Georgiana: Continental Travel   62
 6. Reform   75
 7. Science and Reform: The Royal Society   88
 8. On the Economy of Machinery and Manufactures   103
 9. The Great Engine   123
10. The Ninth Bridgewater Treatise   136
11. Rail, Steam, and the British Association   143
12. The Analytical Engines: Social Life   164
13. Analytical Engines and the Circumlocution Office   190
14. The Great Exhibition   211
15. The Death of Ada: A Family Reunion   225
16. Final Passages in the Life of a Philosopher   241
Epilogue: From the Analytical Engines to the Modern Computer   254

Published Works of Charles Babbage   256
Appendices
  A. From *Passages from the Life of a Philosopher*   261
  B. On the Mathematical Powers of the Calculating Engine   268
Index   277

LIST OF ILLUSTRATIONS

○ ● ○

Charles Babbage by S. Lawrence, 1845. Oil.   *Frontispiece*
National Portrait Gallery.

PLATES

Between pages 16 and 17

1. Old London Bridge by J. M. W. Turner, 1824. Watercolour.
Victoria and Albert Museum.

2. Fore Street, Totnes, *c.* 1860. From Deckemant's *Antiquities of England*, Vol. 12.
Devon County Record Office.

3. John Babbage's house, with shopfront remodelled in the nineteenth century, *c.* 1860. Deckemant's *Antiquities*.
Devon County Record Office.

4. Guildhall, Totnes, *c.* 1860. Deckemant's *Antiquities*.
Devon County Record Office.

5. Totnes from the river Dart by J. M. W. Turner, *c.* 1824. Watercolour.
British Museum.

Between pages 48 and 49

6. Florence by J. M. W. Turner, *c.* 1827. Watercolour.
British Museum.

7. Georgiana Babbage, from a miniature.
The Wills Collection of Photographica.

8. Charles Babbage aged 40. Pen etching published by Colnaghi.
Trinity College Library, Cambridge.

9. Bridgnorth Bridge by J. M. W. Turner.
Victoria and Albert Museum.

10. First Difference Engine. Completed Part, 1832.
Crown Copyright. Science Museum, London.

Between pages 112 and 113

11. South Devon Railway, where it passed through Babbage's farm at Dainton.
Science Museum, London.

12. Charles Babbage aged 48. Drawing by William Brockendon.
Science Museum, London.

13. Mary Anne Hollier. Probably copied from a daguerreotype or a calotype taken *c.* 1842.
    The Wills Collection of Photographica.

14. Augusta Ada, Lady King, later Countess of Lovelace by Margaret Carpenter, *c.* 1835. Oil.
    Crown Copyright. National Physical Laboratory.

15. J. M. Jacquard. Woven silk portrait.
    Victoria and Albert Museum.

Between pages 176 and 177

16. Charles Babbage, *c.* 1850. Photograph by Antoine Claudet.
    National Portrait Gallery.

17. Charles Babbage, *c.* 1860.
    The Wills Collection of Photographica.

18. Mary Anne Hollier, 1856.
    The Wills Collection of Photographica.

19. Henry Prevost Babbage and his family, *c.* 1870.
    The Wills Collection of Photographica.

20. Experimental Carry Column for an Analytical Engine, late 1860s. Replica.
    Science Museum, London.

21. Mill of minimal Analytical Engine under construction when Babbage died. In this photograph the racks stick out in positions they could never have reached in use.
    Science Museum, London.

FIGURES IN TEXT

1. Ground and premises in the occupation of Charles Babbage. *p. 128*
   Public Record Office, London.

2. Plan of Analytical Engine with grid layout, 1858. Redrawn. *p. 243*

TECHNICAL DRAWINGS
Science Museum

Between pages 256 and 257

1. Small multipurpose machine tool, 1858.

2. Difference Engine No. 1. End elevation, 1830.

3. Difference Engine No. 1. Plan and side elevation, 1830.

4. Difference Engine No. 3. Sliding mechanism for sequence control, 1849.

5. First drawing of circular arrangement of new engine, 1834.

6. Plan for part of calculating engine with rack interconnection system, 1835.
7. Superposition of motion, 1836.
8. Two distinct engines with thirty figures each, 1836.
9. Method of carrying by anticipation, 1836.
10. General plan of mill, 1837.
11. General plan of Mr. Babbage's Great Calculating Engine, 1840.
12. Direction of Analytical Engine, 1841.
13. Plan of bolts for store and mill, *c.* 1858.
14. Digit counting apparatus, 1859.
15. Part of mill, 1864.
16. Apparatus for registering time of workmen.

ABBREVIATIONS

———————— ○ ● ○ ————————

BL   The British Library
DCRO   Devon County Record Office
RS   The Royal Society of London
Decline of Science   *Reflections on the Decline of Science in England and on
                                Some of Its Causes*
Economy of Manufactures   *On the Economy of Machinery and Manufactures*
Passages   *Passages from the Life of a Philosopher*

Journals

Econ. Hist. Rev.   *Economic History Review*
Edin. Phil. Jrl.   *Edinburgh Philosophical Journal*
Mem. Astron. Soc.   *Memoirs of the Astronomical Society*
Notes and Records   *Notes and Records of the Royal Society of London*
Phil. Mag.   *The Philosophical Magazine*
Phil. Trans.   *Philosophical Transactions of the Royal Society of
                                London*
Proc. Geol. Soc.   *Proceedings of the Geological Society*
Proc. Inst. Civil Eng.   *Proceedings of the Institute of Civil Engineers*
Proc. Roy. Soc.   *Proceedings of the Royal Society of London*
Quart. Jrl. Geol. Soc.   *Quarterly Journal of the Geological Society*

PRINCIPAL SOURCES

———————————————○●○———————————————

Royal Society—Babbage/Herschel Correspondence.
British Library—Babbage Letters and Papers.
Science Museum—Babbage Drawings, Notations, and Notebooks.
Museum for the History of Science, Oxford—Buxton Papers.
Devon County Record Office—Totnes, Teignmouth, and Ippelpen records.
Periodicals Library, Cambridge—Babbage Papers.

Theses consulted include:

Bruce Collier, 'The Little Engines that Could've', Harvard, 1970.
Walter Lyle Bell, 'Charles Babbage, Philosopher, Reformer, Inventor', Oregon, 1975.
John David Yule, 'Impact of Religious Thought in the Second Quarter of the Nineteenth Century', Cantab. 1976.

PRELUDE

○●○

Child of Two Revolutions

The England into which Babbage was born was almost entirely rural. To the foreign visitor arriving from the Continent the southern counties presented the aspect of a well-kept garden. Even in London everyone lived within a quarter of an hour's walk of the countryside. Eighty years later, when Babbage died, the majority of Englishmen lived in towns and cities, and England was by far the most highly industrialized country that had ever been seen. For a brief period Britain had been the workshop of the world, but the nation was soon to learn that its industrial supremacy had gone for ever, the technological lead, particularly in the new science-based industries, passing to Germany and the rapidly developing United States of America. During his life Babbage was the leading advocate of the systematic application of science to industry and commerce. He played an important part in the development of both pure science and technology, and with a few of his friends made a great impact on British industrial development. They had seen—and Babbage was the first to see and saw most clearly—the dangers for British industry if scientific methods were not generally applied. Although the industrial revolution had been made by practical men, with little direct help from the higher reaches of science, industry could no longer continue to develop satisfactorily without scientific method. Babbage and his friends fought the battle for applied science on a national scale, and lost. The consequences of this defeat remain the subject of active debate.

Babbage was first a mathematician. He made important contributions to pure mathematics and, with two friends, transformed mathematics in Cambridge and thus throughout Britain. At the age of thirty he made a change of direction startling in its suddenness. He began to construct a machine for making mathematical tables: the manufacture of number. Building this machine, his first Difference Engine, led to crucial advances in machine tools and engineering techniques affecting the whole development of precision mechanical engineering. In 1834 Babbage started on a new project, the Analytical Engines. Unique precursors of the modern computer, almost forgotten and then rediscovered in the middle of the twentieth century, they are one of the great intellectual achievements in the history of mankind.

When constructing his first Engine Babbage turned to the general study of machinery and manufacture, developing his doctrine of the union of theory and practice.[1] Later, in his campaigns for the application of scientific method to industrial problems, he had allies among the engineers, including Fairbairn and Whitworth. But Babbage was the central figure, and he continually pursued the application of statistical methods to economic and social problems. What Condorcet was to the French revolution, Babbage was to English reform.

Charles Babbage was very sociable and became a leading figure in English society. A man of immense personal charm and boundless vitality, he mixed in widely varied circles from his early days in Cambridge. Societies and clubs seemed to arise almost spontaneously wherever he moved. His friends tended at first to be liberal, later often moderately radical; though as the reform movement waned and they grew older they became more conservative. He was equally at ease with intelligent working men, country clergymen, men of science, at court, or at the dining tables of the aristocracy.

Babbage knew virtually all the liberal reformers of his time; and as leading man of science was a respected figure among them. His acquaintance on the Continent, in France and Italy, and to a lesser extent in Austria and Prussia, was wide, and he became an important European scientific figure. By chance he became a friend of several of the Bonapartes, who considered European revolution to be their family's profession.

Babbage became a militant reformer, founding scientific societies, launching the 'decline of science' campaign, the scientific counterpart to the struggle for the first reform bill, and participating, both as campaign organizer and candidate, in liberal and radical politics. In the 1820s the scientific movement in England had a radical tincture which it has never quite regained. On the Continent Babbage's friends—Arago in France, von Humboldt in Prussia, the Bonapartes and Menabrea in Italy—played their parts in the liberal movement. Babbage was the child of two revolutions: industrial revolution in Britain, and political and social revolution in France.

London at the end of the eighteenth century was both the capital and the industrial centre of Britain, and also the commercial centre of a great trading network. The ships on the Thames, which fascinated the young Babbage, were continual reminders of the importance of foreign trade, as were the great London docks built by John Rennie to prevent the pilfering of valuable colonial produce. Rennie's sons, also engineers, were to figure in Babbage's life. In the shops could be seen products of Staffordshire's potteries and Manchester's looms. A few of the great steam engines at work were clear evidence of the industrial revolution in progress. As early as 1718–22 John and Thomas Lombe had built a silk mill at Derby, five or six storeys high, employing three hundred hands, and powered by water from the Derwent. It was the first of the modern factories with automatic tools and specialized functions for operatives. But it was not until the 1770s, when power-driven machinery reached the

[1]See note. p. 4. (end of chapter).

woollen, iron and cotton industries, that industry had much effect on the countryside, and then most works were situated remotely to be near a falling stream which would supply the power.

Later in his life Babbage was to make the most detailed study of the operation of factories and their machinery, but during his early years the overwhelming visual impression was of a traditional way of life slowly changing. After the turmoil of the seventeenth century, most of the eighteenth was a long period of peaceful development, scarcely disturbed by distant conflict. Readers of Jane Austen will recall how little even the Napoleonic wars affected everyday life in the English countryside. Nevertheless, by the end of the century it was obvious to the enquiring mind that the industrial revolution was well on its way. And the young Babbage saw something else besides: as a banker's son he saw that most potent of forces, mobilized capital in action. A banker's life is the reduction of all considerations to quantitative terms. This mechanically inclined and mathematically gifted child learned the lessons well.

After the death of Newton, English science lapsed into an age of relaxing mediocrity. There was nothing in the eighteenth century to compare with the great spate of discoveries of the scientific groups of the interregnum and the early Royal Society. But the very dullness can be misleading. The habit of a more scientific approach was spreading into every part of life. A great deal of hard ground-clearing was going on, while the sciences aspired to a Newtonian clarity and perfection. Abstract science in fact developed far more rapidly in the France of the enlightenment and among the encyclopaedists than in the country where the industrial revolution was taking place, suggesting a *prima facie* answer to a question often asked: 'What was the relationship between science and technology during the early phases of the industrial revolution?' It would seem that the industrial revolution owed almost nothing directly to the more theoretical sciences. In a sense the very question is misplaced: it only becomes apposite after the 1830s, when pure science began to develop independently and apart from the inferior applied sciences. In the eighteenth century there was no clear dividing line between science and technology.

During the English civil war and the interregnum, inspired by the ideas of Francis Bacon and the millenarianism of the time, Samuel Hartlib and his associates advanced many proposals for applying science to the requirements of society. But they were premature: neither science nor society was ready. Nevertheless, their ideas were among the principal formative influences on the Royal Society and their interest in the use of scientific methods in agriculture is reflected in the early programme of the Society.

During the eighteenth century the *Académie Royale des Sciences* made repeated declarations of the importance of applying science but did little about it in any organized manner. It was in France after the revolution that the first serious attempt was made to develop science systematically for practical ends.

The metric system was the most obvious result. Napoleon fostered scientific development with an eye to its military usefulness. Thus when Babbage came as a student to look at the contemporary world of science outside Britain it was natural in the first place to look to France. Seeking for the main formative influences on Babbage's scientific views we find in England first the great progenitor, Francis Bacon, then William Petty, Newton, and the Lunar Society—Babbage came to know the grandchildren of its members. But more immediately he was influenced by Condorcet, de Prony, and Napoleon's protégés among the Parisian men of science.

While the Napoleonic wars gave a considerable impetus to technical development in England—the Portsmouth block-making machinery, which played a crucial part in establishing the preconditions for Babbage's Difference Engine, is one example—they also resulted in the establishment by Pitt of a reactionary régime which had a claustrophobic effect. Joseph Priestley's house was ransacked by a mob inspired by the authorities and the Lunar Society ceased to exist. The result was to turn Babbage's generation of young mathematicians at Cambridge militantly liberal.

NOTE

Babbage's repeated insistence on the importance of collecting 'facts' places him squarely in the mainstream of the Baconian tradition. The importance of his work in this field is well illustrated in the report of the secretary to the Smithsonian (Joseph Henry) for 1873, p. 23:

In the report for 1856, is given a plan by the late Mr Charles Babbage, of London, of a series of tables to be entitled the 'Constants of Nature and Art'. These tables were to contain all the facts which can be expressed by numbers, in the various sciences and arts, such as the atomic weight of bodies, specific gravity, elasticity, specific heat, conducting power, melting point, weight of different gases, liquids, and solids, strength of different materials, velocity of sound, of cannon-balls, of electricity, of light, of flight of birds and speed of animals, list of refractive indices, dispersive indices, polarizing angles, etc.

The value of such a work, as an aid to original investigation, as well as in the application of science to the useful arts, can scarcely be estimated. To carry out the plan fully, however, would require much labor and perhaps the united effort of different institutions and individuals, devoted to special lines of research. Any part of the entire plan may, however, be completed in itself ... The Institution commenced about fifteen years ago to collect materials on several of the points of this general plan ...

After an interruption caused by the civil war a series of tables of specific gravities, boiling points, and melting points was compiled from the best authorities. However, as the Institution was short of funds, the tables were printed as a private venture.

I

○●○

Childhood

Charles Babbage was born on 26 December 1791 in his father's house in Walworth, Surrey, five hundred yards from the famous hostelry of the Elephant and Castle.[1] His parents had recently married and come to live near London where his father was working, probably as merchant and banker: the roles were interchangeable at the time. Both Babbage's parents sprang from old Devonshire families. His ancestry can be traced to Totnes, Teignmouth, and the surrounding countryside: yeomen and goldsmiths, a solid well-established West Country background. Walworth was among the fields, a hamlet attached to Newington. Within walking distance or a short ride across London bridge from the City, it was a sensible choice for a country family.

In his autobiography, *Passages from the Life of a Philosopher*, Babbage noted that 'In the time of Henry the Eighth one of my ancestors, together with a hundred men, were taken prisoner at the siege of Calais'.[2] There is also a tradition in the Babbage family, coming through a third cousin of Charles, that four brothers, left behind after the fall of Calais, were expelled from France with the Huguenots in the reign of Queen Elizabeth and returned to Devonshire, two settling in the North of the county and two in the South. Again, another family tradition recorded that 'When William the Third landed in Torbay, another ancestor of mine, a yeoman possessing some small estate, undertook to distribute his proclamations. For this bit of high treason he was rewarded with a silver medal, which I well remember seeing, when I was a boy. It had descended to a very venerable and truthful old lady, an unmarried aunt, the historian of our family, on whose authority the identity of the medal I saw with that given by King William must rest.'[3]

The heartland of the Babbage territory is the triangle defined by Totnes, Dartmouth, and Teignmouth, rich farming country largely formed of the small fields and winding lanes typical of land which has never been open fields. What open fields there had been in Devon were probably enclosed by the end of the fourteenth century.[4] From Totnes one can pass down the winding estuary of

---

[1] Baptismal register of St Mary Newington, 6 Jan. 1792, County Hall, London. The birthday of 26 Dec. was known in the family.

[2] *Passages*, 5.

[3] Ibid.

[4] W. G. Hoskins, *Making of the English Landscape*, 215–16, Penguin, 1970.

the Dart to Dartmouth, perched on the cliff, where Thomas Newcomen lived and worked. To the north, past the fishing village of Brixham, is the expanse of Tor Bay, used during the Napoleonic wars as a naval base. There Charles saw Napoleon on his way to exile in St. Helena. Torquay first emerged as a residence for naval personnel; later becoming a summer resort, where Charles spent several summers with his young family after his marriage. From Torquay the coast continues round the cliffs to Babbacombe and Marychurch; then north again to the mouth of the Teign. On the north side of the estuary is Teignmouth where Charles's father had a house after his retirement. It was Charles's home for about eight years between the ages of fifteen and twenty three.

To the north-west of Totnes, past Dartington, Staverton, and Shinners Bridge, lay Buckfastleigh, Ashburton, and the wild moors. To the north was Exeter, the capital and richest city of Devon. To the west was the great port and naval base of Plymouth. But for centuries the extraordinarily wealthy Totnes was the second town in Devon.

Before Babbage's time the West Country had been a prosperous area for wool, cloth, mining, and primitive industry. One of the effects of the Industrial Revolution was to shift the industrial centre of Britain northwards. The South-West fell into relative decline, and with this decline Bristol lost to Liverpool its ancient position as England's leading Atlantic port. Babbage was to help the young Brunel in a brave but ultimately futile attempt to maintain Bristol's supremacy.

It is curious, and may be more than coincidence, that from Totnes and its port at Dartmouth should come not only Charles Babbage but Thomas Savery and Thomas Newcomen, pioneers of the steam engine. The Saverys were prosperous merchants, an old Totnes family. In about 1614 they purchased the nearby manors of Shilston and Spiddlescombe, and in 1745 we find Benjamin Babbage, grandfather of Charles, concerned with a Mary Savery in land purchase near Harberton.[5] Savery, who was granted on 25 July 1698 the historic patent for 'raising water by the impellant force of fire', was one of the most fertile inventors of his time. But Newcomen was really more important, and his own inventions were made entirely independently of Savery. He was a blacksmith and many people have found it hard to believe that such a man could have been responsible for the remarkable developments of the working steam engine. But we should think of Newcomen as working in iron, copper, brass, tin, or lead, visiting also the Cornish mines to sell tools and his services as a skilled craftsman. In the twelfth and thirteenth centuries Devon and Cornwall had been producing most of Europe's tin, and this prosperous mining is the background to the society that produced Newcomen and Savery. In

[5] Deed in possession of Mrs Elizabeth Babbage: indenture between Benjamin Babbage for the one part; Mary Savery and others for the other part.

Newcomen's time Devon mining was in decline but evidently a strong tradition of craftsmanship remained. Newcomen's engine comes from innumerable unrecorded attempts by the craftsmen of Devon to pump water from the mines, and such work calls forth a wealth of skill and talent which is continually underestimated by academic commentators. Newcomen's work and much other skilled craftsmanship was part of local tradition when Charles Babbage was growing up.

Totnes appears as a small town in the tenth century. Coins were minted there intermittently between the reigns of Edgar and William Rufus. On the hill the castle site was fortified at an early date and earthen fortifications were thrown up round the houses by the river. Stone ramparts replaced the earthworks early in the twelfth century. The lovely Fore Street, still the axis of the town, joined castle to river. Near Totnes are Dartington Hall and Berry Pomeroy, castle of the Seymours.

Babbage owned a field in Totnes under the walls of the ruined castle. It had been in the family since early in the eighteenth century, associated originally with a house they owned at Northgate.[6] He kept the field all his life. If you stand on the walls of Totnes castle looking across to Dartington, Babbage's field slopes down below, still rough grazing. There used also to be an orchard nearby known as 'Babbage's orchard'.

In the reign of Henry VIII Totnes, second only to Exeter in Devon, was a thriving cloth town, full of rich merchants. But it had been slow in developing the 'new draperies' and by the early seventeenth century it was in decline, its place being taken by Tiverton, Crediton, and other new towns; although a guide of 1825 refers to Totnes as noted for its serges.[7]

By the latter part of the seventeenth century the Babbage family was well established. We get glimpses not of great wealth but of comfortable circumstances. For example Charles recorded that 'an ancestor married one of two daughters, the only children of a wealthy physician, Dr Burthogge, an intimate friend of John Locke'.[8] The first definite evidence we have of the family in Totnes is on 18 April 1628 when Roger Babbidge payed a poor rate of 4d. This Roger had at least two sons, Roger and Christopher. Roger was buried in 1665 and in February 1666 we find the widow Babbidge and Roger Babbidge, the son, paying rates on separate houses. The second Roger Babbidge is recorded as having four children, Richard, the eldest, Hester, John, ancestor of Charles, and Sarah. On 19 April 1687 Roger Babbidge and Richard Babbidge were admitted members of the guild. The Babbages were becoming, if they

---

[6] DCRO, Rate books and other documents.

[7] John Hannaford, *The History of Totnes, Its Neighbourhood and Berry-Pomeroy Castle in Devonshire*, 19, Totnes (1825?).

[8] *Passages*, 5.

were not already, part of the commercial oligarchy which controlled old Totnes.[9]

In 1719, as a large deed records, John Babbage—the name had by now assumed its modern form—purchased from the borough a lease of 2,000 years on a house next to another he already owned. Probably John Babbage was living in this latter house. The borough of Totnes was selling a number of leases at the time: raising money and no doubt dividing up the town's property among the leading families. The house whose lease John Babbage purchased still exists. John is described sometimes as shopkeeper, sometimes as goldsmith. Quite likely John's forebears had also been goldsmiths, as sons often continued their fathers' trades, but of this there is no direct evidence, and John was the younger son. From the land tax assessment for 1747 we find that Mrs Babbage—she must be John's widow—owns four houses and Benjamin, her son, another house. Evidently John had made money.

Benjamin Babbage was a leading figure in Totnes. In 1740 we find Mr Benjamin Babbage goldsmith marrying by licence Mistress Margaret Laver: people of consequence. Benjamin became church warden and was mayor of Totnes in 1754. He had four children who lived to maturity: John (b. 1741); Anne (b. 1747), probably the maiden aunt whom Charles described as the family historian; Margaret (b. 1749), who married Joseph Taunton; and Benjamin (b. 1753), father of Charles. From these bare facts and other information provided by Charles we can sketch out a picture of his father's life.

Charles noted that his father had made most of his fortune by his own efforts although he came from a wealthy family. Old Benjamin died intestate in December 1761 when his younger son Benjamin was only seven or eight years old. The elder son, John, became a surgeon. Evidently he was little interested in the trade of goldsmith. A Dr Babbage took over the Dartington Vestry contract for two pounds ten shillings a year in 1770 and was still there in 1790: probably the same man. As elder son John inherited the family property: old Benjamin's houses later appear in his name. Thus when young Benjamin grew up it fell to him to re-establish the prosperity of the family business, which he did with conspicuous success.

Charles also admitted to more disreputable ancestors and relates a curious story about one of them, possibly the brother[10] of his greatgrandfather, John:

Somewhere about 1700 a member of my family, one Richard Babbage, who appears to have been a very wild fellow, having tried his hand at various trades, and given them all up, offended a wealthy relative.

To punish this idleness, his relative entailed all his large estates upon eleven different

---

[9] The story of the Babbages in Totnes has been sketched mainly from records in the Devon County Record Office.

[10] This might from the dates have been another Richard, son of John's brother.

people, after whom he gave it to this Richard Babbage, who, had there been no entail, would have taken them as heir-at-law.

Ten of these lives had dropped, and the eleventh was in a consumption, when Richard Babbage took it into his head to go off to America with Bamfylde Moore Carew, the King of the Beggars.

The last only of thè eleven lives existed when he embarked, and that life expired within twelve months after Richard Babbage sailed. The estates remained in possession of the representatives of the eleventh in the entail.

If it could have been proved that Richard Babbage had survived twelve months after his voyage to America, these estates would have remained in my own branch of the family.

I possess a letter from Richard Babbage, dated on board the ship in which he sailed for America.

In the year 1773 it became necessary to sell a portion of this property, for the purpose of building a church at Ashbrenton. A private Act of Parliament was passed for that purpose, in which the rights of the true heir were reserved.[11]

The profession of Charles Babbage's forebears as goldsmiths is a crucial factor in his background. Although amongst the early country bankers few seem to have been gold or silversmiths,[12] in London the progression from goldsmith to banker was typical; and it was in any case a natural progression. Whether Charles's father actually worked as a practising craftsman is not known, but it is likely enough and would go far to explaining Charles's early competence as a craftsman as well as his broad mechanical interests. In the goldsmith turned banker we have two of Charles's great interests combined: precision manufacture and a quantitative approach to economic and social problems.

We may picture the young Benjamin gradually building up his banking activities in Totnes and the surrounding district, developing and extending his father's business and range of contacts. There is no reason to assume that he formally opened a bank. Indeed that is unlikely, but much could be done without the formality. In particular the cloth trade required capital. But the cloth trade was declining and, like that much greater Devonshire banker, Francis Baring, Benjamin moved to London, the commercial and banking centre of the country. To make the move he must already have been a prosperous man with many connections and a thriving business.

It would have been natural for the Babbages to act as local agents for banks in other towns. Two houses with which they may well have had arrangements were Praeds Bank of Truro, founded before 1774, and the Exeter Bank, which opened its doors on 9 July 1769, first of the banks in Exeter. Among its original partners was William Mackworth Praed. Later Benjamin Babbage was to be a partner in Praeds London bank and it is not difficult to see how the connection

---

[11] *Passages*, 5–6.
[12] L. S. Pressnell, *Country Banking in the Industrial Revolution*, Clarendon Press, 1956.

was made. From late in the 1780s the Industrial Revolution gathered momentum leading to an increasing demand for credit and the opening of new banks. It was a time when bankers could thrive. The careless went to the wall but Benjamin was exceedingly careful.

In about 1790 he married Elizabeth (Betty) Plumleigh Teape. The cautious Benjamin had waited to get married until he had a substantial fortune. His bride also came from an old Totnes family. A James Teape was mayor of Totnes in 1731, 1741, and 1746; a Samuel Teape in 1757. The flagstone of the Teapes' vault can still be seen under the tower of Totnes church; the Babbages also had a vault in Totnes church near the tower before the building was remodelled in the nineteenth century.[13] The church associations of the family were close and played an important part in Charles's thought. It is likely that the Babbages had been a nonconformist family, grandfather Benjamin being the first to conform, a common practice among families on their way up in the world. The stern conscience of his father, an inheritance from the non-conformists, seems softened in Charles by the Teape charm. But the intense dedication to work remained, transferred to mathematical and philosophical pursuits.

It was in 1791 that Benjamin took his bride to London where they bought the house in Crosby Row on the Walworth Road, now in the borough of Southwark. While they were there the row of houses was renamed York Place and then Chatham Place. Later it became Walworth Road. The house was pulled down when Larcome Street was made.[14] In the autumn of 1791 Michael Faraday was born to a poor family near the Elephant. On 26 December 1791, in the recently built terrace house in Crosby Row, Charles Babbage was born.

Charles relates two stories from his early years in Walworth:

Two events which impressed themselves forcibly on my memory happened, I think, previously to my eighth year.

When about five years old, I was walking with my nurse, who had in her arms an infant brother of mine, across London Bridge, holding, as I thought, by her apron. I was looking at the ships in the river. On turning round to speak to her, I found that my nurse was not there, and that I was alone upon London Bridge. My mother had always impressed upon me the necessity of great caution in passing any street-crossing: I went on, therefore, quietly until I reached Tooley Street, where I remained watching the passing vehicles, in order to find a safe opportunity of crossing that very busy street.

In the mean time the nurse, having lost one of her charges, had gone to the crier, who proceeded immediately to call, by the ringing of his bell, the attention of the public to the fact that a young philosopher was lost, and to the still more important fact that five shillings would be the reward of his fortunate discoverer.

I well remember sitting on the steps of the door of the linendraper's shop on the opposite corner of Tooley Street, when the gold-laced crier was making a proclamation

13 *Totnes Times*, Obituary of Charles Babbage, 28 Oct. 1791.
14 Parish Rate Books, The Southwark Room, Southwark Borough Library.

of my loss; but I was too much occupied with eating some pears to attend to what he was saying.

The fact was, that one of the men in the linendraper's shop, observing a little child by itself, went over to it, and asked what it wanted. Finding that it had lost its nurse, he brought it across the street, gave it some pears, and placed it on the steps at the door: having asked my name, the shopkeeper found it to be that of one of his own customers. He accordingly sent off a messenger, who announced to my mother the finding of young Pickle before she was aware of his loss ...

The other event, which I believe happened some time after the one just related, is as follows. I give it from memory, as I have always repeated it.

I was walking with my nurse and my brother in a public garden, called Montpelier Gardens, in Walworth. On returning through the private road leading to the gardens, I gathered and swallowed some dark berries very like black currants:- these were poisonous.

On my return home, I recollect being placed between my father's knees, and his giving me a glass of castor oil, which I took from his hand.

London Bridge was north of Crosby Row; Montpelier Gardens a few hundred yards to the south.

These two stories enabled me to solve the long-standing puzzle of Babbage's place of birth. Although he had himself stated that he was born in London the location was unknown. It was sometimes said that he was born in Teignmouth or Totnes, presumably from the family background. Indeed an exhibition celebrating his birthplace was even financed in Totnes, although a glance at the baptismal register suffices to make a Totnes birthplace exceedingly improbable. After following a number of false trails it occurred to me to enquire where Charles would have been baptized if he had been born halfway between London bridge and Montpelier gardens. The answer was at St Mary Newington. In the baptismal register, there was recorded for 6 January 1792: Charles, son to Benjamin and Betty Plumleigh Babbage. The birthday of 26 December was not in question; the year of birth was evidently 1791.

The only other detail Babbage records from his childhood in Walworth was that his father had a collection of pictures. These included 'a fine picture of our Saviour taken down from the cross. On the opposite wall was a still-celebrated "Interior of Antwerp Cathedral".' In October 1794 a second son, Henry, was born to the Babbages but he died in infancy. In May 1796 another son was born who was also named Henry, but he also died while a small child. In March 1798 a sister was born for Charles. She was named Mary Anne. Brother and sister remained on warm terms for life and she outlived him. In the house in Newington John Babbage, Benjamin's elder brother, a surgeon of Totnes, died in the spring of 1792.[15]

By the turn of the century Benjamin was moving into formal banking

15 *Gentleman's Magazine*, 1792.

partnership with Praed. Presumably to be near Fleet Street where the new bank was to be built, the Babbages moved at the end of 1799 across the river to a small house at 10 George Street, Adelphi. This was the same house designated 10 York Buildings in the survey of London: 'These are three-storey premises over basement with tiled roof and a brick front to the upper portion, the lower portion being stuccoed. The whole appears to date from the latter half of the eighteenth century. The staircase has close strings, turned balusters and panelled dado. The rooms have been redecorated and were in all probability panelled.'[16] George Street reached the Thames near the Duke of York's steps, now separated from the river by the embankment.

The new building for Praed's bank was designed by Sir John Soane.[17] *The Times* for 15 July 1801 notes: 'Monday [13 July] was laid, in Fleet-Street, the first stone of a new Banking House. At the head of the firm is William Praed Esq. Member for St Ives.' On 5 January 1802, it reported, 'That elegant new building just erected in Fleet Street, was last week opened as a Banking House with the firm of Praeds, Digby, Box, Babbage and Co.' The order of names suggests that Benjamin was the junior partner. Almost certainly he had been trading in collaboration with Praed before the opening of the bank: it is hardly likely that business would have waited on completion of the building, or indeed that so splendid a building would have been erected unless it were to house an already thriving and developing business.

In about 1803 Benjamin retired from his partnership and returned to Totnes. Some five years later he purchased 'The Rowdens', a small house with six acres of land high on the cliffs above East Teignmouth. 'The Rowdens' figures in *Passages* and was Charles's home while he was growing to maturity. It had splendid views across Teignmouth harbour and out to sea, and it was situated on the Dawlish road. Charles's room was partly detached from the main house by a conservatory.[18] (The house was burnt down in 1841 and the present house called 'The Rowdens' was built on a different site in 1843.)[19] In 1808 Benjamin became churchwarden of East Teignmouth. The Mackworth Praeds, Babbage's partners in banking, had built Bitton House in Teignmouth, gloriously sited just above the harbour.[20] Teignmouth was not only a pleasant town for retirement, with a mild climate, but also a thriving little port. A guide of 1830 notes: 'The sources of amusement are by no means few. The theatre is generally open in the season; and there is besides a succession of balls, concerts, and promenades every week. ... Teignmouth has an annual regatta, (the first established in the British channel) at which boats and yachts of different sizes

[16] *Survey of London*, xviii, part 2, 81–3 and 135. The house has been destroyed.
[17] The architectural drawings are in the Sir John Soane Museum, London.
[18] *Passages*, 15.
[19] Compare Tithe map for East Teignmouth (DCRO) with ordnance survey map.
[20] 'Bitton House'. Pamphlet in Teignmouth Council Offices.

contend for valuable prizes ... '.[21] However Benjamin maintained his connections with Totnes, and the 1812 poll sheet shows him voting for the Tory, T. P. Courtenay. Both in politics and choosing his way of life Charles was to clash with his father. In the same year that he bought 'The Rowdens' he also purchased a farm in Dainton,[22] which is still there, now called Dainton House. The farm had about one hundred acres in 1844. Charles added to the property and kept it all his life.

While the family had still been living in George Street Charles had suffered a violent fever. His parents had already lost two sons and they decided to send him to the Devon countryside to recuperate, choosing a small school at Alphington, a mile and a half on the Totnes side of Exeter. There were two schools in Alphington[23] at the time and it is not clear to which one he went. Probably the education was similar in both. One was Mr Halloran's Alphington Academy, which for twenty pounds per annum and one guinea entrance gave to sons of gentlemen 'instruction in English, Latin and Greek languages as well as in such branches of education as are necessary qualifications for trade or the sea service'.[24] The latter would include accountancy and navigation, common practice in schools near the ports. Charles was only eight or nine when he went there but that precocious and incisive mathematical mind was already being exposed to subjects which would not usually have been familiar to the products of Oxford and Cambridge.

While he was at Alphington Charles made an experimental inquiry into the existence of ghosts and devils:

I gathered all the information I could on the subject from the other boys, and was soon informed that there was a peculiar process by which the devil might be raised and become personally visible. I carefully collected from the traditions of different boys the visible forms in which the Prince of Darkness had been recorded to have appeared. Amongst them were—
A rabbit,
An owl,
A black cat, very frequently,
A raven,
A man with a cloven foot, also frequent.

After long thinking over the subject, although checked by a belief that the inquiry was wicked, my curiosity at length overbalanced my fears, and I resolved to attempt to raise the devil. Naughty people, I was told, had made written compacts with the devil, and had signed them with their names written in their own blood. These had become very rich and great men during their life, a fact which might be well known. But, after

[21] *The Teignmouth, Dawlish, and Torquay Guide*, by several literary gentlemen, 31, Teignmouth, 1830.
[22] Tithe Map of Ipplepen, DCRO.
[23] BL Add. Ms. 37,199, f 284.
[24] *Exeter Flying Post*, Thursday, 6 Jan. 1791, and 15 Jan. 1795.

death, they were described as having suffered and continuing to suffer physical torments throughout eternity, another fact which, to my uninstructed mind, it seemed difficult to prove.

As I only desired an interview with the gentleman in black simply to convince my senses of his existence, I declined adopting the legal forms of a bond, and preferred one more resembling that of leaving a visiting card, when, if not at home, I might expect the satisfaction of a return of the visit by the devil in person.

Accordingly, having selected a promising locality, I went one evening towards dusk up into a deserted garret. Having closed the door, and I believe opened the window, I proceeded to cut my finger and draw a circle on the floor with the blood which flowed from the incision.

I then placed myself in the centre of the circle, and either said or read the Lord's Prayer backwards. This I accomplished at first with some trepidation and in great fear towards the close of the scene. I then stood still in the centre of that magic and superstitious circle, looking with intense anxiety in all directions, especially at the window and at the chimney. Fortunately for myself, and for the reader also, if he is interested in this narrative, no owl or black cat or unlucky raven came into the room.

In either case my then weakened frame might have expiated this foolish experiment by its own extinction, or by the alienation of that too curious spirit which controlled its feeble powers.

This is a common enough type of experience for an imaginative child, but it was pursued very systematically. The 'too curious spirit' remained with him for life. After worrying whether he had sinned the young Charles concluded sensibly: 'My sense of justice (whether it be innate or acquired) led me to believe that it was impossible that an almighty and all-merciful God could punish me, a poor little boy, with eternal torments because I had anxiously taken the only means I knew of to verify the truth or falsehood of the religion I had been taught.'[25]

Later Babbage came to believe that scientific method pursued to its uttermost limit was entirely compatible with revealed religion and wrote his *Ninth Bridgewater Treatise* to prove the point. He noted with satisfaction that his own parents, while deeply religious, were without a trace of bigotry.[26]

Another story from Alphington shows the characteristic Babbage:

One day, when uninterested in the sports of my little companions, I had retired into the shrubbery and was leaning my head, supported by my left arm, upon the lower branch of a thorn-tree. Listless and unoccupied, I *imagined* I had a head-ache. After a time I perceived, lying on the ground just under me, a small bright bit of metal. I instantly seized the precious discovery, and turning it over, examined both sides. I immediately concluded that I had discovered some valuable treasure, and running away to my deserted companions, showed them my golden coin. The little company became greatly excited, and declared that it must be gold, and that it was a piece of money of great value. We ran off to get the opinion of the usher; but whether he partook of the delusion, or we acquired our knowledge from the higher authority of the

[25] *Passages*, 11–13.   [26] Ibid., 8.

master, I know not. I only recollect the great disappointment when it was pronounced, upon the undoubted authority of the village doctor, that the square piece of brass I had found was a half-dram weight which had escaped from the box of a pair of medical scales. This little incident had an important effect upon my after-life. I reflected upon the extraordinary fact, that my head-ache had been entirely cured by the discovery of the piece of brass. Although I may not have put into words the principle, *that occupation of the mind is such a source of pleasure that it can relieve even the pain of a head-ache*; yet I am sure it practically gave an additional stimulus to me in many a difficult inquiry. Some few years after, when suffering under a form of tooth-ache, not acute though tediously wearing, I often had recourse to a volume of Don Quixote, and still more frequently to one of Robinson Crusoe. Although at first it required a painful effort of attention, yet it almost always happened, after a time, that I had forgotten the moderate pain in the overpowering interest of the novel.[27]

The story reveals formidable powers of concentration.

When Charles's health had quite recovered he was moved to a small school with about thirty pupils in Enfield, then a village north of London. This was the only real schooling, as opposed to tutoring, that he received and from the Enfield days come his schoolboy stories. Things started badly and there were rumours of an attempt to run away:[28]

My first experience was unfortunate, and probably gave an unfavourable turn to my whole career during my residence of three years.

After I had been at school a few weeks, I went with one of my companions into the play-ground in the dusk of the evening. We heard a noise, as of people talking in an orchard at some distance, which belonged to our master. As the orchard had recently been robbed, we thought that thieves were again at work. We accordingly climbed over the boundary wall, ran across the field, and saw in the orchard beyond a couple of fellows evidently running away. We pursued as fast as our legs could carry us, and just got up to the supposed thieves at the ditch on the opposite side of the orchard.

A roar of laughter then greeted us from two of our own companions, who had entered the orchard for the purpose of getting some manure for their flowers out of a rotten mulberry-tree. These boys were aware of our mistake, and had humoured it.

We now returned all together towards the play-ground, when we met our master, who immediately pronounced that we were each fined one shilling for being out of bounds. We two boys who had gone out of bounds to protect our master's property, and who if thieves had really been there would probably have been half-killed by them, attempted to remonstrate and explain the case; but all remonstrance was vain, and we were accordingly fined. I never forgot that injustice.

Even so there is little doubt that the teaching at Enfield formed the foundation of his formal education:

The school-room adjoined the house, but was not directly connected with it. It contained a library of about three hundred volumes on various subjects, generally very well selected; it also contained one or two works on subjects which do not usually attract at that period of life. I derived much advantage from this library; and I now

---

[27] Ibid., 14–15.　　　　[28] *DNB* Captain Frederick Marryat.

mention it because I think it of great importance that a library should exist in every school-room.[29]

Stephen Freeman, the master, was an amateur astronomer[30] and probably first developed Babbage's serious interest in mathematics:

Amongst the books was a treatise on Algebra, called 'Ward's Young Mathematician's Guide.' I was always partial to my arithmetical lessons, but this book attracted my particular attention. After I had been at this school for about a twelvemonth, I proposed to one of my school-fellows, who was of a studious habit, that we should get up every morning at three o'clock, light a fire in the schoolroom, and work until five or half-past five. We accomplished this pretty regularly for several months.

One may reasonably picture Babbage studying mathematics in the early hours: it is difficult to imagine him getting up at 3 a.m. to study classics.

Inevitably word spread and others asked to join, some to work and others to play. Fearing detection Charles refused, but one of the boys, Frederick Marryat, was quite determined to be allowed to join in:

Marryat slept in the same room as myself: it contained five beds. Our room opened upon a landing, and its door was exactly opposite that of the master. A flight of stairs led up to a passage just over the room in which the master and mistress slept. Passing along this passage, another flight of stairs led down, on the other side of the master's bed-room, to another landing, from which another flight of stairs led down to the external door of the house, leading by a long passage to the school-room.

Through this devious course I had cautiously threaded my way, calling up my companion in his room at the top of the last flight of stairs, almost every night for several months.

One night on trying to open the door of my own bed-room, I found Marryat's bed projecting a little before the door, so that I could not open it. I perceived that this was done purposely, in order that I might awaken him. I therefore cautiously, and by degrees, pushed his bed back without awaking him, and went as usual to my work. This occurred two or three nights successively.

One night, however, I found a piece of pack-thread tied to the door lock, which I traced to Marryat's bed, and concluded it was tied to his arm or hand. I merely untied the cord from the lock, and passed on.

A few nights after I found it impossible to untie the cord, so I cut it with my pocket-knife. The cord then became thicker and thicker for several nights, but still my pen-knife did its work.

One night I found a small chain fixed to the lock, and passing thence into Marryat's bed. This defeated my efforts for that night, and I retired to my own bed. The next night I was provided with a pair of plyers, and unbent one of the links, leaving the two portions attached to Marryat's arm and to the lock of the door. This occurred several times, varying by stouter chains, and by having a padlock which I could not pick in the dark.

At last one morning I found a chain too strong for the tools I possessed; so I retired to my own bed, defeated. The next night, however, I provided myself with a ball of

[29] *Passages*, 18–19.   [30] BL Add. Ms. 37,184, f 30.

1. Old London Bridge by J. M. W. Turner, 1824.

3. John Babbage's house, with shopfront remodelled in the nineteeth century, c. 1860.

2. Fore Street, Totnes, c. 1860.
John Babbage's house is on the right.

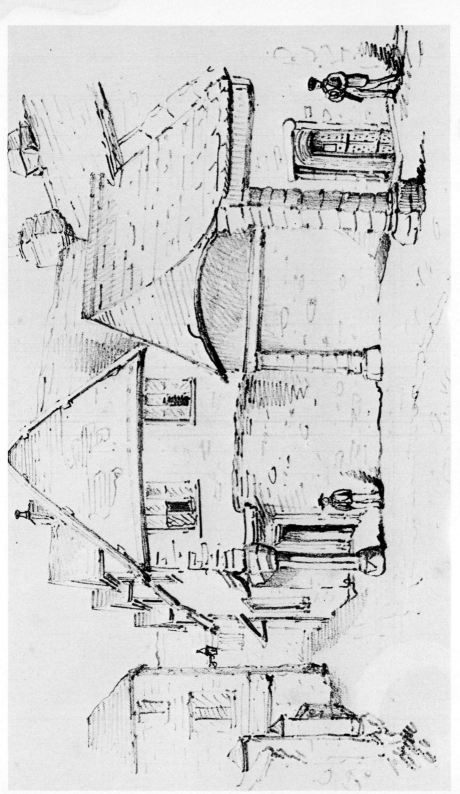

4. Guildhall, Totnes, *c.* 1860.

5. Totnes from the river Dart by J. M. W. Turner, c. 1824.

packthread through a link of the too-powerful chain, and bringing it back with me to bed, gave it a sudden jerk by pulling both ends of the packthread passing through the link of the chain.

Marryat jumped up, put out his hand to the door, found his chain all right, and then lay down. As soon as he was asleep again, I repeated the operation. Having awakened him for the third time, I let go one end of the string, and drew it back by the other, so that he was unable at daylight to detect the cause.

At last, however, I found it expedient to enter into a treaty of peace, the basis of which was that I should allow Marryat to join the night party; but that nobody else should be admitted. This continued for a short time; but, one by one, three or four other boys, friends of Marryat, joined our party, and, as I had anticipated, no work was done. We all got to play; we let off fire-works in the play-ground, and were of course discovered.

Our master read us a very grave lecture at breakfast upon the impropriety of this irregular system of turning night into day, and pointed out its injurious effects upon the health. This, he said, was so remarkable that he could distinguish by their pallid countenances those who had taken part in it. Now he certainly did point out every boy who had been up on the night we were detected. But it appeared to me very odd that the same means of judging had not enabled him long before to discover the two boys who had for several months habitually practised this system of turning night into day.[31]

Marryat later became Captain Marryat, author of *Mr. Midshipman Easy* and something of an artist. He happened to be on St Helena at the time Napoleon died and made a famous death-bed sketch of him. The two boys remained friends when they were grown up. Babbage must have thought highly of Stephen Freeman. They remained in contact for the rest of Freeman's life and Babbage later entrusted his two eldest sons to Freeman's care.

Charles was next sent for a few years to join five other boys 'under the care of an excellent clergyman near Cambridge'. His tutor seems to have been of the same persuasion as the Reverend Simeon, leader of the Evangelicals. Presumably this was Benjamin's reason for sending Charles to Cambridge for tutoring, even if proximity to the university and his tutor's acquaintance with its requirements were further inducements. Excellent though the clergyman may have been he seems to have taught Charles but little. Even the religious instruction seems to have been of doubtful service:

I came into frequent contact with the Rev. Charles Simeon, and with many of his enthusiastic disciples. Every Sunday I had to write from memory an abstract of the sermon he preached in our village. Even at that period of my life I had a taste for generalization. Accordingly, having generalized some of Mr. Simeon's sermons up to a kind of skeleton form, I tried, by way of experiment, to fill up such a form in a sermon of my own composing from the text of 'Alexander the coppersmith hath done us much harm.' As well as I remember, there were in my sermon some queer deductions from this text; but then they fulfilled all the usual conditions of our sermons: so thought also two of my companions to whom I communicated *in confidence* this new manufacture.

[31] *Passages*, 19–22.

By some unexplained circumstance my sermon relating to copper being isomorphous with Simeon's own productions, got by substitution into the hands of our master as the recollections of one of the other boys. Thereupon arose an awful explo;ion which I decline to paint.

However the young Babbage did learn something which he later felt would have made a striking example in his book, *The Economy of Manufactures*:

Mr. Simeon had the cure of a very wicked parish in Cambridge, whilst my instructor held that of a tolerably decent country village. If each minister had stuck to the instruction of his own parish, it would have necessitated the manufacture of four sermons per week, whilst, by this beneficial interchange of duties, only two were required.[32]

It must have seemed ridiculous to send a boy all the way from Devon to Cambridge to receive so inadequate an education. So instead he went to Totnes Grammar School next to the old Guildhall 'under the guidance of an Oxford tutor, who undertook to superintend my classical studies only'.[33] Babbage's classical knowledge must have reached a sufficient standard to see him through Cambridge with little further study. In mathematics he was well able to teach himself from such books as he could obtain. Long after, Babbage's youth in Totnes was not forgotten. In 1871 after his death a Totnesian claimed: 'I have relatives still living who perfectly recollect the studious-looking boy, swinging himself round by a pillar of the old piazza before his father's door and running across to the school turning neither to the right nor the left.'[34] Others remembered Benjamin living at the site which in 1871 was the Commercial Inn, earlier a substantial mansion.[35]

Babbage often complained of ill-health but all the evidence suggests that he was in remarkably good condition for most of his life. Certainly he seems to have been in good physical shape during his youth:

The grounds surrounding my father's house, near Teignmouth, extended to the sea. The cliffs, though lofty, admitted at one point of a descent to the beach, of which I very frequently availed myself for the purpose of bathing. One Christmas when I was about sixteen I determined to see if I could manage a gun. I accordingly took my father's fowling-piece, and climbing with it down to the beach, I began to look about for the large sea-birds which I thought I might have a chance of hitting.

I fired several charges in vain. At last, however, I was fortunate enough to hit a sea-bird called a diver; but it fell at some distance into the sea: I had no dog to get it out for me; the sea was rough, and no boat was within reach; also it was snowing.

So I took advantage of a slight recess in the rock to protect my clothes from the snow, undressed, and swam out after my game, which I succeeded in capturing. The next day, having got the cook to roast it, I tried to eat it; but this was by no means an agreeable task, so for the future I left the sea-birds to the quiet possession of their own dominion.[36]

[32] Ibid., 24.   [33] Ibid., 25.
[34] Obituary of Charles Babbage, *Totnes Times*, 28 Oct. 1871.
[35] Ibid.   [36] *Passages*, 205.

The young Charles was always interested in mechanical contrivances, and one that he constructed in Totnes nearly brought him to grief. Shortly after the attempt to shoot sea birds, while 'residing on the beautiful banks of the Dart', he constantly indulged in swimming in its waters.

One day an idea struck me, that it was possible, by the aid of some simple mechanism, to walk upon the water, or at least to keep in a vertical position, and have head, shoulders, and arms above water.

My plan was to attach to each foot two boards closely connected together by hinges themselves fixed to the sole of the shoe. My theory was, that in lifting up my leg, as in the act of walking, the two boards would close up towards each other; whilst on pushing down my foot, the water would rush between the boards, cause them to open out into a flat surface, and thus offer greater resistance to my sinking in the water.

I took a pair of boots for my experiment, and cutting up a couple of old useless volumes with very thick binding, I fixed the boards by hinges in the way I proposed. I placed some obstacle between the two flaps of each book to prevent them from approaching too nearly to each other so as to impede their opening by the pressure of the water.

I now went down to the river, and thus prepared, walked into the water. I then struck out to swim as usual, and found little difficulty. Only it seemed necessary to keep the feet further apart. I now tried the grand experiment. For a time, by active exertion of my legs, I kept my head and shoulders above water and sometimes also my arms. I was now floating down the river with the receding tide, sustained in a vertical position with a very slight exertion of force.[37]

The episode indicates once again an experimental approach to the truths of revealed religion. Unfortunately the apparatus got out of order and Charles was nearly drowned. The experiment was forerunner to many later experiments more thoroughly planned and carefully executed.

In 1810 Charles secured a place at Trinity College, Cambridge. After an elementary introduction he had taught himself mathematics; and he went to university with the benefit of remarkably little guidance: 'My father, with a view of acquiring some information which might be of use to me at Cambridge, had consulted a tutor of one of the colleges, who was passing his long vacation at the neighbouring watering-place, Teignmouth. He dined with us frequently. The advice of the Rev. Doctor was quite sound, but very limited. It might be summed up in one short sentence: "Advise your son not to purchase his wine in Cambridge."'[38]

[37] Ibid., 206.   [38] Ibid., 25.

2

○●○

Cambridge

Babbage went up to Cambridge in October 1810, his head full of plans for clubs and associations; and he looked forward with delight to having his mathematical difficulties explained. During his years at Totnes grammar school and Teignmouth he had studied such books on mathematics as came to hand—a mixed bag—but now, he felt, he would at last receive competent guidance. He knew of Lacroix's great expository book on the differential and integral calculus and being informed that it cost two guineas wished to purchase a copy in London on his journey. During the Napoleonic wars French books were very difficult to obtain and going to the French bookseller, Dulau, Babbage found to his dismay that the price of the book was seven pounds, a huge price at the time. However after much deliberation the book was purchased.[1]

Arriving in Cambridge, Babbage saw his tutor Hudson, secured lodgings, and spent most of the night reading the precious book. A few days later Babbage approached Hudson, seeking explanation of a mathematical point which was causing him some difficulty. Hudson listened politely and dismissed the question, noting that it would not be asked in the examinations in the Senate House and advising Babbage to concentrate on more elementary topics. A second difficulty was treated by one of the lecturers in a similar manner. When a third difficulty arose which raised a more fundamental question the person Babbage approached not only knew nothing of the matter but sought to conceal the fact. His high hopes of Cambridge had been speedily disappointed and he had recourse instead to the papers of the academies of Petersburgh, Berlin, and Paris which were now available to him in the libraries he was using.

Although Babbage always cultivated mathematics one suspects that more commonplace undergraduate activities played a larger part in his first year of residence. Always sociable he mixed widely in many different circles, the famous charm already at work. His closest friends were a group of ten or a dozen young men who usually breakfasted with him on Sunday after chapel, arriving at about nine and staying until after noon, spending the morning in relaxed discussion. Evidently they considered themselves an intellectual élite. Resolving themselves into a Ghost Club they made a study of the evidence for the supernatural, conducting a considerable correspondence on the subject. At another time they resolved themselves into the Extractors Club with plans to

[1] *Passages*, 34–8.

extract any member from the madhouse should his relatives ever incarcerate him in one. Chess and whist with other groups took up a great deal of his time. The whist four often sat from the end of evening chapel until the sound of the morning chapel bell called them once again to religious duties.

Babbage was very fond of sailing, which he had doubtless learned on the Dart and at Teignmouth with its famous regatta, and kept a light London-built boat on the Cam. Occasionally he took two or three friends on trips down the river. Naturally he chose strong oarsmen to row if the wind dropped or was contrary. The intellectuals of the ghost club contemptuously referred to his boating companions as 'Babbage's Tom fools'. On these occasions an 'aegrotat', certifying that he was unwell and could not attend chapel, hall, or lectures, was procured by his servant and forwarded to the college authorities. Babbage never even saw the piece of paper. The college chef produced a large well-seasoned meat pie, a couple of chickens and other food. This was packed in a hamper with several bottles of wine and a bottle of brandy and Babbage set off with his friends down the River, beyond Ely and into the fens, sailing when the wind was fair and rowing when there was none. Whittlesea Mere, soon to be fertile cornland, was one of their favourite resorts for sailing, fishing and shooting. Sometimes they reached Lynn. After five or six days of adventures and exercise in the open air they returned to Cambridge in splendid health. A booklet of the time, 'Hints to Freshmen', warned: 'A student at the university has no occasion for a numerous acquaintance. To *him* very frequently *amici sunt fures temporis.*'[2] Babbage cheerfully ignored such warnings and moved in more varied circles than most of his contemporaries. He kept detailed accounts[3] during his first two years at Cambridge, still under his father's influence. His allowance was £300 per annum. The entries indicate a steady life with a regular but moderate consumption of wine. There are many entries for the purchase of books. On 3 January 1811 he bought a pair of skates, and an entry on 10 April notes five pounds spent in London, probably on a spree. Not until the beginning of the summer vacation did he return to Devonshire. In July and August there are expenses for balls. Probably it was in Teignmouth that Babbage met Georgiana Whitmore whom he later married.

Babbage returned to Cambridge at the end of October. In November twelve pounds was spent on a 'northern journey', probably visiting the Bromheads. In December he spent seven pounds on eight volumes of the proceedings of the École Polytechnique. Ten pounds went on a spree in London. He bought a dog for two guineas, and eight shillings went on mending his bugle. In April there were bills for a furnace, chemicals and retorts. In May Babbage made his first

[2] Revd. Philip Dodd, *Hints to Freshmen of Cambridge*, 41, London, 1807.
[3] During his first year at Cambridge Babbage spent £300 4s. 8½d.; during his second, £298 14s. 1d., but a few small debts were beginning to accumulate. The account books are in the collection of Mr Alfred W. van Sinderen.

subscription of fifteen shillings to the Analytical Society. In the same month he returned to Teignmouth, and during the summer he became engaged to Georgiana, visiting her family at Dudmaston for the first time.

The Cambridge in which Babbage found himself was slowly recovering after a period of stagnation during the last sixty years of the eighteenth century. In Trinity, which was Babbage's college during his first year, academic standards had begun to rise after a crisis in 1786–7 when a group of younger Fellows had made a stand against the Master and Seniors who then governed the college. Appealing to the crown against current malpractices in the election of Fellows they were able to put an end to the flagrant favouritism which had been commonplace, and election to Fellowships became primarily by merit.

Cambridge remained famous for mathematics. Everyone who sought academic honours was until 1824 compelled to read maths, whatever their interests. But in spite of its renown the subject was in a very poor state until Babbage and his friends appeared on the scene. Oxford mathematics was even worse. The problem derived from one of the most famous quarrels in the history of science: the argument between Newton and Leibnitz on who had first developed the calculus. The atmosphere was dark with accusations of plagiarism. It is now quite clear that the two developments were essentially independent but the quarrel had been bitter and the academics of different countries had taken sides.[4] Moreover the two men had used quite different notations: Newton the famous 'dots' and Leibniz the 'd's. Thus British mathematicians, unswervingly loyal to Newton and trained exclusively in his methods, were often unable even to read Continental mathematical literature. Unfortunately the 'd's were the more powerful method and, partly as a result, Continental mathematics developed rapidly while British mathematicians contributed little of significance. One book, *Principles of Analytical Calculation* (1803), using the Leibnitz notation, had been written by a Cambridge don, R. Woodhouse, but he had done nothing to develop the system in the Cambridge curriculum. Babbage had a copy of Woodhouse's book, and also of Lagrange's *Théorie des Fonctions*',[5] through which he had become familiar with the Continental notations in the first place.

Trinity was the centre of Newtonian scholarship. Richard Bentley, incomparable classicist and Master during Newton's later years, had done much to encourage science and had been the patron of two of Newton's young disciples, Roger Cotes and Robert Smith, whom he attached to the college.

---

[4] The separation between Continental and British science caused by the Newton/Leibniz controversy was further developed between England and France by the difference between Newtonian and Cartesian philosophy—in spite of the fact that Voltaire and de Maupertuis had made universal gravitation more fashionable in Paris than Cartesian whirlwinds. It was a curious chance that Babbage, one of the most Newtonian of nineteenth-century thinkers, should have been the leading Francophile in England.

[5] J. L. Lagrange, *Théorie des Fonctions Analytiques*, Paris, 1797.

Bentley also promoted the second edition of Newton's *Principia* which Cotes edited. The Newtonian tradition remained and in supporting the Continental notation Babbage was taking a bold step. Moreover the leading country of the time for both science and mathematics was France, but to support French science and mathematics after the French revolution and during the wars with France was a political act. William Frend, whose daughter Babbage was later to meet as wife of Augustus de Morgan and acolyte of Lady Noel Byron, had been driven out of Cambridge in 1793 after an academic pseudo-trial in the Senate House, part of a country-wide anti-Jacobin campaign. Although Trinity College had stood aside from the witch hunt, its Fellows well balanced between Whig and Tory, conservative pressure from the University authorities was strong, reflecting the dominant feelings of the establishment.

Undergraduates were not permitted to form societies to discuss politics at the time. In 1811 when some enthusiasts sought to form a branch of the British and Foreign Bible Society they were effectively discouraged by the dons from doing so, partly because the Society was not limited to Anglicans, but basically because of fear that an undergraduate society would soon turn to politics. The evangelical seniors felt it necessary to keep the conduct of the organization entirely in their own hands. Not until 1814 was the Union Society formed by the coalescence of three smaller rival societies and it met for the first time in 1815. At first it discussed both political and literary subjects but in 1817 a meeting was interrupted by the Proctors who commanded the members to disperse and not to reassemble. Whewell, a B.A. and shortly to be elected to a Fellowship at Trinity, pompously replied: 'Strangers will please to withdraw, and the House will take the message into consideration.'[6] However the Union was disbanded and when it was re-formed in 1821 political subjects were excluded from discussion. Not until the time of the reform agitation ten years later was free discussion of political subjects permitted. Thus, as a result of the long-lasting Newton-against-Leibniz controversy and the primacy of French science, the question of the choice of mathematical notation improbably acquired political overtones.

Equally curiously Babbage's move to attack the problem arose from the activities of the Evangelicals whom he had encountered while at school in the person of the Reverend Simeon. At that time Cambridge rocked to a heated controversy about the form in which the bible should be printed. One party desired notes to make the bible more comprehensible; the other scornfully rejected all explanations of Holy writ as profane attempts to improve perfection. The walls of the town were placarded and broadsheets sent to the houses.

One evening in the spring of 1812 Michael Slegg of Trinity, a friend of Babbage's, was drinking with him and discussing mathematics to which both were addicted. The chapel bell rang and Slegg took leave, promising to return

[6] D. A. Winstanley, *Early Victorian Cambridge*, 25, C.U.P., 1940.

for coffee. Babbage took up a broadsheet demanding in exaggerated terms the circulation of the unexplained word of God. The temptation to parody was strong and he sketched out a plan of a society for propagating the gospel of the 'd's while consigning to perdition all supporters of the heresy of dots. It was maintained that Lacroix's work was so perfect that comment was unnecessary. Slegg laughed and took the parody to show to Bromhead, a mutual friend, who thought the joke too good to lose and proposed the formation of a society for cultivating mathematics. A preliminary meeting was held on Thursday in Bromhead's lodgings which constituted itself 'The Analytical Society'. The first proper meeting of the society was on the following Monday, and the first memoir read was by Bromhead 'On Notation'.

The 'Hints to Freshmen' also remarked: 'I have observed, that they are loudest in complaint of the inability of Tutors, who are least qualified to determine their merits.'[7] But even on arrival at Cambridge Babbage was surprisingly well qualified to judge at least the mathematical merits of his tutors. 'You are inclined to throw aside the books recommended by your tutor, and adopt a course of reading better suited to your taste. Beware of the first step to idleness and folly. If you neglect THE STUDY OF THE PLACE, there is little hope of your perseverance in *any* literary pursuit. An Undergraduate at Cambridge is, in ninety-nine instances out of an hundred, a student for the senate-house, or no student at all.'[8] As it happened Babbage was the one in a hundred. Where the more academically inclined of his friends were concerned with Tripos and Fellowship examinations Babbage was unencumbered with such distractions and free to pursue his own inclinations.

The prevailing political opinion amongst Babbage's friends was one of militant Liberalism. To that the Analyticals added belief in mathematical science. This group formed the background to Babbage's scientific career. It is impossible to understand its part in the foundation of scientific societies, in the 'decline of science' controversy, reform of the universities and science generally, unless the continuity from the formative Analytical Society is grasped. For long they continued to refer to themselves as the 'Analyticals' until the accretion of more allies and the loss of some former members to other pursuits rendered the term inappropriate. The compact formed in Cambridge was never forgotten. All things then seemed possible. The Liberals, including Babbage, had a great regard for Continental and particularly for French science. But Britain had defeated France, and while conducting the wars had found the strength to construct an extensive and remarkable set of roads, bridges, docks, and canals. Britain was the workshop of the world and her engineering strength and manufacturing skills were unrivalled. The Analyticals would reform mathematics in Cambridge, and thus in Britain; all science would follow, and with it technology and manufacture.

[7] Dodd, op. cit., 50.   [8] Ibid.

It is worth pausing to glance at the central group of the Analytical Society. John Herschel was the only child of Sir William Herschel, one of the greatest practical astronomers and founder of cosmology. John Herschel himself became one of the nineteenth century's leading astronomers, particularly famous for his survey, made near Cape Town, of the stars of the southern hemisphere. He became Babbage's closest friend. Alexander D'Arblay was the only son of Mme D'Arblay (Fanny Burney). Edward Ryan became part of Babbage's family. While at Cambridge they were courting two sisters from the Whitmore family of Dudmaston in Shropshire, and married them in 1814. Ryan was appointed judge of the supreme court in Calcutta in 1826 and was later Chief Justice of Bengal. Edward Ryan was there at the end to help Babbage sort out his affairs when he was dying. Sir Edward Ffrench Bromhead was a landowner and unfortunately did not pursue science for long. He seems to have had the shrewdest political head of the group. He became patron of the great blacksmith mathematician Green, helping him to get to Cambridge. George Peacock, later Dean of Ely, was one of the leading nineteenth century reformers of Cambridge University. Babbage, Herschel, and Peacock went on to make outstanding contributions to mathematics. Herschel, a year senior to Babbage at Cambridge, was elected president of the Analytical Society. The Society as such functioned for only two years, from the spring of 1812 until the summer of 1814 when Babbage went down.

The flavour of the meetings may be gathered from a letter written by Frederick Maule, one of the founders, headed Edmonton, Jan 16 1813[9] to Babbage, now moved to Peterhouse, possibly to secure rooms in college: 'You will be ready to imagine that I have forgotten the A[nalytical] S[ociety] and their boisterous debate so well described in your letter that I transported myself in thought to the scene of action and heard the damns, the nonsense, the arguments, the objections, etc. with greater personal safety if not with the same clear perception.'

The dons naturally ridiculed the whole venture; and when the Analyticals continued on their course, hinted darkly that they were young infidels and that no good would come of it. Babbage had proposed the translation of Lacroix's smaller work. However the first and only production of the Society was the *Memoirs of the Analytical Society*. Originally intended to include contributions by several members it was in the event the work of Babbage and Herschel. The book is in sections: Preface—Babbage and Herschel; I On Continued Products—Babbage; II On Trigonometrical Series—Herschel; III On Equations of Differences—Herschel. The preface, which was probably drafted by Babbage,[10] gives a brief history of analysis during the previous century and

---

[9] BL Add. Ms. 37,182, f 3. The letter is actually dated 16 Jan. 1812 but this must be an error.
[10] *Memoirs of the Analytical Society*. Babbage made the first drafts of the preface while Herschel

shows a remarkable familiarity with the literature. It focuses attention on a question which Babbage was to take up repeatedly during his life, proclaiming in mathematics

the accurate simplicity of its language ... An arbitrary symbol can neither convey, nor excite any idea foreign to its original definitions. The immutability no less than the symmetry of its notation (which should be ever guarded with a jealousy commensurate to its vital importance) facilitates the translation of an expansion into common language at any stage of an operation,—disburdens the memory of all the load of the previous steps,—and at the same time, affords it a considerable assistance in retaining the results.[11]

The point is reiterated:

The importance of adopting a clear and comprehensive notation did not, in the early period of analytical science, meet with sufficient attention; nor were the advantages resulting from it, duly appreciated. In proportion as science advanced, and calculations became more complex, the evil corrected itself, and each improvement in one, produced a corresponding change in the other.[12]

One can already anticipate from this the great importance Babbage was later to attach to his own 'mechanical notation',[13] which he was to develop for describing the Calculating Engines. Problems of nomenclature and language were being discussed in other sciences during this period, corresponding to the rapid advances taking place. A major reform of chemical nomenclature was introduced in France near the end of the eighteenth century.

The Memoirs were printed in Cambridge by J. Smith, Printer to the University, at the end of 1813,[14] and financed by ten active members of the Society. The contents were over the heads of the senior Cambridge mathematicians. Edward Ffrench Bromhead, in an undated letter, probably early in 1814, to Babbage on receiving his copy of the 'Memoirs of the A.S.' wrote: 'They [the Memoirs] are too profound to do us any good, & not one mathematician in 10^∞ can understand them ... The true faith will never flourish till a book has been published in *English*, in *Octavo* [i.e. cheaper], on the *plan* of Woodhouse's Anal[ytical] Calcul[us], & in a *compact* and *tangible* shape. Whoever does this will be the father of a new English school. Herschel will *not*

---

was still busy with examinations. The strong historical bias as well as the style have the Babbage flavour. Moreover he was at the time far more widely read in Continental mathematics even than Herschel, whose reading extended after he had finished with examinations. Babbage saw the first and part of the second memoir through the press, and when he had returned to Devonshire in June 1813 Herschel saw the remainder of the memoirs and the preface to completion.

[11] Ibid., Preface, i.   [12] Ibid., xvi.

[13] C.B., 'On a Method of Expressing by Signs the Action of Machinery', *Phil. Trans.* 1826, 116, 250–65.

[14] *The Memoirs of the Analytical Society* were published anonymously. Babbage alone had wanted boldly to attach the names of the authors to the papers.

do this. Try it yourself.'[15] Similar problems arose later after Babbage had gone down. In 1816 Bromhead was writing from Thurlby Hall, Newark, Notts, to Babbage who was by then settled at 5 Devonshire Street, Portland Place: 'I have only just returned from Cambridge ... Very many thanks for your kind letter and invaluable paper, which will bear a hundred readings ... I did not find a single soul, even among Senior Wranglers, Herschel excepted, who understood a word about it. One very high man, a fellow, observed that $f\alpha(x)$ could not $= f(x)$, as then $\alpha(x) = x$ QED.'[16] This blunder is a schoolboy howler: such was the level of Cambridge mathematics when the Analyticals launched their campaign.

Unconcerned with studying for honours or seeking a fellowship, and certainly not bothering to attend maths' lectures (a Cambridge tradition followed by many of his eminent successors), Babbage commenced translating the smaller work of Lacroix, only to lay it aside. Some years later Peacock called on him in Devonshire Street to say that both Herschel and himself were convinced that the change from the dots to the 'd's required the translation of an eminent foreign work into English. This led to the completion of the translation as a joint work by Babbage, Herschel, and Peacock,[17] and to its publication some months later: technical publication was a great deal more rapid then than it is today.

Progress of the Leibniz notation was slow, even though the tutors of the largest Cambridge colleges adopted it. Thinking and reasoning in a new language is difficult and requires of all but the most energetic minds a direct interest. Babbage decided to provide the incentive by preparing a volume of problems with worked solutions. Such a book would be invaluable for students. He used a work by Hirsch and communicated his idea to Herschel and Peacock. Once again it appeared as a combined production of the three men.[18] In Cambridge the book was celebrated in bad verse:

> *Then* we hope to see problems and questions which balk all us
> Solved by the aid of the Integral *Cal*culus
> When civilized, savage, Oxmanians and Hottentots
> Shall set up the d's instead of those rotten dots
> When Peacock and Babbage from Cam to the Indus
> Shall glance in their glory from booksellers windows ... [19]

In a few years the change was completed. Sixteen years after the Analyticals had commenced battle, Babbage in Rome read in Galignani's newspaper: 'Yesterday the Bells of St. Mary rang on the election of Babbage as Lucasian Professor of Mathematics'.[20] The Lucasian professor was elected by the

[15] BL Add. Ms. 37,182, ff 13 & 15. [16] Ibid., f 46.

[17] S. F. Lacroix, *Sur le Calcul Différentiel et Intégral*, Tr. C. Babbage *et al.*, Cambridge, 1816.

[18] C.B. *et al. Examples to the Differential and Integral Calculus*, 2 vol. Cambridge, 1820.

[19] John Herschel to Babbage, 4 March 1823, RS f 183. [20] *Passages*, 29.

masters of the Colleges; not at the time the best set of men, perhaps, to select for intellectual distinction; but Cambridge had given its official blessing.

Mathematics was by no means Babbage's only scientific study at Cambridge. Although he did not attend many maths lectures he attended lectures on chemistry. Having a spare room—life could be spacious in those days for the comfortably off—he turned it into a private chemical laboratory. John Herschel worked with him until he established a rival laboratory of his own, and they both occasionally assisted Smithson Tennant, Professor of chemistry, with his experiments. Babbage respected and liked Smithson Tennant and hoped to be introduced by him to London scientific society. Unfortunately whilst crossing a drawbridge at Boulogne a bolt was displaced and the drawbridge collapsed, flinging him to the bottom of the moat and killing him instantly. After taking up residence in London Babbage soon abandoned practical chemistry but the knowledge he had gained was later to prove valuable.

If Babbage was not concerned with examinations or a Fellowship it is natural to enquire what he did plan to do after university. Probably when going to Cambridge he had expected to try for honours, possibly also for a Fellowship, and drifted into the position where they were out of the question through sheer distaste for routine studies. Probably also his mind was not made up. Science and mathematics simply did not appear as a practical possibility as a vocation: rather they were a gentleman's hobby. Herschel, for example, moved to the Inns of Court after Cambridge to read for the bar. It must be assumed that Babbage had decided not to follow his father into banking: certainly there is no evidence of such an intention. The one plan of which we have direct evidence seems curious in retrospect: Babbage planned to enter the church. He was in love and this may explain his intention: settling down in a quiet country parsonage with Georgiana may have appealed. If the plan had been realized the country would have had, in place of the Charles Babbage we know, yet another eccentric mathematical clergyman.

Through Georgiana and her family Babbage had met a remarkable man who had a profound influence on his life: Lucien Bonaparte, a younger brother of Napoleon.[21] Lucien had quarrelled with his powerful brother and was living

---

[21] Lucien was born at Ajaccio in 1775, third of the Bonaparte children. Educated in France he had returned to Corsica and been there for a month when the revolution began. Only fifteen he joined enthusiastically and energetically in the popular movement, soon becoming an effective orator. When in April 1793 Corsica under the old nationalist leader Paoli declared for separation from France, Lucien was head of the delegation seeking assistance from the Jacobin clubs of Marseilles and Paris. It must have been a heady experience for the youth to address a huge meeting of the revolutionary club in Marseilles. Sickened by the sight of a guillotine in action he declined to accompany a Marseilles delegation to Paris and remained to await their return. A few days later his mother arrived with the rest of her family, gaining impeccable status as refugee patriots.

A letter from Herschel at the time gently mocks Lucien's influence on Babbage: 'Citizen,

quietly in exile in Worcestershire, not far from the Whitmores in Dudmaston. He was regarded with deep suspicion by the British Government who thought that the quarrel with Napoleon had been faked and that he was a spy. Few people of quality visited the Bonapartes, so the Whitmores and their friends were particularly welcome. We may thus picture Babbage, Edward Ryan, Wolryche Whitmore and some of his sisters, whenever they were in Dudmaston driving frequently to the estate of Thorngrove which Lucien had purchased. Lucien Bonaparte was a most romantic figure. He had rejected a kingdom for love, standing up to the Emperor of most of Europe when no other man in France dared do so. He had participated in the French revolution with great bravery; as president of the Council of Five Hundred he had been the architect of 18 Brumaire and many people thought he had a better political brain than Napoleon himself. Lucien carried with him an aura of power: the mark of a man who has lived at the centre of great events. To the young Babbage he represented the best aspects of the new France: the revolution without the terror; a Bonaparte without bonapartism; the Republic against the Empire; and the vitality of Paris's intellectual society.

There have always been a number of puzzling features in Babbage's life. He became a leading society figure, and while much may be attributed to his personality it seems to require some further explanation. He was part of the Duke of Somerset's circle, but then so were other men of science such as Fitton. Then there is a militancy in Babbage's approach to the organization and application of science which stands alone in England. The range of his Continental associations is also remarkable. In all this the influence of Lucien Bonaparte can be seen. From the Prince of Canino, to give Lucien the title which he later acquired and which Babbage accords him, Babbage learned his militant approach to science, which had a sweep and cutting edge quite foreign to the English scientific *milieu*. From the Prince of Canino Babbage gained a European view of political events; he later gained social connections among the Continental aristocracy; and he acquired a sense of style.

In a fit of enthusiasm Napoleon appointed the mathematician Laplace minister of the interior. One can but applaud the sentiment inspiring the appointment but Laplace proved quite impractical and Lucien was appointed in his place. Lucien Bonaparte gave splendid entertainments that year in his mansion, the Villa Le Plessis Charmant. Quarrels between the brothers continued and Lucien was sent as ambassador to Spain. He then returned to Paris to be appointed member of the tribunate and soon Grand Master of the newly created Legion of Honour. He was wealthy and assembled a magnificent collection of paintings as well as becoming known as a generous patron of the arts.

Your letter of the 20th reached me ... I remain *citizen*, Yours sincerely, Herschel'. 1 July 1812, f 2, RS.

Lucien was a republican and as Napoleon moved towards dictatorship the fraternal quarrels became violent. Napoleon wanted him to contract a dynastic marriage, which Lucien refused to do. Instead he married clandestinely a widow, Madame Jouberthon. There was a sharp break and Lucien went to Italy. There he bought the estate of Canino from the Pope. In 1810, however, after further disagreement with his brother, he was sufficiently alarmed to sail for the United States. Captured by two British ships he and his family were taken to Malta and then to England. The British Government suspected that he had really been on a mission to bring the United States into the war on the side of France, and so he was kept under light surveillance. Although he had lived unostentatiously at Canino, at Thorngrove he lived in style to maintain his position while in captivity. He had a showy carriage and servants in splendid liveries. The young Babbage had never seen anything like it and Lucien's personality was formidable.

French science had been developing rapidly well before the revolution. Lavoisier was making important advances in chemistry. After the revolution some men of science, including Lazare Carnot and Gaspard Monge, who were republicans, immediately took leading parts in the economic and military administrations. Others, including Lavoisier himself and Condorcet, were killed, unable to live down their associations with the old régime. The newly founded *grandes écoles* served as model institutions for both teaching and research, giving a great impetus to science. Napoleon himself was an enthusiastic patron. He took a whole team of men of science with him on the expedition to Egypt and established the Institute of Egypt as a major instrument of government. When he returned to Paris, with him in the leading carriage was not only general Berthier but the men of science, Monge and Berthollet. The appointment of Berthollet and Laplace to well paid positions as senators enabled them to found the leading private scientific society of the age, the Society of Arcueil. Babbage was later to become a close friend of several former members. In the meantime he learned of the best aspects of French scientific society from Lucien. It was at Thorngrove that Babbage really grasped the possibilities of organizing science and applying it to the improvement of society. He began to dream of raising science to a leading position in England, comparable to the position it had been accorded in Napoleon's Paris, for the benefit of the whole nation. For Babbage, deeply in love with Georgiana, visiting the Bonapartes who had fled from France for love, bringing with them an aura of European power and Parisian style, it was all incomparably romantic. For the rest of his life Babbage saw France *couleur de rose*.

3

○ ● ○

Marriage: Early Years of a Philosopher in London

On 2 July 1814, shortly after coming down from Cambridge, Charles Babbage married Georgiana Whitmore in Teignmouth. They went for their honeymoon to Chudleigh, a charming village a few miles inland. Much of Chudleigh had been destroyed by fire a few years earlier but the village had been quickly rebuilt. Probably they stayed at the old coaching inn where many of the homeless villagers had taken refuge after the fire. Then for economy they moved into a farm. The marriage was to be very happy and while it lasted provided a secure and stable basis for Charles's life. However Benjamin Babbage did not approve. He had not married until he was well off and thought it highly improvident for Charles to marry before he was earning a good living.

From Chudleigh Charles wrote on 1 August to Herschel:

To be a little serious however I will tell you the events of the last few days. I am married and have quarrelled with my father. He has no rational reason whatever; he has not one objection to my wife in any respect. But he hates the abstract idea of marriage and is uncommonly fond of money.

I cannot go into the Church for this will not accord sufficient propriety (for a curacy is all I should get). I am therefore thinking of getting some situation connected with the mines where though I might get but a very little at first I might have opportunities of turning my chemical knowledge to advantage. However I should be glad to get any employment where I should have some future prospects though it might not produce much at first. If you hear of anything in any way that you think likely to suit me I should be glad to hear of it. I do not much care what it is only that it should not require very much bodily exertion as I was never well able to bear much of that. I believe I sent you what I have done relative to functions. I have added the following theorems . . .[1]

On 7 August a slightly shaken Herschel replied: 'I am married and have quarrelled with my father—Good God Babbage—how is it possible for a man calmly to sit down and pen those two sentences—add a few more which look like self justification—and pass off to functional equations.'[2]

Under the stress the reticent Charles for once let his pen flow, expressing his feelings. The letter deserves to be quoted *in extenso*:

August 10th 1814
Dear Herschel,
    Your letter of 7th August has just reached me but the other to which you allude has not yet arrived.

---

[1] Babbage-Herschel correspondence, f 25, RS.   [2] Ibid., f 28.

My opinions concerning Cambridge are in many respects similar to your own, but there are two reasons for which I shall always value a university education—the means it supplied of procuring access to books—and the still more valuable opportunities it affords of acquiring friends. In this latter point of view I have been singularly fortunate. The friendships I have formed while there I shall ever value; nor do I consider my acquaintance with yourself as one of the least advantages.

You must indeed have been surprised with my letter. I did not mention many circumstances because as we had rarely talked on such subjects, I thought they might be uninteresting. Now however I shall lay aside all reserve and give you an account of my present situation. For although in the commerce of the world we must use its currency and pay and receive what we have so emphatically styled *Hum*; yet between ourselves we shall, I hope, ever preserve that confidence and candour which has hitherto marked our friendship. You seem to have a great horror at my having quarrelled with my father; were he such a man as from the slight knowledge I possess I conceive yours to be; it would indeed be to me a subject of the deepest regret; but the case is different and you must know a little of his character before you judge. My father is not much more that sixty, very infirm, tottering perhaps on the brink of the grave. The greater part of his property he has acquired himself during years of industry; but with it he has acquired the most rooted habits of suspicion. It is scarcely too much to assert that he *believes* nothing he *hears*, and only half of what he sees. He is stern, inflexible and reserved, perfectly just, sometimes liberal, never generous, is uncultivated except perhaps by an acquaintance with English Literature and History. But whatever may be his good qualities they are more than counterbalanced by an accompaniment for which not wealth nor talents not the most exalted intellectual faculties can compensate—a temper the most horrible which can be conceived.

Seeking the happiness of no earthly being he lives without a friend. A tyrant in his family his presence occasions silence and gloom; and should his casual absence afford opportunity for the entrance of any of the more pleasing sensations they are invariably banished at his return. Tormenting himself and all connected with him he deserves to be miserable. Can such a man be loved? It is *impossible*.

Religion or sense of duty might possibly induce one to bear with him, and difficult nay almost impossible as is the task I see it executed in the conduct of my most excellent mother. She does so because she thinks it right. I too do what I think is right and should esteem it treason to myself to sacrifice my own happiness in the *vain* attempt of pleasing such a character. I will only observe that this is not the hasty sketch of irritated feelings. I am not angry with, I pity such a being. Ryan will confirm parts at least of this though he has only seen the fairer side.

I will now say something of my wife as it certainly is no secret. She was a Miss Whitmore the youngest of eight sisters of a very good family in Shropshire. She has some fortune and will have more about five thousand in all. My father has no objection to her, or her fortune or her family, this he has said. Why then you will ask have we quarrelled? This I cannot tell. I know as little as yourself what reasons he assigns. Everybody blames him. He allows me £300 during my life simply because he promised it to me. So that I now have about £450 a year with which I can live very comfortably for the present; but as I do not expect a shilling more from my Father I am looking for some employment where I may be able ultimately to make some money. I will now mention the circumstances which made me think of getting some employment connected with the mines.

About a month since I met at dinner at my Father's a Mr Champernowne, a gentleman of great respectability and considerable fortune who lives in the neighbourhood; in the course of conversation I found that he had travelled much and studied mineralogy under Werner but understood nothing of chemistry. He was rather pleased with me and invited me to see his collection of minerals which is a very good one. This I did the next day; shortly after he invited me to dinner where I met the only scientific men in the neighbourhood. These of course are by no means of the first rate. He then mentioned a tour which he and a friend were going to make to a part of Devon where there are mines and they invited me to accompany them.

As soon as I was married and knew my father's determination I wrote to Mr Champernowne and stated the case that circumstances were now altered and begged that if in his intended tour he met with anything that would suit me that he would recommend me to it. He returned me a very polite answer and said that if he met with such a thing he would. On meeting him a few days since at Exeter he invited me and my wife to spend a few days at his seat which is near Totnes and said he would write to me to appoint a time when he returned. This is very kind considering I had only been acquainted with him for a month; it certainly was rather *cool* to make such an application however coolness is necessary to a man of the world as well as to a philosopher. If he does not succeed in finding out anything I shall advertise in some of the country papers for some such situation. If any one applies to me in consequence I must give references for character and respectability; for the latter I shall refer them to Mr C[hampernowne] and as to the former I shall desire them to apply to *you* for my chemical knowledge and mathematical skill; or perhaps I shall send them to *Tennant*: what think you of this last application; would it be *too cool*?

However as I before remarked coolness is necessary, and I am determined to get on, if I have health I have not the least doubt I shall; and my Father will, if I get any money myself, very possibly forgive me for its sake; as he likes it better than any thing on earth. I applied to Mr Lane (who is the only person on whom my father places the least confidence and who is his best friend if friendship can exist with such a man as my father) we consulted what was best to be done he approved of my plans and offered to convey my letter to my Father and use all the influence he possessed with him.

One evening I took my former tutor (an Oxford man) over with me in a chaise to Teignmouth and was married by him the next morning and came to Chudleigh. We have very comfortable lodgings in a farm house in a most romantic country; here I shall stay until I find some situation. We live considerably within our income and I shall be able to pay all my Cambridge debts which are not £90 at next October. You will probably be better able to account for some apparent inconsistencies in my character when I inform you that I have been engaged for the past two years and of course my spirits fluctuated with the fluctuation of my prospects.

I wished to have taken orders when I had expectations of possessing my father's fortune. I should then have had more leisure for philosophical pursuits but without these prospects it would be madness to enter a profession where I have no hopes of getting on.

As it now is I shall not have much time to spare but whatever little I may have I shall always devote to those studies. Chemistry I admire, very much and would always pursue but Analysis was always my favourite amusement and will probably occupy most of my idle hours at it is much less expensive.[3]

[3] Ibid., f 29.

Then Charles launched once again into functional equations.

With a little care £450 per annum was sufficient to live on comfortably. Young and deeply in love, the newly married couple looked to the future with confidence. Dartington, where they stayed with Arthur and Louisa Champernowne, was a lovely place in which to spend part of their honeymoon. Dartington Hall[4] had been a royal estate, one of the first great houses in the country to be built without fortifications, shortly before the Wars of the Roses began.[5]

The Champernownes kept a simple household. They had only a curricle drawn by an old horse, two men and two or three female servants. But Arthur Champernowne spent a lot of money collecting paintings. On the walls were Titian's *Noli me Tangere* and Rubens's *Horror of War* and *Triumph of Caesar*. He also collected geological specimens and was a member of the geological society. He was involved in mining activities in Devon but apparently no opportunities opened for Charles. Nor can his advertisement, if it was in fact placed, have led to anything. By October the Babbages were considering moving nearer to London. Charles asked John Herschel if Slough, where Sir William Herschel had his home and telescopes, would be convenient. Herschel was not encouraging: 'I have repeatedly heard my mother declare there is not to be found so expensive a place in the kingdom. But the worst of the story is the society, which owing to the vicinity of the court is in general peculiarly unpleasant.'[6] In the autumn Charles and Georgiana were staying at 31, Arundel Street, Strand,[7] near his father's former house in George Street. By February 1815 Herschel was writing to Charles at 46, Lucas Street, Brunswick Square.[8] In the middle of September 1815 Charles 'became possessed of' a small house at 5, Devonshire Street, Portland Place: perhaps the name 'Devonshire' attracted them. This house was to be their family home.

Once settled in London Babbage wasted no time in launching himself on the scientific scene and was soon invited to give a course of lectures on astronomy at the Royal Institution. This series of lectures, of which the drafts of nine remain,[9] forms an interesting popular history of astronomy, and Babbage's detailed knowledge of the subject is impressive. John Herschel's aunt Caroline advised Babbage, and from the Herschels he had the guidance of the outstanding astronomers of the age. The lectures could have formed the basis

---

[4] Anthony Emery, *Dartington Hall*, O.U.P., 1970.

[5] The great hall and the principal associated buildings were planned and built in the last twelve years of the fourteenth century by Henry Holland, Earl of Huntington, half brother of Richard II. An Elizabethan residential block was added to one end of the hall and it is in this block that the Babbages would have stayed. While Totnes had lost its former importance, Dartington had quietly mouldered away. When the Babbages stayed there Champernowne had recently removed the slates from the roof, intending to replace them after inspecting the timbers.

[6] Babbage/Herschel correspondence, f 31, RS.

[7] Ibid., f 32.   [8] Ibid., f 36.   [9] BL Add. Ms. 37,203.

for a good popular book. They were read during the course of 1815. Following a brief history of early astronomy they discuss in succession: the instruments used in astronomy; the appearances of the heavens; the shape of the earth; the appearance of the moon, eclipses, and tides; the structure of the sun and its effect on the earth; the inner planets; planetoids and the development of the telescope; the outer planets; scientific theory in general; and the comets.

The most interesting lecture is the tenth, in which Babbage discusses the relation between theory and experiment in science:

The indiscriminate zeal against hypothesis so generally avowed by the followers of Bacon has been much encouraged by the strong and decided terms in which [hypotheses] have been reprobated by Newton.

But the language of this great man must be qualified and limited by the exemplification he has himself given of his general rules and it should be remembered that they were particularly directed against the vortices of DesCartes which were purely fictitious and were the prevailing doctrine of the time.—A very learned and acute writer has observed that:

'The votaries of hypothesis have often been challenged to show one useful discovery which was made that way.'

In reply to this challenge it will be sufficient on this occasion to mention—The theory of Gravitation and the Copernican system—Of the former I shall presently endeavour to prove from a sketch of its history that it took place entirely by conjecture and by hypothesis suggested by analogy . . .

The Copernican system offers however a still stronger case inasmuch as the only evidence which the author was able to offer was the advantage it possessed over every other hypothesis in explaining with beauty and simplicity all the phenomena of the heavens—In the mind of Copernicus therefore this system was nothing more than an hypothesis, but it was an hypothesis conformable to the universal law of nature always accomplishing her ends by the simplest means.

Nor is the use of hypothesis confined to those cases in which they have subsequently received confirmation—it may be equally great where they have completely disappointed the expectations of their authors. Indeed any hypothesis which possesses a sufficient degree of plausibility to account for a number of facts will help us to arrange those facts in proper order and will suggest to us proper experiments either to confirm or refute it.

The last paragraph comes quite close to the modern idea that science advances through successive stages of theory and refutation.

Babbage's political interests show when he turns aside to remark that: 'in the midst of [his] profound enquiries Newton had leisure for other pursuits. When the privileges of the University were attacked by James II he appeared as one of its most strenuous defenders—and he made a very successful defence before the high commission court. He was also a member of the convention parliament in which he sat until it was dissolved.'

Bromhead wrote cheerfully to a mutual friend, John William Whittaker at St John's College, Cambridge: 'Babbage does something more than astonish Albermarle Street, having made some beautiful discoveries in the Calculus of Functions which are now before the R. S.'[10] On 7 December 1815 the following certificate was read at the Royal Society:

Charles Babbage of Devonshire St. Portland Place, Esq. a gentleman well acquainted with various branches of useful science particularly the mathematics, being desirous of becoming a member of the Royal Society, we whose names are undersigned do of our own personal knowledge recommend him as worthy of that honour and as likely to become a useful and valuable member of this society.

WM HERSCHEL
J. F. W. HERSCHEL
JAMES IVORY
PETER M. ROGET
I. HAWKINS BROWNE.[11]

After the certificate had hung in the meeting room and been read at the requisite twelve successive meetings, Charles Babbage was elected a member of the Royal Society on 14 March 1816. In the spring of 1817 he twice dined at the Royal Society Club. In a couple of years he had established his position in London scientific circles. It was essential to belong to the Royal in order to play a leading part in the development of science in England, although membership was not restricted to practitioners of science. It was more like a good club. Lord Byron, whose daughter Ada Augusta was to play so important a part in Babbage's life, was a member. Indeed reform of the Royal Society, restricting its membership to men of science, and turning the Society into an effective headquarters for the promotion of English science, had been an early objective of the Analyticals.

Between 1815 and 1820 Babbage carried out much important mathematical work. This included fundamental work on algebra, which was continued by George Peacock, and a series of highly original papers on the theory of functions, usually thought of as a Continental subject. These papers were soon in the hands of the French mathematician Cauchy[12] and the brilliant young Swedish mathematician Abel[13]. One feels that Babbage would have gone on to make more important contributions to the theory of functions, as well as to

[10] Edward Ffrench Bromhead to John William Whittaker, Thurlby Hall 1815. Archives of St John's College, Cambridge. Whittaker (*c.* 1790–1854) was 13th Wrangler in 1814 and one of the founders of the Astronomical Society.

[11] Royal Society Archives.

[12] A. L. Cauchy to Babbage, 9 March 1820: Babbage had sent copies of his papers on the theory of functions to Cauchy, who wrote: 'L'importance du suject que vous y traittez prouve que dans la patrie de Newton il existe encore des géometres qui travaillent au progrès de l'analyse'. BL Add. Ms. 37,182, f 235.

[13] Christian Hansen to Babbage, 21 April 1857; BL Add. Ms. 37,197, f 183.

other branches of pure mathematics, had his interest not been concentrated on the calculating engines. However at this time there were several outstanding mathematicians. In his development of the Analytical Engines, those great precursors of the modern computer, in his development of operations research and in his systematic application of mathematics and science to industry, Charles Babbage was unique.

Meanwhile on 1 March 1815 Napoleon had landed at Cannes with a small detachment of the Guard. What part Lucien played in organizing the return from Elba is difficult to determine, but he apparently knew of his brother's plans. He travelled secretly to Paris, engaged in various abortive covert negotiations, and retired to Switzerland. He then openly returned to Paris where he was installed in the Palais Royal. Having left Napoleon at the height of his power, Lucien rallied to his support at the time of crisis. The régime and the constitution it planned were far more liberal than those of the Empire in the great days of its power; full of progressive ideas and worthy sentiments. By contrast the régime imposed on France after Waterloo and Napoleon's abdication was far more reactionary than the régime in power before the Hundred Days. Soon the Napoleonic legend was replacing the memory of historic reality, while the dream of restoring the revolutionary movement against the oppressive governments of the kings remained to inspire Babbage's generation in Europe.

When Charles Babbage settled in London great civil engineering works were in progress. John Rennie was constructing the Waterloo and Southwark bridges and Regent Street was being built. At the same time in the entrance to Plymouth sound John Rennie was building the enormous breakwater to protect ships from the south-westerlies. A mile long, it contains more than 3,500,000 tons of stone. England had emerged from the French wars victorious, a dominant power, pulsing with energy, but with hardly the rudiments of a state organization to cope with the rapidly increasing problems presented by industrialization and the associated growth of the large towns. There was a tiny civil service, no police force, no means for regulating the factories. The law was suited to the eighteenth century. The two universities, just beginning to recover from a long period of stagnation, were still exclusively Anglican; parliamentary representation was anachronistic and the Whigs a demoralised permanent opposition. What state positions did exist were, in spite of the vaunted reforms of the younger Pitt, still mainly given in patronage, which was in the hands of the ruling Tory party.

Still haunted by memories of the French revolution, the propertied classes went in fear of an English Jacquerie, and the slightest sign of popular discontent filled them with alarm. In 1816 a sharp rise in the price of bread combined with heavy unemployment led to nationwide agitation. Then at the opening of parliament early in 1817 a missile smashed through the Prince Regent's coach windows. This was interpreted as an assassination attempt: perhaps the missile had been a bullet. Parliament panicked. There were wild reports of plots to seize the Bank of England and the Tower, incite the army to mutiny and launch a Jacobin revolution. Habeas Corpus was suspended and, by the standards of the time, though hardly according to modern ideas, a reign of terror imposed. Even the recently formed Cambridge Union Society was suppressed. One magistrate declined to sanction the formation of a local mineralogical society on the grounds that mineralogy led to atheism. William Cobbett, the radical journalist, was driven out of the country by the threat of prosecution for £20,000 of unpaid stamp duty on his enormously successful *Register*. But fundamentally Britain was in a very strong position. It had emerged from the French wars far richer and stronger than it had been in 1793 and national wealth was increasing rapidly. Indeed Rennie's magnificent bridges seemed the embodiment of solidity and strength.

What was Babbage's part in all these excitements? He early gained a reputation as a liberal or radical, though always a moderate radical, to use terms then coming into use. No doubt he took part in elections, both in London and Cambridge, and he was from an early date associated with the more respectable liberal circles in London. But there is no reason to suppose that he took much direct part in political activities at the time. On the other hand during this period scientific work came to be seen as a form of political action. Science would dispel ignorance and superstition and was essential for the rational organization of society. This attitude is closely associated with the name of Jeremy Bentham. The Benthamites, or Utilitarians, were themselves influenced by French ideas, but their approach was characteristically English. And even if Babbage took many of his ideas directly from France he too belonged in the English tradition. There was a deep-seated feeling that England was a free country, the cradle of democracy. That everything in the social organization needed changing and improving was common ground among the liberals. But there was also the confidence that England could take the lead in developing a rational democratic society on a scientific basis.

One does not have to look far to find relations between Babbage's immediate circle and the leading Utilitarians; and the influence of the Benthamites was predominant. John Herschel and other of Babbage's friends were reading for the bar, while Bentham and his allies were seeking to reform the legal system, placing it on a rational basis. Babbage was greatly interested in classical economics, as appears quite clearly in his later book, *On the Economy of Machinery and Manufactures*. One personal link was Henry Bickersteth, later Lord Langdale. He became a friend of Babbage soon after the latter had come to live in London. Bickersteth married rather late in life Lady Jane Elizabeth Harley and their only daughter, Jane Frances, who became Countess Teleki, was a close friend of Babbage in his last years. Bickersteth was in turn a friend of both Sir Francis Burdett and Jeremy Bentham. These radical associations held up Bickersteth's progress during his early years at the bar but he later became a famous Master of the Rolls. Later Babbage was a close friend of the Grotes. The liberal intellectuals in London formed a fairly small circle.

During these five years Babbage continued to seek paid employment to supplement his limited income. Early in 1816 the Professorship of Mathematics at Haileybury, the East India College, became vacant. Babbage applied and was rejected. He had called on one of the directors who was quite frank : 'If you have interest you will get it; if not you will not succeed,'[14] In the summer of 1819, after Playfair had died and Professor Leslie succeeded him, Leslie's own chair of mathematics in Edinburgh became vacant. Babbage applied, backed by a glittering list of recommendations. Bromhead's father warned that 'if you fail it will be because you are not a Scotsman'.[15] Babbage did indeed fail to obtain the post. However he met Dugald Stewart and had a pleasant tour of the country, returning from Edinburgh through Glasgow, Liverpool, Shrewsbury, Worcester, and Bristol, finally joining his family in Torquay.

In this period Babbage was unable to secure a paid post, and this was when he badly needed one. The reason is clear: he was a liberal at a time when the assignment of posts was in the hands of Tory patronage. Fifty years later Macaulay's nephew, George Otto Trevelyan, wrote:

... for the space of more than a generation from 1790 onwards, our country had, with a short interval, been governed on declared reactionary principles. We in whose days Whigs and Tories have often exchanged office, and still more often interchanged policies, find it difficult to imagine what must have been the condition of the kingdom, when one and the same party almost continuously held not only place but power, throughout a period when, to an unexampled degree, 'public life was exasperated by hatred, and the charities of private life soured by political aversion' [from Lord Cockburn's *Memoirs of his Time*]. Fear, religion, ambition, and self-interest,—everything that could tempt and everything that could deter,—were enlisted on the side of the dominant opinions. To profess Liberal views was to be excluded from all posts of emolument.[16]

Fortunately the Babbages had a sufficient private income provided they lived carefully. Georgiana's father died in the late summer of 1816 making more funds available. And a slow decline in prices made what money they had go further. Children were born: at least eight in all between 1815 and 1827, of whom three survived to maturity. Benjamin Herschel was first on 6 August 1815, named for propriety Benjamin after Babbage's forebears but always called Herschel. Next came a second son, who was named Charles, followed by Georgiana, the only daughter. Two sons died in infancy. Then came Dugald Bromhead and Henry Prevost.

[14] *Passages*, 473.   [15] BL Add. Ms. 37,182, f 125, Bromhead to Babbage, 21 July 1819.
[16] G. O. Trevelyan, *Life and Letters of Lord Macaulay*, 144.

In the autumn of 1815 Benjamin Babbage sold the Rowdens to Sir James Nugent and moved to Totnes. In 1823 he was to purchase the lease of another house in Teignmouth and live there for his last few years.[17] On 28 December 1815 Babbage visited his parents in Totnes, and again on 27 April 1816 when his father was very ill. That summer Herschel, who was staying with Sir W. Walson Dent in Dawlish, visited Babbage and Georgiana in Torquay.[18] In 1817 Babbage and Adam Sedgwick, the Cambridge geologist, planned a geological tour together through Devon and Cornwall but university duties prevented Sedgwick from going.[19] However in 1820 Babbage and John Herschel were to make a similar trip.

Georgiana shared Charles' social life when her growing family gave her time. For example, Babbage wrote to Herschel: 'We have just returned from Lincolnshire where we spent a short time with Bromhead after termination of the Cambridge gayeties. Our wandering steps are now directed Westwards and if it shall be perfectly convenient to Sir W[illiam] and Lady Herschel and yourself we will accept your hospitality for a day or two ... We propose leaving town the 4th August ... We spend a week near Bath and other visiting places before we reach Torquay.'[20] In the event the trip West was interrupted by Babbage's visit to Edinburgh. Two or three months in the summer were usually spent in Devonshire. Georgiana's brother, Wolryche Whitmore in Dudmaston, and E. Isaac at Boughton, married to another of the Whitmore girls, were often visited. Babbage's old friend Edward Ryan, now his brother-in-law, was living in Gower Street in North London.

It was probably in 1819 that Babbage made his first visit to Paris with John Herschel. This was the occasion of the fifty-two eggs. Babbage recorded:

[17] East Teignmouth parish records, DCRO.
[18] A letter of 2 September 1816 from John Herschel to John William Whittaker describes the visit:

'I went the other day to Torquay and spent a day or two with Babbage who is well and occupied with his functions. I like his enthusiasm on that subject and as we rambled through the noble and romantic scenes which that neighbourhood affords—climbing the rocks with boyish eagerness, and talking over analytics all the while, my brain became once more warmed with the speculations which used to give me such delight, and I swore to return again to those sources of enjoyment I had allowed myself too long to lose sight of.

By the way Babbage I think will soon receive an accession to his earthly blessings—if appearances can be trusted—young B. is a hearty, ugly, sprawling brat & his father seems to think him a pattern baby.

I will send you the Appendix to the Lacroix as soon as you give me notice they are near the end of my Ms.'

Whittaker, conveniently on the spot in Cambridge, was seeing the translation of Lacroix through the press. When Whittaker was away from Cambridge Wilkinson, also at St John's, took over the tiresome task of superintending the printing. Mss. St John's College, Cambridge.
[19] BL Add. Ms. 37,182, f 93.
[20] Babbage/Herschel correspondence, RS, f 117.

On reaching Abbeville, we wanted breakfast, and I undertook to order it. Each of us usually required a couple of eggs. I preferred having mine moderately boiled, but my friend required his to be boiled quite hard. Having explained this matter to the waiter, I concluded by instructing him that each of us required two eggs thus cooked, concluding my order with the words, 'pour chacun deux.'

The garcon ran along the passage half way towards the kitchen, and then called out in loudest tone— 'Il faut faire bouillir cinquante-deux oefs pour Messieurs les Anglais.' I burst into such a fit of uncontrollable laughter at this absurd misunderstanding of chacun deux, for cinquante-deux, that it was some time before I could explain it to Herschel, and but for his running into the kitchen to countermand it, the half hundred of eggs would have assuredly been simmering over the fire. A few days after our arrival in Paris, we dined with Laplace, where we met a large party, most of whom were members of the Institut. The story had already arrived at Paris, having rapidly passed through several editions.

To my great amusement, one of the party told the company that, a few days before, two young Englishmen being at Abbeville, had ordered fifty-two eggs to be boiled for their breakfast, and that they ate up every one of them, as well as a large pie which was put before them.

My next neighbour at dinner asked me if I thought it probable. I replied, that there was no absurdity a young Englishman would not occasionally commit.[21]

On this trip Babbage and Herschel met many French scientific men. Among those who became particular friends of Babbage's were Arago, Biot (whose son, an orientalist, was later to translate *On the Economy of Machinery and Manufactures* into French), and Fourier. Laplace lived at Arcueil, a village a few miles from Paris, where Berthollet also had a house. Many others had made the pilgrimage to Arcueil before Babbage and Herschel, including for example the Scots physicist Sir John Leslie in July 1814 and Mary Somerville in 1817. Most of Babbage's closest friends among the French men of science were members of the Arcueil Society[22] or closely associated with it; so also was his Prussian friend, Alexander von Humboldt. If French science and scientific institutions served as models for the developing science of Europe, the men of Arcueil formed the most important compact scientific group of the time.

The two senior figures were Laplace and Berthollet. Laplace was the great scientific figure of the group. His *Méchanique Céleste*, in five volumes which appeared between 1799 and 1825, was an attempt to apply Newtonian theory to the Universe. He was a protégé of the encyclopaediste d'Alembert. By a fortunate chance Laplace was Napoleon's examiner in his final examination at the *École Militaire* in 1785. The association was later renewed at the first class of the Institute. After his six weeks as Minister of the Interior, Napoleon wrote: 'Mathematician of the highest rank, Laplace was not long in showing himself an extremely poor administrator. From his first actions I realized that I had

---

[21] *Passages*, 195–6.
[22] Maurice Crosland, *The Society of Arcueil*, Heinemann, 1976. Babbage knew Mary Somerville well after he moved to London. Both were friends of Wollaston.

deceived myself. He sought everywhere for subtleties, had only problematic ideas, and carried the spirit of the "infinitely small" into administration.'[23] Nevertheless experience at that level of government was inconceivable for men of science in England at the time. And Laplace's wife had become lady-in-waiting to Napoleon's sister Elisa. As the leading theoretician it was natural that Laplace should be a friend of Sir William Herschel, the leading practical astronomer. When visiting England Laplace went to see Sir William in Slough as well as going to Babbage's house in Devonshire Street.

Berthollet was a chemist, deeply involved in the industrial application of chemistry. He introduced the use of chlorine for bleaching cloth, which made the industrial process far more rapid. In 1794 he was one of the instructors of a group of men brought to Paris to study the extraction of saltpetre and the manufacture of gunpowder, urgently needed for the war effort. During the Terror he analysed a sample of brandy, supposedly poisoned for use in a plot. Summoned before the Committee of Public Safety for giving an unwanted answer and declaring the brandy wholesome, he defied Robespierre by drinking a glassful. Berthollet and Monge, who shared Napoleon's experiences in Egypt, became closely attached to him. In 1807 Napoleon gave Berthollet 150,000 Francs to clear his personal debts, support crucial to the Arcueil Society. The whole group was equally interested in scientific theory and industrial application. Although the two senior members were loyal to Napoleon the younger men were mainly good republicans. Arago and Poisson were among the students of the École Polytechnique who refused to send an address of congratulation to Napoleon when he proclaimed himself emperor. Thus the group's predominant political opinions were not unsympathetic to Babbage's own.

Dominique Francois Jean Arago was both man of science and a political figure. In 1830 he was elected permanent secretary of the Académie des Sciences. In the same year he was elected deputy for the Eastern Pyrenees, joining the extreme left wing opposition to Louis Philippe. In 1848 as Minister of War and the Navy in the provisional government he abolished flogging and secured the end of slavery in the French colonies. Babbage remained in touch with Arago for the rest of his life. Another close friend was Jean-Baptiste Biot who carried out a wide range of scientific work, publishing hundreds of papers. Babbage saw Biot many times and recalled the last occasion on which he did so: 'The last time during M. Biot's life that I visited Paris I went, as usual, to the Collège de France. I inquired of the servant who opened the door after the state of M. Biot's health, which was admitted to be feeble. I then asked whether he was well enough to see an old friend. Biot himself had heard the latter part

---

[23] *Oevres de Napoleon Ier à Saint Hélène*, 'Consuls Provisoire', para III, Correspondance, 392. Quoted by Crosland, op. cit.

of this conversation. Coming into the passage he seized my hand and said "My dear friend, I would see you even if I were dying."'[24]

The Prussians set up an institute to train men for the engineering corps in imitation of the École Polytechnique. Both bodies also trained civilians. Later similar institutes were established widely in Germany, gradually dividing into technical universities and polytechnics for educating highly skilled craftsmen. These institutions were of crucial importance in the rapid development of Germany's technically based industries. In England, where state intervention was widely deplored by the establishment, no comparable development took place, leaving weaknesses in technical education which are still far from being eliminated. Justus von Liebig, prince of organic chemists, chose Paris for his education and studied with members of the Arcueil Group. Later Liebig, established the greatest school of chemistry in the nineteenth century. It became one of Babbage's cherished aims to introduce French concepts of scientific organization into England.

The Frenchman whose work had the greatest direct effect on Babbage's own work was Gaspard François de Prony, director of the École des Ponts et Chausées and France's leading civil engineer. He had survived the revolutionary period thanks to the protection of Lazare Carnot and on the establishment of the École Polytechnique became professor of analysis. De Prony was associated with Laplace and Berthollet in planning the metric system but he declined to accompany Napoleon on his Egyptian venture, his eye fixed on the directorship of the École des Ponts et Chaussées where he had himself been trained. The previous director, Antoine de Chezy, died while Napoleon was in Egypt and de Prony got the post. Although he was made a member of the Legion of Honour when it was formed, none of the plum appointments, such as Senator, handed out by Napoleon came to him.

It was probably on this trip that Babbage was first able to see the great numerical tables of de Prony which were to have such an influence on Babbage's work. De Prony had been commissioned during the Republic to prepare a set of logarithmic and trigonometric tables to celebrate the new metric system. These tables were to leave nothing to be desired in precision and were to form the most monumental work of calculation ever carried out or even conceived. In particular the logarithms of the numbers from 1 to 200,000 were to be calculated. With customary methods of calculation the work would have been impossible: there simply were not enough computers—that is to say people carrying out calculations—to complete the work in a lifetime. Chancing on a copy of the de luxe edition of Adam Smith's *Wealth of Nations* in a second-hand bookstall, de Prony opened it at random and discovered the chapter on the division of labour. This was the crucial idea, and on this basis he drew up his plans.

[24] *Passages*, 198.

The calculators were divided into three sections. The task of the first section was to plan the overall strategy of the project, investigating the various analytical expressions for each function to be tabulated and selecting those best suited to numerical calculation by simple steps. In this section were half a dozen of the best mathematicians in France. This élite had little or nothing to do with the actual numerical work. Its labour completed, the formulae which it had chosen were delivered to the second section.

The second section consisted of seven or eight competent mathematicians. It was their job to convert the formulae into sets of actual numbers: an exceedingly laborious task. This group then distributed the sets of numbers among the members of the third group, receiving from them in due course the completed calculations.

The third section was formed of sixty to eighty people, nine-tenths of whom knew no mathematics beyond simple addition and subtraction, the only two operations they were required to carry out. It is interesting that these people were found more often correct in their calculations than those with a more extensive knowledge of mathematics. Anyone with experience of trying to add up modern milk sheets will agree with this finding.

The tables thus calculated occupied seventeen large folio volumes. Babbage was enormously impressed. Later he was to base the logical structure of his Difference Engine on the methods of organization used by de Prony to calculate his tables.

Although Babbage had joined the Royal Society and its central importance was not in question, the reformers felt the need for other scientific societies to complement the Royal. As early as 1815 Babbage was thinking of forming a 'First Class of the Institute'[25] on the French model. Bromhead suggested asking eminent men to permit their names to be inscribed as members of the First Class of the Institute whenever it should be formed. Evidently it had been discussed among the Analyticals, as Bromhead wrote: 'You are our active centre of force in this business, do not relax or we are undone. Do not quarrel with the R.S. or use any intemperate language which may reach them. Try to become one of them and be all things to all men.'[26] At the time Babbage was also thinking of starting a new scientific journal as a vehicle for publishing their work.

In 1819 Babbage's friends in Cambridge formed the Cambridge Philosophical Society and Peacock inscribed Babbage's name as a member. Sir Joseph Banks, the formidable President of the Royal Society, objected, as he did to all new scientific societies in England, seeing them as threats to the Royal. Bromhead wrote to Babbage: 'Woodhouse has left the Cam. Phil. Soc. in the lurch, I suppose from fear of Sir Joseph.'[27] A year later he wrote that the Analytical

[25] BL Add. Ms. 37,182, f 38.   [26] Ibid., f 52.   [27] Ibid., f 201.

Society would always be called in history the parent of the Cambridge Philosophical Society, just as the meetings in Oxford were allowed to have been the origin of the Royal Society.[28]

The first national essay in scientific organization in which Babbage was concerned was the formation of the Astronomical Society. John Herschel's diary for 1820 gives a glimpse of the convivial atmosphere in which it began:

Sat. Jan. 1. Dined with Peacock and Babbage at Provost Goodall's at Eton, and met Col Thackeray, Vice Provost Roberst, Capt. Roberts R.N. etc.

Sun. Jan 2. Peacock and Babbage left Slough after spending a few days here.

Wed. Jan 12. Dine at the Freemason's Tavern to meet Dr Pearson and other gentlemen to consider of forming an astronomical society.

Sat. Jan 15. To attend Committee of ye Astronomical at Geol Soc. Rooms at 10 o'clock. Evening. Mr Lowry's to tea to meet Wollaston, Babbage, Gompertz, Perkins and Col Fairman.

Tues Jan 18. Spent morning at Dr Pearson's. Babbage came about 1. Read over and arranged address for circulation with the notices of formation of ye Astronomical Soc. Dined and returned with Dr P and Babbage to the meeting of the Ctee in the Evening.[29]

The dinner of 12 January at the Freemason's tavern, Great Queen Street, Lincoln's Inn Fields became legendary. It was probably after this dinner that it was resolved to form the Astronomical Society. Present were: Charles Babbage, Arthur Baily, Francis Baily, Captain Thomas Colby, Henry Thomas Colebrooke, Olinthus G. Gregory, Stephen Groombridge, John Herschel, Patrick Kelly, David Moore, Revd. William Pearson, James South and Peter Slawinsky. Babbage was elected one of the first Secretaries of the Society, a post which he held from 1820 to 1824. He carried out the duties conscientiously, attending regularly at the council meetings. John Herschel became the first foreign secretary. However the presidency of the Society presented a problem. The first person approached, the Earl of Macclesfield, declined the position.[30] The Duke of Somerset, an amateur mathematican and later a close friend of Babbage, was approached and accepted. But Somerset soon yielded to the influence of Sir Joseph Banks and changed his mind.[31] Sir Joseph had earlier made violent attempts to stop the foundation of the Geological Society, seeing all such bodies, concerned with specific scientific disciplines, as tending to weaken the Royal Society of London itself. Finally old Sir William Herschel was persuaded to accept and, on the promise that the position would be purely

[28] 7 March 1821, Ibid., f 323.

[29] *History of the Royal Astronomical Society, 1820–1920*, Ch. I, Royal Astronomical Society, 1923.

[30] Francis Baily to Babbage Gray's Inn, 12 February 1820. BL, Add. Ms. 37,182, f 221.

[31] Francis Baily wrote to Babbage: 'Sir Joseph tells the Duke of Somerset that our society will be *the ruin* of the Royal Society: no mean compliment to us, but not very respectful to that learned body.' BL, Add. Ms. 37,182, f 237.

titular, was elected first president of the Astronomical Society. Sir William's enormous scientific prestige would seem to have made him the ideal choice for the post from the start, and one notes with surprise the preference, even among that group of staunch reformers, for an aristocratic figurehead.

Shortly after the formation of the Astronomical Sir Joseph Banks died and the presidency of the Royal Society became vacant. There were two candidates, Wollaston and Humphry Davy. Babbage, Herschel, and some friends campaigned for Wollaston. Edward Ffrench Bromhead wrote to Babbage in June 1820 urging caution and giving sound advice: 'I am sure Wollaston will not succeed, and if we cannot carry a friend of our own, our policy is to turn the scale, and at least avoid making an enemy of the future president. If we cannot succeed now, we should not shut ourselves out from the vacancy *after* the present—The Royal Society wants revolutionizing.'[32] Babbage and Herschel were in almost daily contact so that we have no letters recording their discussions on the subject. It is from Bromhead, living in the country, that we get the clearest statements showing how the old Analyticals regarded themselves as a militant reforming group. Humphry Davy did indeed become the new President, and the decisive battle was deferred for a decade.

[32] Ibid., f 270.

4

○ ● ○

Science in Action: Start on the First Engine

The 1820s opened with Babbage and Georgiana still living tranquilly in Marylebone, south of recently enclosed Regent's Park. Babbage had found neither employment nor regular means of making money and his father missed no oportunity of reminding him of this failure. But Babbage was supremely confident in his abilities, and family life was both lively and contented. Georgiana was filling his home with children. She made an excellent wife and was little interested in lavish entertaining. Everything speaks well of her. The surviving correspondence gives glimpses of continuing modest entertainment, the occasional grand ball and frequent visits to friends and relatives in the country. These were Babbage's happiest years. They had one or two servants, lived very comfortably, travelled when they wished, and sufficient remained to help finance Babbage's researches. The children grew up surrounded by their father's experiments. We find a parcel of type from the Difference Engine being used as a toy at one time; and a model of Babbage's pretty if Heath-Robinsonish device for carrying letters in little carriages on wires suspended from church steeples and high posts was set up between the drawing room and workshop.

Commonly mathematics is a young man's subject and after settling in London Babbage was looking round for other fields of activity. He was reading widely, particularly in political economy, and was abreast of all the latest work in the physical sciences. After considering several possibilities he finally embarked on the construction of a calculating engine: the first Difference Engine. Government finance was secured on what was for the time a large scale and the project became well-known to men of science and in political circles. Babbage's social contacts widened and he began to dine with the great.

Before he started to work in the field, mechanical calculators had been very simple hand-operated devices. Apart from some work in antiquity by Hero of Alexandria, the first digital calculator was a simple wooden device made by Wilhelm Schickard of Tübingen, which he called a calculating clock. It was burnt and is known only through correspondence between Schickard and Kepler.[1] The first calculators made for sale were constructed in the 1640s by the young French philosopher Blaise Pascal and after that many simple digital

[1] For detailed references to early calculators see: Brian Randell, *The Origins of Digital Computers, Selected Papers*, Springer, 1973.

calculators were designed. An intractable problem was always the 'carry' system, for carrying numbers from the units to the tens, the tens to the hundreds, and so on. In the 1660s Samuel Morland devised a neat little calculator[2] with an extra set of wheels on which the carries could be stored or hoarded for subsequent manual transfer by the user. Some other machines used forms of automatic mechanical carry. Later Babbage was to use both hoarded and several forms of mechanical carry system as well as combinations of the two. All the mechanical calculators before Babbage's Difference Engine required continual intervention by the user, making calculation not only slow but liable to human error.

During his life Babbage was to prepare many detailed plans for calculating engines. These were of two main classes: the Difference Engines, so called because they were designed to compute tables of numbers according to the method of finite differences, and then automatically to print the tables as they were computed; and the Analytical Engines, which were versatile, programmable automatic calculators. The first Difference Engine was Babbage's great practical engineering project, vastly more complex than any previously conceived mechanical calculator. The advances in machine tools and machining techniques developed in making the first Difference Engine were to have a far reaching effect on the subsequent development of precision engineering. But compared with the Analytical Engines, the Difference Engine was a straightforward logical system. It is the Analytical Engines which are of greatest interest to the history of science as complex logical systems, precursors of the modern digital computer. An Analytical Engine, like a computer, is an abstract system. But in practical computers the abstract system is designed to be capable of being realized in physical form, and the abstract structure is itself greatly affected by the constructional technology in which it is designed to be made. Owing to their peculiar nature Babbage's work on the calculating engines can be followed at several different levels: the abstract, logical structure; the devices designed to effect each part of the engines; manufacturing methods for constructing components for the engines; mathematical methods for preparing problems for calculation; and both programming and microprogramming. Further, as Babbage studied political economy, statistics, operations research, and innumerable commercial and industrial problems, he was constantly thinking of new applications for his beloved engines, and was beginning to open up perspectives which have come more fully into view with the developments of modern computing. Babbage was one of the great polymaths and he pursued his ideas vigorously at the several different levels, sometimes almost independently of each other, but where appropriate bringing different levels of consideration together.

His first Difference Engine was intended for making tables although it could

[2] Morland's calculator with a hoarded carry can be seen in The Science Museum, London.

6. Florence by J. M. W. Turner, c. 1827.

8. Charles Babbage aged 40.

7. Georgiana Babbage, from a miniature.

9. Bridgnorth Bridge by J. M. W. Turner.

10. First Difference Engine, completed part, 1832.

also be used for other purposes, such as solving equations. Babbage envisaged its use in calculating mathematical tables of many classes but he rested his public case primarily on the production of tables for navigation. At that time navigational tables were full of errors which continually led to ships being wrecked. It would only be necessary to save a few vessels from destruction to justify the cost of the project. For Britain, the leading seafaring country, the case for constructing a Difference Engine was very strong.

Joseph Henry of the Smithsonian later spelt out the importance of tables which a difference engine could have produced. After listing squares and cubes, square-roots and cube-roots, he continued:

This class of tables involves only the arithmetical dependence of abstract numbers upon each other. To express peculiar modes of quantity—such as angular, linear, superficial, and solid magnitudes—a larger number of computations are required. Volumes without number of these tables also have been computed and published at infinite labour and expense. Then come tables of a special nature, of importance not inferior, of labor more exacting—tables of interest, discount, and exchange; tables of annuities and life insurance, and tables of rates in general commerce. And then, above all others, tables of astronomy, the multiplicity and complexity of which it is impossible to describe, and the importance of which, in the kindred art of navigation, it would be difficult to over-estimate. The safety of the tens of thousands of ships upon the ocean, the accuracy of coast surveys, the exact position of light-houses, the track of every shore from headland to headland, the latitude and longitude of mid-sea islands, the course and motion of currents, direction and speed of winds, bearing and distance of mountains, and, in short, everything which constitutes the chief element of international commerce in modern times, depends upon the fullness and accuracy of tables.[3]

Henry then expatiated on the huge numbers of errors in virtually all the sets of tables then in use.

Babbage gave several descriptions of the origin of his idea of constructing a Difference Engine:

The first idea which I remember of the possibility of calculating tables by machinery occurred either in the year 1820 or 1821: it arose out of the following circumstances. The Astronomical Society had appointed a committee consisting of Sir J. Herschel and myself to prepare certain tables; we had decided on the proper formulae and had put them in the hands of two computers for the purpose of calculation. We met one evening for the purpose of comparing the calculated results, and finding many discordancies, I expressed to my friend the wish, that we could calculate by steam, to which he assented as to a thing within the bounds of possibility.[4]

That is how Babbage in 1834 recalled the genesis of the idea. Years later as he relates in his autobiography, a friend reminded him of an earlier notion: 'One evening I was sitting in the rooms of the Analytical Society, at Cambridge, my head leaning forward on the table in a kind of dreamy mood, with a table of

[3] Report of the Smithsonian, 1873, 166.
[4] C.B., *History of the Invention of the Calculating Engines*, 10, Buxton papers.

logarithms lying open before me. Another member, coming into the room, and seeing me half asleep called out "Well, Babbage, what are you dreaming about?" to which I replied, "I am thinking that all these tables (pointing to the logarithms) might be calculated by machinery."[5] There is no real contradiction between these stories as an idea may occur fleetingly in one's mind on several occasions before being taken up in earnest. In any case making tables for the Astronomical Society led Babbage to start designing calculating engines.

The making of numerical tables, whether of logarithms, trigonometric, or other mathematical functions, seemed to Babbage just the sort of operation to which machinery could be successfully applied: machinery, he held, is always useful when great accuracy is required and particularly where the same process is repeated in almost endless succession. 'I wish to God these calculations had been accomplished by steam.' The idea persisted, and one spare evening Babbage sketched the outlines of a design. In a few hours he satisfied himself that machinery could be made to compute tables by the method of finite differences, the mechanical analogue of De Prony's system, and even to print directly the tables computed. The excitement of the investigation made Babbage ill. Recommended by his doctor to rest and refrain for a while from work on calculating engines he went to stay with the Herschels in Slough.[6] From such incidents we see how completely an intellectual pursuit could absorb Babbage, sometimes leading to incipient breakdown. Babbage seems to have enjoyed good physical health until his last years, the origin of his complaints being almost entirely nervous.

Babbage considered two main methods for storing numbers. One was to store each digit of a number on a straight metal strip or rod. He derived this method from a mechanism used to control the striking of the hours in clocks. The second was storing the digits on toothed wheels as in the early hand calculators of Pascal, Leibniz and others. The latter is more straightforward because when counting passes from 9 to 0 a carry must, as every schoolchild learns, be conveyed to the next place. Counting proceeds 9, 0, 1, 2, . . . while the wheel continues turning in the same direction. A strip on the other hand requires a mechanism for the return to zero. Later Babbage deemed either method practicable although a calculator using metal strips for storage would require more parts than one using toothed wheels. In all his detailed plans numbers were held on columns of toothed wheels and in the early plans each wheel had ten teeth. His own lathe lacked necessary facilities, and he got a workman to make the required toothed wheels. What is extraordinary is not that his lathe should have lacked facilities but that he had a lathe at all and could use it: by no means a common skill for a gentleman at the time. So worried was Babbage that somebody might steal his idea that he had several parts executed by different workmen, assembling the model himself. This is a fear from which

[5] *Passages*, 42.   [6] Babbage biography (fair copy), 2, Buxton papers.

many inventors suffer. Usually they need not worry, for if an invention has any real originality few understand the point or are even remotely interested.

But Babbage's friends included a very remarkable set of men who rapidly grasped at least part of the significance of his plans, thus adding to his illusion that the nation would welcome calculating engines with open arms. Important among his friends was William Hyde Wollaston. Discoverer of palladium, Wollaston was both an outstanding man of science and a successful entrepreneur: a man after Babbage's heart. Wishing to discuss several projects he was considering, Babbage visited Wollaston's house at 1 Dorset Street, the house which Babbage later bought and where the great work on the Analytical Engines was carried out. Wollaston poured cold water on the other plans, but encouraged Babbage to build a calculating engine.[7] Those who really understood the importance of the project were very few, however, and if Babbage could have foreseen the practical and moral difficulties he would later encounter through sheer lack of comprehension it is doubtful whether he would ever have ventured upon constructing his Difference Engine.

By 1822 a model with six figure-wheels worked satisfactorily, tangible proof that the principle was correct. Central to his plan was to have the machine not only calculate but also print the results of calculation: the possibility of careless error had to be eliminated at each stage in preparing tables. Thus the machine was to have two major connected parts: the first for calculation, the second to make stereotype plates of computed results. Babbage still had to decide whether to start the project in earnest. It was a drastic departure from his earlier work and although he was convinced of the fundamental importance of the subject he was tempted to return to pure mathematics and particularly to his beloved theory of functions. He announced the project by publishing his proposals in the form of an open letter dated 3 July 1822 to Sir Humphry Davy Bart., President of the Royal Society. The document made a number of points, referring to 'the powers of several engines which I have contrived'. For Babbage the terms 'engine' and 'machine' were synonymous. He went on to mention 'a machine for multiplying any number of figures (m) by any other number (n), I have made several sketches ... I have also thought of principles by which, if it should be desirable, a table of prime numbers might be made, extending from 0 to ten millions.'

Then came the main proposal for a Difference Engine, and he also advanced a concept later to prove of crucial importance in the transition from Difference Engine to Analytical Engine: 'Another machine, whose plans are much more advanced than several of those just named, is one for constructing tables which have no order of differences constant.' The idea of making a table of prime numbers, of mathematical interest but of no practical interest at all at the time, had been discussed by Babbage in a paper at the Astronomical Society where

[7] *Passages*, 42.

he referred to the formula $x^2 + x + 41$ for generating prime numbers. Davies Gilbert sent a copy of the open letter to Robert Peel who seized on this point. Born into an industrial family but moving amongst the aristocracy he was deliberately distancing himself from all detailed, not to say sordid, questions of technology and industry. Peel's forebears knew their cotton but young Robert had been educated as a gentleman. Confronted with the question of a calculating engine he recoiled fastidiously with a classical quotation and flippant remark: '"Aut haec in nostros fabricata est machina muros. Aut aliquis latet error." I should like a little previous consideration before I move in a thin house of country gentlemen, a large vote for the creation of a wooden man to calculate tables from the formula $x^2 + x + 41$. I fancy Lethbridge's face on being called on to contribute.'[8] The quotation comes from the second book of the *Aenaeid* and the 'wooden man' refers of course to the Trojan horse. It may be translated: 'It is an engine designed against our walls or some other mischief hides in it.' Babbage was soon, with reason, deeply suspicious of Peel's attitude to science. When the question of Babbage's engine later confronted the Duke of Wellington, a superb military engineer, he grasped the essentials immediately.

However Babbage had shown his six figure-wheel model to a large number of people, demonstrating its operation. The open letter had circulated widely and a copy reached the Lords of the Treasury who on 1 April 1823 referred it back to the Royal Society, requesting 'The opinion of the Royal Society on the merits and utility of this invention.' Wasting no time, on 1 May the Royal Society reported to the Treasury: 'Mr. Babbage has displayed great talent and ingenuity in the construction of his Machine for Computation, which the Committee think fully adequate to the attainment of the objects proposed by the inventor; and they consider Mr. Babbage as highly deserving of public encouragement in the prosecution of his arduous undertaking.'[9]

In June Babbage had an interview with the Chancellor of the Exchequer (Mr Robinson, later Lord Goderich, and then the Earl of Ripon). Babbage was ecstatic. He wrote to John Herschel:[10]

Dear Herschel,
I had some conversations this morning with the Chancellor of the Ex[r] who treated me in a most liberal and gentlemanlike manner. He seems quite convinced of the utility of the machine and that it ought to be encouraged. At present he is to procure for me £1000 and next session, if I want more to complete it he is willing that more should be granted or that I should have a committee of the house if a larger sum were wanted than that fund could be charged with.
He acquired the knowledge of the Astronomical medal two days since from Gilbert so that you got it for me just at the fortunate moment—Mr Brougham had given him

[8] Peel to Croker, 8 March 1823: *Correspondence and Diaries of John Wilson Croker*, i, 262–3, John Murray, London 1884.
[9] *Passages*, 29.   [10] Babbage/Herschel corr. f 184, RS.

very just views on the subject of the machine and the manner in which he spoke of it was more gratifying than the grant itself as I have liberal people to deal with. I shall not be annoyed about pence and the particular mode in which I may think it right to distribute them and I shall, I hope, be able to bring the thing to perfection or at least to a good practicable working state and that in a few years we shall have new (but not patent) stereotyped logarithmic tables as cheap as potatoes. Mr Ryan expects you on Monday night when I hope to tell you more of my interview with the nation's purse-bearer.

It was quite delightful to see [Davies] Gilbert's joy when I told him of my success. He proposed that we should have a frolic in consequence of the good news which is to consist in an excursion to see the great telescope.

Give our kind regards to your mother and believe me dear Herschel

most sincerely yours,

C BABBAGE

Devonshire Street
27th June 1823

Unfortunately by a serious omission no minute was made of the meeting between Babbage and the Chancellor.[11] The project was unprecedented. Neither administrative machinery nor the law was adequate to running efficiently the primitive industrial capitalism of the time, let alone to aiding the systematic application of science and technology to industry and commerce which Babbage was already beginning to envisage. Everything hung on the chance understanding of individuals. But, apart from a handful of practitioners, science and technology were unknown territory to the English gentleman. The exceptions were outstanding military and naval officers.

The project to design and construct the first Difference Engine was now launched. One room of the house in Devonshire Street had long served as a workshop. A second room also became a workshop while a third was converted into a forge. On Marc Brunel's recommendation Babbage employed an engineer named Clement who worked from home, having just one good lathe in his kitchen at 21 Prospect Place, Southwark when Babbage took him up: now his business thrived. A first-class workman and competent draftsman, Clement had worked for Henry Maudslay, as did so many of the leading engineers of the time. Indeed, nearly the whole of mechanical engineering in Britain was developing from Maudslay's methods and workshop practice. Maudslay's 'Lord Chancellor', an accurate micrometer which he had made with the utmost care, was, so to speak, the highest court of appeal in precision engineering of the time. Under Babbage's technical inspiration, and funded by his project, Clement's workshop became the leading centre for the development of precision machine tools. The school of workmen trained in working on the Difference Engine was to have a major effect on the development of light

[11] *Passages*, 71.

engineering in England. These developments are not easy to describe. They concern such questions as the precision with which parts are made and the forms of gear teeth, and require quantitative discussion. Indeed there is room for a great deal of research on this topic. However, from what we know it is clear the developments were of great importance for industry. The development of high-speed textile machinery, for example, depended on such techniques. Photomicrographs show that a tool ground to the required profile was used to cut the teeth of the number wheels in the Difference Engine, but the teeth do not have the modern form of the involute.

The Difference Engine was designed to calculate tables by the method of finite differences. A set of numbers placed on the machine at the start of a sequence of calculations is subjected to a series of additions. This is a very simple procedure of which examples are now commonly learned during the first year of secondary school: it is a commonplace of school mathematics that a table, say of square numbers, can be calculated by successive additions. Establishing the precise sequence of additions required to calculate a particular table efficiently may require considerable mathematical knowledge, but the basic principle is elementary. Rather than give yet another description of the method, a popular account written in the 1860s by Babbage for his autobiography is printed as an appendix.

Babbage's object was to produce tables accurately, which required more than mechanical reliability as there was the possibility of errors by the operators. From its inception the Machine was designed to print as well as to calculate its results: otherwise errors would certainly be introduced during typesetting. Let us look in a little more detail at this printing which was just one of many problems to be solved in making the Engine.

His first plan was to use moveable type. Ten boxes were each to hold 3,000 pieces of one of the ten digits. The types would be fed one at a time as instructed by the calculating part of the Engine. Obviously an operator might accidently place a number in the wrong box. Babbage removed this source of error by cutting notches in places characteristic of each number. When a box was correctly filled a thin wire could pass through the aligned notches but if one figure was wrong the wire would not get through. A second notch was common to all the digits so that an assembled block of type could be tied together. A trial box worked satisfactorily and a tied block of type survived the rigorous test of being used as a toy for several years by Babbage's children.

A second method of printing used ordinary type fixed to the rim of thin wheels. On instruction from the calculating part, each of a coaxial set of wheels rotated to the required digit. As succeeding numbers were composed the row of type-wheels would descend, impressing the tabular numbers into a plate of soft plaster composition which set hard in a few hours. The plaster mould could then be used to cast type. The plan was simple and effective. Similar print wheels have been used as output devices in modern computers, but usually with an ink ribbon in direct impact printing.

In designing the Difference Engine Babbage faced many subtle and difficult mechanical problems. They were not only questions of detail, problems which being clearly stated merely required solution. Rather it was a matter of approaching creatively entire classes of problem. Starting with an investigation of printing machinery he proceeded to an exhaustive study of mechanical devices, and particularly machinery used in manufacture, in all the mechanical industries in the country. A creative engineer of Babbage's calibre could readily see many alternative uses for a device originally designed for some specific purpose. Initially his principal motivation had been the study of mechanical devices in themselves. But Babbage was a banker's son and his interest soon extended beyond mechanical devices and even production machinery to more general commercial and industrial problems. In 1832 he was to publish some part of these studies in his epoch-making book *On the Economy of Machinery and Manufactures.*

Babbage's visits to industry became celebrated: charming, courteous, very knowledgeable, and remarkably patient, he was an excellent and appreciative observer and guide. In 1823 he took Georgiana on a tour through England and Scotland, studying industry and staying in Edinburgh with Dugald Stewart. When he began working on the Difference Engine Babbage was beginning to move in the higher reaches of society. In 1823 he was elected to the Royal Society Dining Club.[12] After the Decline of Science controversy he was later to resign. He became an intimate of the rather intellectual circle around the Duke of Somerset. On some of his industrial visits Babbage was accompanied by the Duke's elder son, later M.P. for Totnes. The Duke and Duchess were good friends of Babbage's and we note again the Devon connections. Edward Augustus Seymour, the eleventh Duke, belonged to the Seymours of Berry Pomeroy, a family which acquired the ruined castle of Totnes and rectorial rights. In 1829 the eleventh Duke also acquired the Templer estate at Stover with the Stover canal, Haytor granite quarries, and railway. The Duke sought Babbage's advice on several technical problems arising on his properties and we may be sure that the subsequent developments in Totnes did not take place without many suggestions from Babbage. A new stone bridge across the Dart, designed by Marc Brunel at the Duchess's request, enabled the Duke to lay out several new streets on the other side of the river. A hundred houses were built, mostly small neat buildings designed for the families of working men.

In the early 1820s great engineering works were still comparatively rare. The occasional steam engine remained an object of wonder. Although steam-power was beginning to be used in ships, not until the spread of the railways, starting in the 1830s, did the new engineering become a commonplace. Indeed

[12] Sir Archibald Geike, *Annals of the Royal Society Dining Club*, Macmillan, 1917.

apart from clocks, watches and simple locks and fastenings, mechanism was still a rarity in everyday life. When great works of civil engineering got under way they caught the public imagination in a manner somewhat analogous to space travel in the present age. One device which aroused interest was the diving bell. It had been invented by John Smeaton, famous for building the Eddystone lighthouse. He had used a clumsy wooden apparatus in 1770 when working on Ramsgate harbour and John Rennie had developed it into a solid, well engineered vehicle in 1813 while he was carrying out extensive repairs to the East pier head, also at Ramsgate. Later Rennie used his diving bell when working on the Plymouth breakwater. Visiting Plymouth in 1818 Babbage and two friends descended to a depth of about twenty feet.[13] Babbage noted the more obvious phenomena: changing pressure on the ear drums, and how to relieve it; the light which filtered through the water, sufficient for delicate operations. He measured the divers' temperatures, both before and after descent; also the direction of a compass needle. Signals were communicated to the people above by heavy hammer blows on the side of the hull. In 1826 Babbage wrote an article on the diving bell in the *Encyclopaedia Metropolitana* and described a submarine with an open bottom he had invented. Compressed air would permit a vessel to remain under water for several hours. By having an open vessel a diver could attach explosives to ships at anchor and destroy even an ironclad ship, which then seemed the ultimate in strength. Of all the engineering works of the time the Thames tunnel attracted most attention, first as an heroic undertaking and then during its long cessation as a joke. The Difference Engine was by no means the only project which was to run into difficulty and exceed its estimated cost. But where other projects could point to immediate public utility, it required considerable knowledge and intellectual effort to comprehend the significance of the Difference Engine. Small though it was when compared with the great steam engines and civil engineering works, Babbage's Engine was technically in many ways the most remarkable piece of machinery of them all. The project to construct the Thames tunnel forms an instructive counterpoint to Babbage's work on the Difference Engine.

Brunel's diary for 5 March 1824 contains an interesting entry: 'Waited on the Duke of Wellington by appointment, the object of which was to have the plan of the mode of proceeding explained to him. His Grace made many very good observations and raised great objections; but after having explained my Plan and the expedients I had in reserve, His Grace appeared to be satisfied and to be disposed to subscribe'.[14] And subscribe he did. Moreover his support continued when he became Prime Minister.

Political history has stigmatized Wellington as a reactionary, but from the point of view of the history of British technology the replacement of the

---

[13] *Passages*, 208–11.   [14] Paul Clements, *Marc Isambard Brunel*, 96, Longman, 1970.

competent, progressive and far-sighted Wellington as Tory leader was a serious setback, which was to have a profound effect on the fortunes of Babbage's engines.

The contract to supply the tunnelling shield was won by Henry Maudslay. What a tiny group it was! Maudslay's workshops owed their early growth to constructing the block-making machinery for the Portsmouth naval dockyards to Brunel's designs. The Portsmouth machinery was in turn the important precursor to Babbage's engine, in using machine tools to make equipment with interchangeable parts, though the precision of Babbage's work was incomparably greater. The main succession of workshops in developing lathes, precision machine tools, and machining techniques runs through Maudslay, Babbage and Clement, and then Joseph Whitworth, who worked on Babbage's engine. If one considers the development of precision engineering in Britain in the middle of the nineteenth century the question is usually: from which branch of the Maudslay tree did it come?

The tunnel was started at the beginning of 1825 but there were many difficulties. Brunel did not know that gravel dredging had made deep holes in the bed of the Thames. Brunel's son, Isambard Kingdom Brunel supervised day to day work with boundless energy and courage, gaining invaluable experience as an engineer. The river broke in and a diving bell was borrowed from the West India Dock Co. to inspect the damage. The holes were filled but there were further inundations. Several men were killed and funds ran out. Wellington offered assistance with all his authority as prime minister. He addressed a meeting at the Freemason's Tavern saying, 'there is no work upon which the public interest of foreign visitors has been more excited than it has been on this tunnel'. The Duke of Somerset and other grandees were present; Wellington was splendid; but only £9,660 of further funds were subscribed. The great tunnel was bricked up and the Brunels took a holiday. Years later at Wellington's insistence Parliament sanctioned a loan of £270,000 from public funds. A new shield was made, this time by the Rennies, and in 1836 work was started again. On 12 January 1842 Brunel wrote to Babbage with satisfaction:

The shield is now out of service, in an attitude, however, to demonstrate that it has been the chief agent in accomplishing of a subaqueous structure, twelve hundred feet in length, after numberless difficulties of the most formidable character.

Any day you may please to come with a friend, *let me know* that I may be in the way.[15]

On Saturday, 25 March 1843 the tunnel was ceremonially opened. Today the underground runs through Brunel's tunnel. After work on Babbage's Difference Engine stopped in 1833 there was to be no new subscription of public funds. But then, only very recently has calculating machinery become as glamorous as tunnels.

[15] Ibid., 144.

While making designs for the Difference Engine, Babbage found great difficulty in ascertaining from ordinary drawings—plans and elevations—the state of rest or motion of individual parts as computation proceeded: that is to say in following in detail succeeding stages of the machine's action. This led him to develop a mechanical notation which provided a systematic method for labelling parts of a machine, classifying each part as fixed or moveable; a formal method for indicating the relative motions of the several parts which was easy to follow; and means for relating notations and drawings so that they might illustrate and explain each other. As the calculating engines developed the notation went through many editions, becoming a powerful but complex formal tool. Although its scope was much wider than logical systems, the mechanical notation was the most powerful formal method for describing switching systems until Boolean algebra was applied to the problem in the middle of the twentieth century. In its mature form the mechanical notation was to comprise three main components: a systematic method for preparing and labelling complex mechanical drawings; timing diagrams; and logic diagrams, which show the general flow of control.

The curious phenomena of electromagnetism were beginning to be discovered during this period, and inevitably Babbage took an interest. In the spring of 1825 Gay Lussac visited London and described Arago's experiments with rotating discs. Plates of copper and other substances set in rapid motion in a magnetic field and under a magnetized needle caused it to deviate from its direction, finally dragging it round with them. At this time John Herschel was secretary of the Royal Society and had rooms in Devonshire Street. Babbage and he carried out some quite extensive experiments in Babbage's house.[16] They tried the effect with discs of many different substances using Babbage's lathe: only metals and graphite showed the effect and they concluded that the conductivity of the disc was the important point. They also tried the effect of rotating a powerful compound horse-shoe magnet about its axis of symmetry. However they did not solve the problem of electromagnetic induction: later their friend Michael Faraday did.

Babbage was still troubled by his father's nagging complaints that for all his talents he would never receive proper financial reward. Moreover it was necessary to educate the growing family and provide for their future. For some time Babbage had considered working in assurance and in 1824 he had been invited by the founders to organize a new life company which was being formed called The Protector.[17] The company had impressive backing, including that of the Marquis of Lansdowne, Lord Abercromby, a clutch of directors of the

[16] C. Babbage and J. F. W. Herschel, 'On Electrical and Magnetic Rotations', *Phil. Trans.* 1826, 116, 494–528.
[17] cf. Add. Ms. 37,183. f 133, BL.

Bank of England and the East India Company, and sundry other eminent and respectable gentlemen. Babbage was offered £1,500 a year, with liberty to practice as an independent actuary, and apartments over the establishment. His friend Francis Baily FRS who had practised as an actuary advised Babbage that he could easily make £1,000 per annum from free-lance actuarial work. How easily an educated man could pass from one profession to another at that time!

Babbage calculated the requisite tables for the new company from the actual mortality rates of people insured by the Equitable Society, established in 1762. Subscriptions had been raised and all arrangements were complete for opening The Protector Life Assurance Company at offices in Frederick's Place, Old Jewry, when the plan was abandoned, two of the directors having proved awkward. Babbage received a hundred guineas in return for his 'valuable services', which just about covered his expenses. He was well respected for the part he had played and made valuable political friends through the venture. The plan was not completely given up and another opportunity arose in due course for Babbage to run a life assurance company. However after experiencing the amount of work involved he felt obliged to choose between the Difference Engine and assurance: the Difference Engine won.

But it would have been out of character for Babbage simply to drop the subject. He continued his studies, expanding the field of interest, and wrote a book called *A Comparative View of the Various Institutions for the Assurance of Lives*. It was at once an analysis of the life assurance offices and a consumer's guide. If only consumer's guides were as well written today! The opening has the authentic Babbage ring: 'In exposing the disgraceful practices which prevail at some assurance offices . . . I feel that little more is requisite than by rendering those practices known, to make them universally condemned.'[18] But he was strongly in favour of life assurance as contributing to social stability. The actuaries, he said, are usually men of skill and ability, but the public should know that their advice 'is frequently neutralized by passing through the ordeal of a board of directors, far too intent on profit, and who in their joint capacity, esteem it no degradation to sanction measures, which they would be very sorry to be considered as acting upon in their character as individuals.'[19]

Babbage then gave an excellent simple guide to the quantitative basis of life assurance, discussing the mortality tables on which assurance premiums were calculated. The figures of the Equitable Society suggested that there had been little change in mortality rates since its foundation. If this is correct it indicates that life expectancy among the section of the population taking out assurance policies (for those of sufficiently mature age to take out assurance) was little changed between the 1760s and the beginning of the nineteenth century.

Assurance organizations were established on three bases: mutual, or co-

---

[18] *Assurance of Lives*, ix.   [19] Ibid., xi.

operative; private; and offices which offered what would now be called with-profit policies. Babbage speaks warmly of the Equitable and Amicable mutual offices as 'great institutions... unfettered by the partial and contending interest of any proprietory. Founded alike for the mutual benefit of all who choose to become members, they have pursued their separate course, diffusing comfort and security over a multitude of families: they form, as it were, a portion of the public.'[20] A marked contrast is provided by private offices paying a commission to agents. The agents were not employed by the company but were the solicitor, broker, or other expert consulted by the person taking out insurance. There was plenty of opportunity for corruption there. Babbage gives an example of a lawyer who had advised a clergyman to take a highly disadvantageous policy on which the lawyer received a commission.

In contemplating with scorn the mercenary agent who betrayed, for so trifling a sum, the confidence reposed in him by his client ... ought not some portion of our indignation to be reserved for those who tempted him to this breach of trust?[21] ... It will naturally be enquired who authorized the practice we have been reprobating, and whether the long lists of respectable names, displayed at the head of many of these institutions, are placed there only to beguile the unwary, and to lead them to suppose that the same honourable principles, which govern the directors in their private capacity, will be adhered to when they act together as a body. There are many persons thus situated, whose known integrity or high rank make it impossible to suppose, that they are aware of a practice thus carried on in their name.[22]

In spite of his reservations about some private insurance companies Babbage was strongly in favour of private enterprise. There was plenty of room for mutual and co-operative organizations, and the Government should only step in where private or co-operative organizations could not cope. An excellent example of such a situation was provided by the calculating engines: there was no profit in them for an individual but great potential benefit for the public weal.

After his study of life assurance the next book Babbage published was a Table of Logarithms of the natural numbers from 1 to 108,000. The tables were compared with previously calculated sets of logarithmic tables. Some part of the comparison was carried out by the then Lieut. Col. Thomas Frederick Colby of the Royal Engineers, who pioneered the ordnance survey, and to Colby the book was dedicated. It became the standard set of logarithmic tables for many years, although at the time Babbage no doubt hoped it would soon be superseded by tables prepared directly by the Difference Engine.

During these years Babbage saw much of his university friends. He was particularly close to John Herschel. Following an unhappy love affair of Herschel, in the late summer of 1821 Babbage and Herschel visited Northern Italy, meeting men of science and engaging in strenuous climbing on Monte

[20] Ibid., 14.   [21] Ibid., 137.   [22] Ibid., 140.

Rosa. When Herschel was thinking of getting married Babbage gave him advice on settlements, and when there was a threat of breach of promise action once again Babbage was called on for counsel and advice. It was probably during this period that Babbage's beloved younger sister, Mary Anne, married Henry Hollier. Herschel became one of the trustees of her marriage trust. John Guest, the famous coal owner and industrialist, was also a trustee, presumably for the groom's side. Babbage helped his brother-in-law, Wolryche Whitmore, in his election campaigns in Bridgnorth.

In 1826 Babbage took Georgiana to Paris. Whilst there he was able to compare the proofs of his book of tables with the great tables of M de Prony which were still in manuscript. They dined with Laplace and on 16 August Babbage was sitting next to Laplace at the Observatory when a message arrived through the great French semaphore telegraph from Gombart in Marseilles announcing the discovery of a new comet. At another dinner Babbage first met the Belgian statistician Quetelet. Georgiana must have enjoyed exploring Paris and Babbage was very good at combining work with pleasure.

5

○ ● ○

The Death of Georgiana : Continental Travel

After the visit to Paris, Babbage continued working on the Difference Engine. He was also engaged on the proofs of his logarithmic tables, and making his classic study of manufacturing techniques and organization. Although he was very systematic in his work, in his day to day life minor details tended to be forgotten. Colby, who was helping Babbage with the logarithmic tables, wrote to Edward Ryan on 21 November 1826: 'It is unfortunate that our excellent friend is not enough a man of this lower world to have his address at the Post Office that letters may follow him without delay ... Babbage [has] been rambling to Birmingham[1] ...' On that trip he met a very intelligent clockmaker who not only understood the mechanical notation but even suggested an improvement. Babbage was delighted. He was always respectful of skilled craftsmen, and hoped the notation would one day come into general use for describing complex machinery. It would certainly have introduced some order into a large class of patent applications which is still in an indescribable muddle. That the notation did not come into widespread use, or at least that some such notation was not generally adopted, was a result of the replacement of mechanical equipment by electromechanical and later by electronic systems for many purposes.

Old Benjamin Babbage had been a sick man for many years. In February 1827 his condition became acute and on the 27th of the month he died, at the house he had leased in Teignmouth. He was in his 74th year. He was buried in East Teignmouth church as he had directed in his will[2], 'in as private a manner as possible and with as little expense as may be consistent with decency': an old nonconformist formula.

Georgiana wrote to John Herschel:[3]

Mr Babbage died the *Tuesday* following the same day Charles left Town ... To feign sorrow in so happy a release would in *me* be hypocracy. Of the affairs I as yet know but little, but I do not imagine the event will make any material difference to us, during Mrs B's life, and no great accession at *any period*, but I do not speak with any certainty and I have always felt an indifference on the subject that with *my family* almost astonishes myself. It does not proceed from carelessness but the little difficulty I find in accomodating my wants and wishes to my circumstances. Naturally active I am an

---

[1] BL Add. Ms. 37,182, f'373.   [2] P.R.O.
[3] Babbage/Herschel corr. f 353 RS.

excellent wife to a poor man but never should have been sufficiently fond of stile or company to make a proper wife to a rich man.

Yours affec^{tly}

G. BABBAGE.

There was little love lost between Georgiana and old Benjamin.

Herschel wrote to enquire whether Charles wished to be a candidate for the Savilian chair of mathematics which was vacant at Oxford, and Georgiana replied[4] on 12 February 1827:

It is seldom that I cannot say a priori what Charles *would* like or would not, but with regard to this professorship I cannot judge. Before his father's death he would have taken it had it been offered or at least a certainty that if he declared himself a candidate he would have it, but whether this event will make him *more* or *less* desirous of obtaining it, without more knowledge of our affairs I cannot say. One great inducement in endeavouring to procure an addition to our income was his father's always *despairing* his abilities (if I may use the term) and saying C's abilities would never procure him anything. This made dear C feel more keenly the fruitlessness of his endeavours. This trial is now past. I think it would be good for his family but my boys must make their own way, for whatever harrassed their father never would turn to my happiness or theirs ...

In another letter, addressed to Herschel, Georgiana wrote[5]:

My dear Mr H.

I have this morning received a letter from dear Charles. He says, 'I shall leave this place Monday. If on my arrival in Town I find reasonable hope of success I will write the requisite letters. The additional income of the Professorship although not in the present circumstances so *immediately necessary* is by no means undesirable as the increase to our own income from my father's death is small and far less than the world will believe' ...

However the appointment had in effect already been settled. Baden Powell was to get the Chair and Babbage did not apply.

Apart from a number of small bequests, Benjamin had divided his property between his wife, Betty Plumleigh, and Charles who was also the sole executor. His wife received £9,000 of stock for her absolute use and the interest on a further £10,000. On her death the £10,000 capital was to be divided among Charles's children. Betty Plumleigh also received the lease on the house in Teignmouth, all Benjamin's silver, china, glass, household goods and furniture, liquors, coal, and other goods about the house, and his square piano-forte. She had the use for life of a diamond necklace, king-pin, and locket, probably heirlooms, which were afterwards to go to Charles. William Doidge Taunton, Thomas White Windeatt, Catherine Celeste Moutier, Mary Windeatt, Sarah Windeatt, and May Taunton each received nineteen guineas. The Tauntons and Windeatts were Charles's cousins. The Totnes Bluecoat school also

---

[4] Ibid., f 210.   [5] Ibid., f 357.

received nineteen guineas. Benjamin's plate, library, and the residue of his property went to Charles. The value of the legacy was about £100,000. It included the farm at Dainton, and the Castle Meadow in Totnes. Charles was now comfortably placed. He possessed sufficient resources to support his family and with care his scientific studies for the rest of his life.

At the time his two elder sons, Herschel and Charles, were at the small school in Enfield under Stephen Freeman which Babbage had himself attended with Marryat. Freeman had charged reduced fees for the two boys, although he had not mentioned the fact to Babbage, because of the latter's financial circumstances. The boys were now withdrawn and sent to Bruce Castle School, newly opened by the Hills in Tottenham. Freeman was deeply upset. He was very proud of Babbage as his most successful former pupil and also valued the connection for the school. The timing was particularly unfortunate as Freeman's wife had recently died and his daughter, who had taken over the domestic side, felt the withdrawal of both the boys as a reflection upon her management. If Herschel alone had been taken away she would have understood, but to lose little Charles at the same time was a blow. However Bruce Castle had special attractions for Babbage, and he did what he could to comfort the Freemans.[6]

Thomas Wright Hill and his sons had previously run a very successful school at Hazlewood near Birmingham. The school which Babbage had visited early in 1827 or possibly earlier was based on the attempt to apply Utilitarian principles to education[7]. After studying the Hazlewood system, Jeremy Bentham had been so impressed that he abandoned his own attempts in that direction. Bentham, Grote, Joseph Hume, and many other leading radicals were advising parents to send their children to Hazlewood. It was as much the progressive thing to send your sons to Bruce Castle in 1827 as to send children to Dartington Hall School in 1935.

Bruce Castle, which was based on the well-developed Hazlewood system, was run as a self-governing institution by a headmaster, teachers, and committee of boys. While the headmaster retained the right of veto, it was never in fact exercised. The committee had a Chairman, Secretary, Judge, Magistrate, Sheriff, and Keeper of Records, all boys. While the head appointed the Attorney General, the Judge nominated the Clerk and Crier of the Court, and the Magistrate his two constables. It was a model Utilitarian state in miniature. Discipline was unrelenting and punctuality rigidly enforced. Vacations were regarded as a regrettable necessity 'interrupting that regularity, which becoming more and more exact from day to day attains to such a degree of perfection before the end of the session, as to contribute exceedingly to the comfort of the master, and the improvement of the scholar.'[8] Little wonder that

---

[6] Stephen Freeman to Babbage, 11 May 1827, BL Add. Ms. 37,184, f 30.
[7] M.D. and R. Hill, *Public Education*, 2nd. ed. London, 1825.   [8] Ibid., 72.

a former pupil was later to complain that the system worked 'at too great a sacrifice. The thoughtlessness, the spring, the elation of childhood were taken from us, we were premature men.'[9]

On 10 April 1827 Thomas Wright Hill wrote from Hazlewood to Babbage. After discussing the Calculating Engine the letter continued: 'My sons will be delighted with the care of your family—Rowland will be in London on Saturday and will seize an early opportunity for calling in Devonshire Street. Hazlewood was very fortunate in obtaining your good opinion. It will be the pride of Bruce Castle to preserve it.'[10]

Babbage always discussed his technical interests with his visitors and it is typical that Thomas Wright Hill should mention the Calculating Engine. At that time Babbage and his friend Colonel Colby were also carrying out studies on the postal system. Although he may well have heard of these studies earlier this was probably the first occasion on which the young Rowland Hill, later the creator of the penny post, encountered an operational research approach to the postal system. It would also have been his first encounter with the concept of a uniform postal rate, an immediate corollary of Babbage's theory of the cost of verifying prices, a theory which he was later to discuss in his book *On the Economy of Machinery and Manufactures*.

In July 1827 young Charles died. Soon after came the heaviest blow. Georgiana became very ill. On 4 August Babbage took her and the children to Boughton to the house of her sister, Harriet Isaac. By the end of August Georgiana was dead[11]. A newly born boy also died. The effect on Babbage of the loss of Georgiana was devastating. After a year's tour of Europe he was to return fully able to work: indeed he seemed to fling himself into all his activities with a peculiar passion, seeking to still the grief caused by his loss. His famous charm was unaffected; even his gaiety and sense of humour recovered; but his family life was gone, and there was an inner emptiness. In his public controversies there is a new note of bitterness of which there was no trace while Georgiana was alive. Far more than the difficulties over the Calculating Engines, far more than any public battles and disappointments, the loss of Georgiana left Babbage a changed man.

His mother, who had moved to Devonshire Street, was alarmed at the effect. John Herschel suggested travel, offering to accompany him. Babbage went to stay with the Herschels in Slough. Herschel attempted to reassure Babbage's mother but she remained very worried, writing to John Herschel on 8 September 1827: 'you give me great comfort in respect to my son's bodily health. I cannot expect the mind's composure will make hasty advance. His love was too strong and the dear object of it too deserving ...'[12]

---

[9] *DNB* Rowland Hill.   [10] BL Add. Ms. 37,184, f 4.
[11] It is possible that Georgiana may have died at the beginning of September.
[12] Babbage/Herschel corr. f 215, RS.

Babbage and Georgiana had been planning a visit to Ireland together before Georgiana died. Now Babbage and John Herschel made a short trip, visiting Colby and Trinity College, Dublin. After he had returned to England Babbage decided on a long tour. He planned to cross Europe and then travel to the Far East. He wished to travel alone but, to calm his mother's fears, although he could not bear the idea of taking a servant, he took one of his workmen named R. Wright. Later Wright was to be his chief workman during the second phase of work on the Analytical Engines. Babbage's eldest son, Herschel, stayed during the school holidays with Babbage's mother and his only daughter, Georgiana, while the two younger boys, Dugald and Henry, lived with their aunt and uncle, the Isaacs at Boughton. John Herschel offered to look after the work on the Difference Engine being done by Joseph Clement.

Babbage was then able to make detailed arrangements for his journey. Aware of the importance of novel manufactures and scientific techniques he took with him a selection of samples to show to the people he met. The stomach pump had recently been invented. Babbage took the parts of an instrument which could be used either as stomach pump or syringe for cupping, neatly packed in a small box. Diffraction gratings, which split up light into pretty colours, made attractive presents and Babbage took several gratings: a steel die ruled by Mr Barton, and a smaller piece of steel ruled with not quite the same perfection. He also had a dozen large and a dozen smaller gold buttons stamped by Barton's dies with lines of varying spacing between four thousand and ten thousand lines per inch. The smaller piece of steel was kept in Babbage's waistcoat pocket accompanied by a gold button, which was softer and kept in a sandalwood case to avoid damage.[13] Accompanied by his son Herschel, Babbage made a last visit to the Thames Tunnel where they met the younger Brunel. A flow of liquid mud was staunched under his direction. Shortly afterwards the tunnel was flooded: six men were drowned while Brunel escaped with difficulty by swimming. Babbage purchased a dozen copies of the description of the tunnel, six in French and six in German: more beads for the natives. A day or two later he left for the Continent.

Babbage crossed to the Low Countries and visited Louvain where he met the rector of the university. The old university, which had been one of the finest in Europe in the sixteenth century with 6,000 students, had been suppressed by the French. It had then been re-established by the King of Holland, but had only three hundred students when Babbage was there. Next he went to Liége, a manufacturing centre which was becoming the Birmingham of the Low Countries. The university at Liége was housed in a modern building erected by the King of Holland in 1817. Babbage continued to the old fortified town of Maastrich, which had famous quarries, and then to Aachen where he no doubt

[13] *Passages*, 372–4.

saw the tourist attractions. He then went to the commercial free town of Frankfurt, one of the great trading cities of central Europe. The houses of the wealthy merchants and bankers on the quays facing the Main in the new town were palaces. The old town was a maze of tiny streets of wooden houses with overhanging gables. In Frankfurt he met the eldest son of the coachmaker to the Tsar. This young man had been travelling through Western Europe making drawings of all the most interesting carriages he could discover. Some were selected for elegance; others, including the Lord Mayor of London's, for inelegance. The two travelled to Munich together and from this intelligent young man Babbage learned the structure and form of every part of a carriage. He made detailed notes of critical points, and by the time they reached Munich he knew enough to design a carriage himself. The young Russian was going to Moscow and begged Babbage to accompany him. However, Babbage regretfully declined since he was anxious to press on through Italy to Turkey and the East. In the Bavarian capital of Munich the stomach pump aroused interest. Dr Weisbrod, the king's physician, had an exact copy made by the chief surgical instrument maker. From Munich the route lay through Innsbruch, Verona, and Padua to Venice.

Babbage made a tour of the factories of Venice.[14] At one where the celebrated Venetian gold chains were made he sought to buy as samples a few inches of each type of chain manufactured but was told they were only sold in longer lengths. He then produced a hardened steel die for the proprietor to see in the sun. A room was darkened and with a single lamp the diffraction effects were even more impressive. Babbage was rewarded by being allowed to purchase short pieces of all the types of gold chain. On his return to London these pieces of chain were duly measured providing data illustrating the relative contributions of the cost of raw material and the amount of labour employed to the value of commodities, which he used in his book on *The Economy of Manufactures*. A few years later he was interested to hear that the factory had taken up his idea and was selling short lengths of gold chain mounted on black velvet as souvenirs.

After Venice, where Babbage must have learned that the route to the Far East was closed, he visited Parma, Reggio, and then Bologna where he spent several pleasant weeks at the university. He also got to know a shopkeeper who made barometers and thermometers as a sideline, who proved technically very

[14] He was accompanied on these visits by Captain G. Hutchinson who was in charge of the gun foundries in India. Hutchinson was returning to England and in Vienna he saw the gun foundry, which also bored the barrels and was 'infinitely superior to anything I had seen of the kind'. He was fortunate in meeting the nephew of Mr Reichenbach (deceased), who had designed the Vienna foundry, and purchased from him a set of plans for the foundry—there was very little military secrecy in those days. As a result Hutchinson was able to introduce many improvements into the machinery which he had ordered in England for installation in India. G. Hutchinson to Babbage, 5 Sept. 1832, BL Add. Ms. 37,187, f 114.

knowledgeable. Babbage was planning to make a presentation of his scientific toys and instruments to friends and professors at the university, and he enquired whether a humble instrument-maker would be acceptable company in such a gathering. He was glad to find that the instrument-maker was held in high regard and would be warmly welcomed. Babbage never forgot that his own forebears had been goldsmiths and remained proud of the background.

In Bologna a method he knew of making holes in glass was demonstrated, leading to an incident he later described:[15]

Finding myself in the workshop of the first instrument maker in Bologna, and observing the few tools I wanted, I thought it a good opportunity to explain the process to my friend [who was showing me the city]; but I could only do this by applying to the master for the loan of some tools. I also thought it possible that the method was known to him, and having more practice that he would do the work better than myself.

I therefore mentioned the circumstances of my promise [to show my friend the trick], and asked the master whether he was acquainted with the process. His reply was, 'Yes; we do it every day.' I then handed over to him the punch and the piece of glass, declaring that a mere amateur, who only occasionally practised it could not venture to operate before the first instrument-maker in Bologna and in his own workshop.

I had observed a certain shade of surprise glance across the face of one of the workmen who heard the assertion of the daily practice of his master's, and, as I had my doubts of it, I contrived to put him in such a position that he must either retract his statement or else attempt to do the trick.

He then called for a flat piece of iron with a small hole in it. Placing the piece of glass upon the top of this bit of iron, and holding the punch upon it directly above the aperture, he gave a strong blow of the hammer, and smashed the glass into a hundred pieces.

I immediately began to console him, remarking that I did not myself always succeed, and that unaccountable circumstances sometimes defeated the skill even of the most accomplished workman. I then advised him to try a larger[16] piece of glass. Just after the crash I had put my hand upon a heavier hammer, which I immediately withdrew on his perceiving it. Thus encouraged, he called for a larger piece of glass, and a bit of iron with a smaller hole in it. In the meantime all the men in the shop rested from their work to witness this feat of every-day occurrence. Their master now seized the heavier hammer, which I had previously just touched. Finding him preparing for a strong and decided blow, I turned aside my head, in order to avoid seeing him blush—and also to save my own face from the cloud of splinters.

I just saw the last triumphant flourish of the heavy hammer waving over his head, and then heard, on its thundering fall, the crash made by a thousand fragments of glass which it scattered over the workshop.

I still, however, felt it my duty to administer what consolation I could to a fellow-creature in distress; so I repeated to him (which was the truth) that I, too, occasionally failed. Then looking at my watch, and observing to my companion that these tools were not adapted to my mode of work, I reminded him that we had a pressing engagement.

[15] *Passages*, 381–2.
[16] The thicker the glass to be punched, the more certainly the process succeeds.

I took leave of this celebrated instrument-maker, with many thanks for all he had shown me.

The man was not invited to Babbage's evening demonstration.

Babbage continued his journey through Florence, where he met the Grand Duke of Tuscany, and by March he was in Rome. He stayed in the Piazza del Popolo. By chance he read in Galignani's newspaper that he had been elected Lucasian Professor of Mathematics in Cambridge. Babbage did not seek honours or academic positions for himself unless they had practical use, although he was very anxious that proper respect should be paid to science. If the election had been made during his father's lifetime, he would have been glad to accept, but only, he asserted, because it would have brought his mother pleasure to see his father happy. Babbage protests too much: Georgiana's letter[17] written after Benjamin's death shows that Babbage cared greatly what his father thought, even if he would not admit it. Babbage had just drafted a letter of refusal, when two of his friends, the Revd. Mr Lunn and Beilby Thompson who were staying in Rome, came to congratulate him. They pointed out that a refusal would be hurtful to the friends who had worked to secure his election. The point was fair. Even though he was anxious to avoid any distractions that might interfere with his work on the Difference Engine when he returned to England, the letter was not sent. Instead he accepted the post and held it for more than a decade.

In Rome, either then, or later in the year on his return from Naples, Babbage met Charles Lucien Bonaparte (later the Prince of Musignano), who had been a child when Babbage was visiting his father in Thorngove. Charles Lucien, eldest son of the Prince of Canino, now well on his way to becoming a noted ornithologist, had returned to Europe from the United States. He introduced Babbage to his sisters, Lady Dudley Stuart and the Princess Gabrielli. Later in Bologna Babbage met another sister, the Princess d'Ercolano. It was probably during this time in Rome that Babbage met Louis Napoleon, then living with his mother in the castle of Arenenberg in Switzerland but visiting Rome each year. Louis Napoleon was at that time merely the younger son and Babbage does not seem to have taken to him as he did to his elder brother, Napoleon Louis, and their cousins, Lucien's children.

In April or May Babbage arrived in Naples, exempt, thanks to his connections with the Duke of Tuscany, from having his luggage inspected by the customs. Vesuvius was the obvious subject of scientific interest there. At that time there was very little understanding of volcanic activity or the internal movements of the earth. The volcano was in a state of moderate activity and Babbage took apartments in the Chiaja. From his bedroom he could watch the eruptions through a telescope while lying comfortably on his bed, but he was

[17] Georgiana Babbage to John Herschel, 12 Feb. 1827, Babbage/Herschel corr. f 210 RS.

determined to look down the throat of the volcano. After a fortnight the eruptions became more regular and the chief guide deemed the time suitable for the trip. Babbage rode as far as possible and then had himself carried in a chair to save his strength. After exploring the top of the mountain he had a few hours sleep in a small hut and rose before dawn to complete his survey of the edge of the crater and then to enjoy the magnificent view as the sun rose.

At that time the bottom of the main crater was a flat eliptical plain. In one corner of the plain there was an embryonic volcano from which the eruptions came. The chief guide prudently declined to enter the crater. Babbage and one companion, assisted by a rope, descended to the plain, about 570 feet below the lowest point on the edge of the crater, his faithful but heavy barometer strapped to his back. He then proceeded methodically to make a tour and survey of the fiery domain. The plain was criss-crossed with ditches, or fissures, which glowed red hot below a depth of two feet. He measured with a tape a base line of 340 feet, took his elevations and remeasured the base line. By that time the marker stick, which was only inserted two inches into a small crack, was in flames. Pressure and air temperature were measured and he collected a few mineral specimens. After observing and noting the time of the eruptions from the small live crater, Babbage approached closer and closer in the quiet intervals. After a few attempts he succeeded in actually lying down on a projecting rock and looking into the sea of molten lava. He watched it bubbling slowly until his watch indicated that the next eruption was due. On returning to Naples the thick boots he had been wearing fell to pieces, destroyed by the heat.

Babbage had received recognition early on the Continent and he was a foreign member of the Royal Academy of Naples, the capital of the Bourbon Kingdom of the Two Sicilies. During his residence the government appointed a commission of members of the Royal Academy to report on the hot springs of the island of Ischia off the bay of Naples, and Babbage was appointed a member of the commission. They spent three of four pleasant days on the island studying the springs and it may be that the experience turned Babbage's mind to the possibility of the industrial application of the heat in thermal springs. Several of the springs, which were quite shallow, were excavated. As they got deeper the temperature of the springs rose, and it occurred to Babbage that if they had bored a few feet deeper they might have reached boiling water which could have been used as a source of power. Later he was to propose the industrial application of thermal power in the hot springs near Volterra in Tuscany.[18] There had recently been an earthquake on Ischia and Babbage was surprised to see how localized the region of severe damage was.

Although Vesuvius and Ischia were the most dramatic phenomena, the whole region was affected by the associated movements of the earth. Babbage

[18] See ch. 9; also *Handbook for Central Italy*, 1st ed., 128, John Murray, 1843.

made a detailed study of the temple of Serapis in Pozzuoli near Naples.[19] The temple was right on the coast and it had been buried. From the strata in which it was embedded and encrustation on the marble columns he was able to estimate the sea level at various earlier dates. He took a large number of samples back to London : Faraday analysed the chemicals ; Professor E. Forbes identified the shells. Babbage read a paper to the Geological Society on his observations together with a theory of the movement of isothermal surfaces within the earth. He sought to prove that large tracts of the earth's surface subside through the ages, whilst other portions rise irregularly at various rates. An abstract was published at the time although the complete paper was not published until 1846. When he read his paper John Herschel was at the Cape, and Babbage considered that the only people who understood his theory were Dr Fitton and de la Beche, founder of the geological survey. A few years later John Herschel published a similar theory, possibly stimulated by reports of Babbage's paper from friends. When he republished his paper as a separate pamphlet in 1847 Babbage was pleased to note that 'in 1838 Mr Darwin published his views on those subjects, from which, amongst other very important inferences, it resulted, that he had, from a large induction of facts, arrived at exactly the same conclusion.'

The study of the evolution of the earth's surface was making very rapid advances at that time and was closely related to the question of the evolution of living forms. The most important work was carried out by Babbage's friend Charles Lyell. In 1828 Lyell visited Naples and continued to Sicily where he climbed round Mount Etna making extensive surveys. The positions of fossil-bearing strata enabled him to make the first reasonable estimate of the time during which there had been life on earth ; and it proved to be far longer than anyone had imagined. The discovery was of profound scientific and theological importance and an essential foundation for Darwin's theory of the Origin of Species. In 1858 it was Lyell, together with Joseph Hooker, who presented Darwin's and Wallace's papers on natural selection to the Linnean Society. In July 1830 Lyell published the first volume of his *Principles of Geology*, an early historical account of geological development. It created a sensation.

The battle of Navarino fought on 20 October 1827 had closed the routes through Turkey to the East and led Babbage to prolong his stay in Italy. He now started on his return journey, first to Rome and then crossing to Ancona on the Adriatic, the principal naval station of the Papal States. He continued to Bologna and Florence where he stayed for several months. While staying there

[19] Abstract in *Proc. Geol. Soc.* 11, 72, 1834. This was the first public mention of Babbage's celebrated theory of the Isothermal Surfaces within the earth. Later John Herschel published an identical theory. Probably Herschel had glanced at the report of the paper which Babbage had read to the Geological Society in 1834 and then forgotten about it. As Babbage later wrote to Sedgwick, Herschel was *incapable* of plagiarism (17 Aug. 1868, BL Add. Ms. 37,199).

Babbage dined frequently at the table of the Comte St Leu, Louis Bonaparte, formerly the King of Holland. He saw the Princess d'Ercolano again and he met Napoleon Louis. Napoleon Louis had married Princess Charlotte, daughter of Joseph Bonaparte, King of Spain. Babbage noted that Napoleon Louis and Charlotte 'reminded me much of a sensible English couple, in the best class of English society.'[20] Napoleon Louis had his own workshop and lithographic press in a room at the top of his father's palace. After Napoleon Louis had died during an abortive revolutionary march on Rome, Charlotte lived with her father. They later lived for a time near Regent's Park in London where Babbage visited them and occasionally received them in his own house in Dorset Street.

In Florence Babbage was again welcomed warmly by Leopold II, the Grand Duke, who asked him if he could suggest any methods for aiding the progress of science in Italy. To several people who had raised the question of forums of discussion Babbage had suggested periodical meetings of men of science from the whole country and now he again made the same suggestion. At the Duke's request he drew up a minute of the discussion. However on consideration the Duke concluded that the time for such a meeting had not yet arrived. Eleven years later in 1839 Babbage was delighted to receive an invitation from the Grand Duke to meet the scientists of Italy, then due to assemble in Pisa, and saying 'The time has *now* arrived.'[21]

After leaving Florence Babbage travelled to Venice[22] and then by steamboat to Trieste. Further north he visited the caves of Adelsburg in Styria and purchased six specimens of the creatures which live in the dark waters of the caves. All of them died on the journey home, but he had them preserved in spirits and sent as specimens to various universities. This is a typical example of the chance origin of so many exhibits in nineteenth century museums and laboratories. Not all collectors were as scrupulous and exact as Babbage and many modern museums are still littered with actual fakes. His route then passed through Laybach, Gratz, the capital of Styria with its own university, and through Bruck to Vienna. The great Imperial capital justified a long stay and, taking advantage of his lessons in coach design from the son of the Tsar's coachmaker, Babbage designed his own carriage there. It was a strong, light, four-wheeled calèche in which he could stretch out and sleep full length. A lamp could be used for cooking and a large shallow drawer permitted plans and dress coats to be stored without folding.

Babbage heard that von Humboldt was in Töplitz and went there in the hope

[20] *Passages*, 203.   [21] Ibid., 431.
[22] Babbage went to Venice to visit a sick friend, probably George Everest. On arrival in Venice he found that Everest had migrated to Vienna whither Babbage followed him, only to find that Everest had returned to Milan. Thus Babbage missed an intended visit to see Lucien Bonaparte again on his estates.

of finding him, only to learn that he had returned to Berlin. Thither Babbage followed him, visiting Prague and Dresden, and taking rooms in Unter den Linden. The next morning he was invited to breakfast with Humboldt and a few scientific friends. There he learned that a major congress of German men of science was to take place within a few weeks. Babbage's review of the congress, published in the *Edinburgh Journal of Science* in April 1829,[23] and a report on a subsequent congress, may reasonably be taken as the first in a sequence of events leading to the formation of the British Association for the Advancement of Science, scene of the great scientific controversies of nineteenth-century Britain.

About eight years earlier Lorenz Oken had prepared a plan for an annual meeting of German botanists and people interested in medicine. This led to the formation of the Deutsche Naturforscher Versammlung which held its first meeting in Leipzig in 1822. Subsequent meetings were held in Halle, Würzburg, Frankfurt on Main, Dresden, Munich, and now in 1828 in Berlin. Its range of activities had gradually expanded until it included all departments of science, including pure mathematics. The movement associated with these congresses played a crucial part in mid-nineteenth-century scientific developments in Germany.

Von Humboldt was busy organizing the Congress and arranged for two young friends, Dirichlet and Magnus, to be Babbage's guides in Berlin. Amongst other activities Babbage was making a collection of signs employed in map-making, possibly stimulated by his friend Colby's work on the ordnance survey. He was pleased to be given by Von Bach a map shaded on the basis of its contour lines, at that time a novelty. A vital question in arranging such a congress was naturally which establishment should feed the delegates. Babbage was put on the select committee chosen to sample the possible hostelries, Humboldt remarking that an Englishman always appreciated a good dinner.

Alexander von Humboldt was a very important scientific figure in Babbage's time, and Humboldt's life reads like a sketch of the scientific movements of the first sixty years of Babbage's life. Meetings with such men as von Humboldt and Arago, both on the Continent and in England, had a very stimulating effect on Babbage, reinforcing his militancy. Alexander had been born in 1769 in the Prussia of Frederick the Great, the son of a Prussian officer. He and his elder brother Wilhelm joined liberal circles at an early age. After a wide-ranging scientific education, experience in mining, and travelling in Europe, Alexander followed Wilhelm to Paris in 1798. At that time the Spanish colonies in Latin America, accessible only to government officials and the Catholic church, were virgin territory for scientific exploration. During a visit to Spain Alexander von Humboldt's social position secured access to high places and the progressive Prime Minister obtained for him a Royal permit of exploration. Humboldt

[23] *Edin. Jrl. Sci.* X, 225–34.

made a legendary five year tour through South and Central America collecting a vast amount of information. It was this enormous increase in the amount of scientific data, obtained by Humboldt and many others, that led Babbage to urge the systematic presentation and study of what he was to call 'the constants of nature and art'. On his return to Paris von Humboldt became closely associated with the Society of Arcueil and a friend of the young Arago.[24] Short of money, he eventually returned to Berlin, where he enjoyed enormous prestige.[25] Babbage drew the conclusion that the social position accorded to men of science was of great importance for the development of science itself and afterwards was repeatedly to insist on the point. In 1848 during the revolution in Berlin, at the head of a mass parade for liberty and national unity was the single figure of old von Humboldt. Today the close association between the scientific movement and the liberal revolutionary movement, an association which formed the European backdrop to much of Babbage's scientific life, is often overlooked.[26]

The Berlin congress opened on 18 September 1828 in a large theatre. Among those present were Gauss, Berzelius, and Oersted. The galleries and orchestra were filled with people from the highest ranks of society while royalty and foreign ambassadors filled the side boxes to hear the opening address by von Humboldt. Babbage was immensely impressed. Compared with any scientific meeting which had ever taken place in England the meeting was on a huge scale; and compared with the position accorded to the sciences in England the importance that seemed to be attached to them in Prussia was in the sharpest contrast. He left determined to take steps towards launching a similar organization in Britain. Leaving Berlin he travelled in a leisurely manner to the Hague. There he sold his highly successful calèche and returned to England.

[24] When King Friedrich Wilhelm III arrived in Paris in 1814 following the allied armies he attached himself to von Humboldt, using him as guide and dispensing funds. Only Arago declined the royal patronage, refusing to show the King round the Observatoire in spite of Humboldt's repeated requests. Before leaving for London Humboldt arrived at the Observatoire accompanied by a friend to say farewell, carriage at the door. He asked if they could see round the Observatoire and Arago agreed. The conversation soon became political, Arago denouncing foreign kings and their disgusting behaviour in Paris. Embarrassed von Humboldt drew him aside asking him to be more discreet as he was speaking to the King of Prussia. Arago had guessed.

[25] He spent much time in the liberal salons, which had been established in imitation of the pre-revolutionary salons of Paris, particularly the salons of Rahel Levin and Bettina von Arnim, among the first women writers of Prussia.

[26] In Germany as in England the subsequent generation of men of science was much more apolitical; in contrast with the earlier generations of Priestley, von Humboldt, Arago, and Babbage, all more directly influenced by the Enlightenment, and concerned with the application of science to help organize society.

6

○ ● ○

Reform

After returning to England late in 1828 Babbage became engaged in an extraordinary range of activities: he wrote a polemical book, *On the Decline of Science*, and led the scientific reform movement; he continued his studies of industry and wrote *On the Economy of Machinery and Manufactures*; he organized several election campaigns for liberal candidates and himself twice stood for election to the newly reformed parliament. Despite his interest in a host of other technical matters, the Great Difference Engine continued to be his principal preoccupation. Before discussing the progress on that celebrated Machine we shall consider his more general interests. To carry out his many activities Babbage leased Dorset House, number 1 Dorset Street, Manchester Square, near the house in Devonshire Street. It was at Dorset House that his soirées began. At first they were quiet affairs for friends and family, with his old mother, his eldest son Herschel, and his beloved daughter, little Georgiana, acquiring social graces. Soon the parties became famous, invitations were much sought, and they became a feature of the London season.

At this time the 'era of discussion' began. Parliamentary reform was in the air. All the changes which the liberals had cherished for so long, which the pioneers of the Analytical Society had worked for in science since 1812, now appeared as practical possibilities. The whole country was astir: not only the middle classes backed by the developing working class, but also the poor in the countryside. Innumerable different interests, hopes and aspirations became concentrated on the single demand that parliament should pass a major Reform Bill.[1] The existing system of political representation belonged to the eighteenth century. Shifting and growth of population had made the electoral system an anachronism. The growing industrial towns were without representation and the rotten boroughs a scandal. The movement for reform, which had been developing slowly and in many different ways for decades, in 1830 sprang into life quite suddenly.

The middle classes, a heterogeneous group, were to be enfranchised by the Great Reform Bill: property, so to speak, received the vote, and the movement for the Bill was organized and remained under the control of the middle class; but to push the House of Lords into accepting the new franchise much broader

---

[1] For a general discussion of the Reform movement and a description of the turbulent events preceding the passing of the Act see: Michael Brock, *The Great Reform Act*, Hutchinson, 1973.

popular forces were in action. In the countryside a movement of arson and machine-breaking put fear into the propertied classes, reminding them inevitably of the *jacquerie*. In the towns monster demonstrations took place and the tiny army and local police could certainly not have put down determined mass action. But England was far from revolution. Although there were many vested interests at stake there was no *ancien régime* to oppose the change: England had had its revolution in the seventeenth century.

The parliamentary struggle was led by the whigs. It has often been held that they misjudged completely the significance of the Reform Bill, seeing it as a conservative measure whereas in fact it was the first step on the path towards modern democracy. But after all they were the immediate beneficiaries of the Bill: following decades of almost uninterrupted tory rule and patronage the great whig houses became once again the scenes of ministerial receptions and their owners the holders of ministerial office.

In 1829 and again in 1831 Babbage was chairman of the London committee to elect William Cavendish as Member of Parliament for Cambridge University. Having a great deal more free time since his family had been broken and dispersed, and no doubt encouraged by much that he had encountered on his Continental tour, Babbage began to take a more direct interest in politics. The 1829 by-election aroused a great deal of interest at the time. Cavendish, later Duke of Devonshire, had waived the privilege then attached to rank of being able to take his M.A. after two years residence. Instead he had entered into open competition and distinguished himself by becoming second wrangler and senior Smith's prize man. The combination of birth and superior knowledge was exactly suited to university representation.

On 3 June 1829, at the British Coffee-House in Cockspur Street, a meeting presided over by the Earl of Euston of members of the Cambridge Senate who were resident in London resolved to nominate Cavendish as candidate[2] and then elected a campaign committee of which Babbage was chairman. The elections of the time were a source of much entertainment[3] and often of profit to the active participants. Babbage remarks that his two most active lieutenants in the Cambridge contests were not neglected.[4] One shortly became a Master in Chancery and the other secured a place in India producing £10,000 a year. The latter must have been Thomas Babington Macaulay. Babbage was an efficient and energetic campaign organizer, commencing his duties early in the morning, continuing until midnight and entering into the spirit of the campaign with zest. A small number of men on the committee did all the real

[2] *Passages*, 260. See also reports in *The Times*: 23 and 28 Nov. and 6, 13, 21 Dec. 1832. *Morning Chronicle*, 10 and 13 Dec. 1832.

[3] For a description of the atmosphere of elections without a secret ballot see for example *The Pickwick Papers* or *Dr. Thorne*.

[4] *Passages*, 264.

work, but committees to support parliamentary candidates were generally large. They had to include, of course, persons of weight; then there were those who wished to appear to have weight; some merely liked to see their names in the newspapers; others again would vote for the opposing candidate unless their names were included; and inclusion on a list might establish a claim upon the political party; young barristers and solicitors were numerous, well aware of the opportunities for professional advantage. The committee was a place for worldly men, but a few honest simpletons were often included to give the cause a high moral tone. When questionable practices were being considered the innocents were kept out of the way so that at a later date they could affirm that no questionable action had even been proposed.

On the matter of 'pairing', then standard practice during elections, Babbage wrote later:

About a dozen years had elapsed after one of the elections I had managed, when the subject was mentioned at a large dinner-table. A supporter of the adverse political party, referring to the contest, stated as a *merit* in his friends that they had succeeded in outwitting their opponents, for on one occasion they had got a man on their side who had unluckily just broken his arm, whom they succeeded in pairing off against a sound man of their adversaries. Remembering my able coadjutors in that contest, I had little doubt that a good explanation existed; so the next time I met one of them I mentioned the circumstance. He at once admitted the fact, and said, 'We knew perfectly well that the man's arm was broken; but our man, whom we paired off against him, had *no vote.*' He then added, 'We were afraid to tell you of our success.' To which I replied, 'You acted with great discretion.'[5]

University elections had a special character. A large part of the electorate was clerical, and attached to the list of electors which was kept in the committee room Babbage placed signs indicating the books each voter had written, the nature of his preferment, the source whence derived, the nature of his expectations, source whence expected, chance of promotion for the impediment, the age of the impediment, and state of his health. Armed with this knowledge Babbage was not above inspiring in some newspaper a paragraph regretting the alarming state of health of the eminent divine who blocked some voter's path. In appropriate cases hope of preferment could easily decide the oscillation of even a cautious voter. He claimed that this worked particularly well because eminent divines were known to take to their sick-beds on the approach of university elections to avoid the bore of being canvassed.

The lengths to which they went to secure the votes of this small, privileged electorate were quite remarkable. Babbage located one elector in Berlin and employed a friend to write to him. Even at that time of slow travel the voter actually returned to England and voted for Cavendish. On another evening Babbage noted on one list the name of Minchin, an acquaintance of his when

[5] Ibid., 263.

they had both been undergraduates at Trinity. It was believed that Minchin had gone to India. Undeterred Babbage pursued his enquiries, discovering that Minchin, now a barrister, was returning to England on the *Herefordshire*. It was inconceivable that the ship would return in time, or that Minchin would receive a letter; and perhaps even more absurd to expect that a barrister would retain the liberal principles of his youth. However Babbage directed three letters to the most likely ports, begging Minchin, if he had not altered his political principles, to call at the committee room in Cockspur Street. At midnight on the first day of polling, sitting alone, Babbage was thrilled when the door opened and Minchin entered wrapped in a huge box coat for travelling on the roof of a coach and, shaking Babbage's hand, pronounced unchanged liberal principles. He had received Babbage's letter from the boat which brought the pilot. Abandoning wife and children to make their own way to London, he had returned on the pilot's boat and got on the roof of a coach about to start for London. Minchin was promptly packed off to Cambridge with an old friend of Babbage's, John Elliot Drinkwater.[6] Cavendish was finally elected with 609 votes, against 432 votes for his opponent, George Banks, a rising young lawyer.

The following year in 1830 the death of George IV on 26 June led to a general election, which was held at the end of July and the beginning of August. During the polling news broke of a revolution in Paris. After a brief struggle the Bourbon Charles X was expelled and replaced by the bourgeois Louis Philippe. If the news of French revolution stirred the whole of England with memories of the previous revolution in France, the relative moderation of current events served to convince the English middle classes that peaceful change was possible, encouraging them to conjure up a popular mass movement in support of the demand for parliamentary reform. Party lines were not clearly drawn in those days and there were no sure figures corresponding to today's estimates of party strength. Nevertheless it soon became clear that the whigs had been greatly strengthened. In November the Duke of Wellington declared sharply against reform. He was defeated, resigned, and was replaced by Grey, the consistent reformer.

The demands for change had been given powerful backing by the rising in the countryside known in the towns by the name of Captain Swing, its mythical leader. The movement spread through more than twenty counties and was on a formidable scale matching the distress in the countryside. The position of the poor countryman had changed drastically for the worse. The common lands on which so many depended for survival had been enclosed. When hard times came the poor farmworker's position was such that he could hardly fail to resist. In 1830 the movement developed on an unprecedented scale, with machine breaking, firing of ricks, threatening letters, often signed 'Swing', angry groups

[6] Ibid., 267–8.

demanding higher wages or extracting immediate succour in the form of money or food, and attacks on particularly unpopular justices and overseers. Nor were the farmers always unsympathetic: indeed they frequently used the riots as excuses to demand for themselves reductions in tithes and rents. The first threshing machine was destroyed near Canterbury in Kent on the night of 28 August 1830. As it spread the propertied classes suspected central conspiracy where there was only spontaneous action and local organization. Late in the year, when he was running the campaign to elect John Herschel as president of the Royal Society, Babbage received a letter from Footscray in Kent giving some indication of the effect of the action in the countryside and the atmosphere in the country while the campaign was being fought: '... in the present state of our neighbourhood it would be most unjustifiable in me to *think* of being about from home. You in London (*except the conspirators who are there*) can form no idea of the effect the ceaseless fires are producing.'[7]

Called to power Lord Grey established a highly aristocratic government, rewarding his followers. This government launched the special commissions which tried the 'Swing' rioters. Local action to stop the riots was sometimes conciliatory, sometimes repressive, but the judicial action was severe. There were eleven executions and 481 men and women were transported to Australia. Neither luddites nor chartists nor early trade unionists paid such a price: a measure of the fear which the risings had caused. The tories were pleased enough to see such firm action but at the same time Grey prepared to launch a far more drastic reform bill than anyone outside a tiny circle suspected. He had in fact been planning such a reform for more than a decade, although few beside Lord Holland were aware of his intentions. On 1 March 1831 Lord John Russell introduced the Bill, which had been prepared in great secrecy by a small commission of four, into the Commons. The tories were stunned; the radicals and the great mass of the people delighted: battle lines were drawn. Macaulay made his reputation in the Commons; petitions and letters poured in from all over the country. On 21 April the Commons refused supplies and the following day the King dissolved parliament. The next election took place in conditions of intense excitement, fought out almost entirely on the single question of the Reform Bill.

The two liberal candidates for Cambridge University, Cavendish and Palmerston, who had adroitly crossed the floor, were confronted by Henry Goulburn and William Yates Peel. The two latter were mediocrities when compared to Palmerston and the academically brilliant Cavendish, the two sitting members. Babbage, whose conduct of Cavendish's earlier campaign had been both efficient and successful, was once again asked to chair Cavendish's London committee. As on the previous occasion he worked from morning till midnight. Catching the national mood the campaign was quite different from

[7] BL Add. Ms. 37,185, f 357.

the previous one. It was marked by a delightful squib which was apparently Drinkwater's idea. One evening a cab drove up before the office of the *Morning Post* with copy claiming to come from Mr. Goulburn's committee:

The Whigs lay great stress on the academical distinction attained by Mr Cavendish. Mr Goulburn it is true was not a candidate for university honours; but his scientific attainments are by no means insignificant. He has succeeded in the exact rectification of a circular arc; and he has likewise discovered the equation of the lunar caustic, a problem likely to prove of great value in practical astronomy ...[8]

Pure nonsense! Cambridge was delighted. However the grey mass of country clergy, terrified of reform, dominated the vote: W. Y. Peel and Goulburn were elected. When the results were announced the successful candidates were hissed, and there were cries of 'Hurrah for the lunar caustic!'[9] The undergraduates took the horses from Cavendish's carriage and drew him out of town amid deafening acclamations. Those resident fellows who had voted tory against academic distinction were reported as hanging their heads in shame. One can see why Robert Peel felt no debt of gratitude to Charles Babbage. But nationally, fought on the slogan of 'the Bill, the whole Bill, and nothing but the Bill', the result of the election was a triumphant success for reform.

When he agreed to manage Cavendish's second campaign Babbage had made the reservation that in the event of his brother-in-law, Wolryche Whitmore's seat at Bridgnorth being contested, he should have three days off to help the campaign there. The contingency occurred and he went by mail to Bridgnorth. Two other brothers-in-law, Edward Ryan and Isaac the Banker, also helped at some Bridgnorth elections. Whitmore had a considerable influence on the development of Babbage's political views, and the situation at Bridgnorth was singular. Returning two members, it was the pocket borough of the senior branch of the family, the Whitmores of Apley Park. In two hundred and fifty years they survived two reform acts and fought twenty two contested elections, losing only two. Wolryche Whitmore was head of a junior branch of the family, the Whitmores of Dudmaston, and was famous as the whig member of a tory family: perhaps he too had been influenced by Lucien Bonaparte during his stay at Thorngrove. Nonetheless his cousin at Apley, Thomas Whitmore, the sitting member and an impeccable tory, supported cousin Wolryche: family before party. The local tories were furious and twice put up a third man. But in vain: the Whitmore hold was too strong to break. Later, however, in the reformed parliament of 1833, Wolryche Whitmore moved to sit for Wolverhampton. An early advocate of reform of the corn laws he was well supported by the West Midlands industrialists.

When he arrived in Bridgnorth Babbage found that the opposing candidate

---

[8] *Passages*, 272.   [9] *The Times*, 27 May 1831.

had withdrawn at the last minute and both his brother-in-law and a leading iron-master, Mr Foster, were to be returned unopposed. A chance meeting gave Babbage an opportunity which led to liberals standing for the Shropshire county seats. With four hours to spare Babbage was waiting at the hotel for the mail to take him back to his committee rooms in Cockspur street when he met a group of Foster's leading supporters, including some of the largest iron-masters and manufacturers in the county. They were pleased at Foster's success, particularly as the manufacturing interests of the district had been utterly neglected by the county members. Babbage urged Foster's supporters to field their own candidates, and retired to another room to draw up placards and an address to freeholders, while the group estimated the number of placards required. Within two or three days every town and village in Shropshire was reading his placards and soon three liberal candidates were in the field. On his return to London Babbage arranged for £500 to be sent to Shropshire by the Patriotic Fund which had been instituted to assist supporters of the Bill. As expected no liberal was elected but the ground was prepared for the first election after the passage of the Reform Bill.

The results of the general election, the way in which it had been conducted and the obvious demonstrations of popular feeling, were not enough: following procrastination in the Commons, on 8 October the Lords threw out the Reform Bill. The King refused to create enough new peers to force it through, and Grey chose the stratagem of introducing a no less efficient bill in its place. In London crowds broke windows in the houses of the Duke of Wellington and a few other tory peers. In Birmingham, a monster meeting of some 100,000 people was called to maintain pressure but to observe the law. In Nottingham a crowd burned down the castle, property of the Duke of Newcastle who was strongly against the Bill, and then burned a few factories for good measure. There were demonstrations throughout the country, and major riots lasting three days in Bristol, in which the mansion house was sacked and several public buildings burned down. The bishops, who had voted against the Bill despite Grey's warning, were attacked and mitred figures replaced the Guys on November the fifth bonfires. A more cheerful note was sounded by Sydney Smith, later a familiar figure at Babbage's Dorset Street parties, who likened the Duke of Wellington to Dame Partington attempting to sweep back the Atlantic Ocean of reform with a mop, and the country rocked with laughter. The middle classes pressed strongly for action but in the main trusted Grey, while Grey in turn worked closely with the decisive Political Unions.

On 12 December Lord John Russell introduced a new Bill greatly improved in detail but unaffected in substance. It passed slowly through the Commons but on 7 May 1832 the Lords voted to postpone the crucial first clause disfranchising many boroughs. The cabinet asked the King to agree to create sufficient peers to pass the Bill: otherwise it would resign office. The resignation was accepted. The tories proposed a government led by Morris Sutton, who was Speaker, and the Duke of Wellington; but the country now thought of him only as the decisive obstacle to reform. As a matter of fact the Duke had grasped that reform could no longer be stopped and was by then quite prepared to see a bill through parliament, but he was distrusted and there was a widespread suspicion that he was planning a military coup. Then decisively the middle class leaders launched a financial attack: they would refuse to pay taxes while promoting a run on the banks, until the Bill was passed. Orders were countermanded; buyers declined to make purchases. There was a heavy demand for metal from the Bank of England stimulated by the slogan: To Stop the Duke, Go for Gold. The City of London was frightened; the country and Commons angry: no tory administration could stand. The whig ministers had never given up their seals and the King asked Grey to resume office. Grey demanded and received a promise of the creation of sufficient peers to secure his majority. Opposition in the Lords collapsed and the Great Reform Bill was passed.

A new age seemed to be dawning. The correspondence of the time between radicals sparkles with these hopes. For instance in January 1832 Babbage had received a letter from a friend in Brighton: '... all the Fitzclarences forthwith are PEERS!!! How much better it is to be a *bastard* than a *philosopher* in England, at PRESENT, and perhaps for TWO YEARS longer. But a mighty change is at hand.'[10] The radicals of all degrees expected that a reform act would indeed prove a prelude to radical reform. They were to be bitterly disappointed. Babbage decided to stand in the general election of 1832 which followed the enacting of the Reform Bill. Like so many others his hopes were now concentrated on the reformed Commons. The prospects seemed favourable as the newly enfranchised electorate was formed of the middle classes for whose interests he spoke: he stood on a Benthamite platform, but among the philosophic radicals he was one of the moderates, furthest from working-class radicalism. This appears for example in his scepticism about the secret ballot.

Several constituencies were mooted, including Tower Hamlets. Then he was invited to stand for the new North London constituency of Finsbury and accepted. The campaign took place during the latter part of November and early December 1832. His committee rooms were on Holborn Hill. Already a national figure, the publication of his book *On the Economy of Machinery and Manufactures* in the summer had made him the unrivalled theoretician of the industrial application of science in the country. His campaign was sufficiently important to attract good coverage in the press.[11] Refusing to conduct a

[10] BL Add. Ms. 37,186, f 232.
[11] An interesting excerpt from the *Mechanics Magazine* was reprinted in *The Times* on 8 Dec. 1832:
'Political matters do not, generally speaking, come within the province of this journal, but

personal canvass we find him addressing meetings at the Canonbury Tavern, the Crown Tavern in Clerkenwell Green, and the gracious hustings on Islington Green. He stood as the candidate of science, confidently believing in reason and reform.

One subject of acute controversy was the 'taxes on knowledge', that is to say stamp duty and duty on paper. In 1832 there was a duty of 4d. a copy on newspapers and *The Times* cost 7d.; 15d. on almanacs; 3s. an edition on pamphlets; and 3s. 6d. an item on advertisements. All the radicals were strongly opposed to such taxes: the crusading popular press played a vital part in the political radical movement and an unstamped alternative press had developed whose publishers risked prosecution. But taxes on knowledge also had wider educational implications and the demand for a national educational system at all levels was central to the theories of the philosophic radicals, as it had been to Condorcet and others during the revolutionary period in France. Babbage was in favour of general education because it would make people happier, and, he declared, 'give me but knowledge for [the people] and I fear nothing'.[12] That was the political issue. But he also made a statement of fundamental importance about scientific and technical education:

All were aware of the rapid strides to perfection the manufactures had made in this country, and he believed the manufactures were in a continual state of advance, and that it was consistent with the national character and superiority that the country should keep thus ahead of other nations. This could only be maintained by a far and wide diffusion of knowledge; and therefore among the advocates of education for the people he numbered himself one of the most strenuous supporters.[13]

Wise words! but to the country's cost they were little heeded. At that time Babbage still had confidence in the 'rapid strides to perfection' of the country's manufactures. Within a few years his confidence had gone.

when the interests of science happen to be mixed up with them why should we be silent? We look for a great deal of good to science, as well as to every other important interest of the country, from the return to Parliament of a gentleman of Mr. Babbage's eminence in the scientific world, tried independence of spirit and very searching and business-like habits; and therefore we take the liberty to say to every elector of Finsbury who is a reader of this journal and a friend to the objects it has especially in view—Go and vote for Mr. Babbage. If you are an inventor, whom the iniquitous and oppressive tax on patents shuts out from the field of fair competition, and are desirous of seeing that tax removed—Go and vote for Mr. Babbage. If you are a manufacturer, harassed and obstructed in your operations by fiscal regulation—and would see industry as free as the air you breathe—Go and vote for Mr. Babbage. If you are a mechanic, depending for your daily bread on a constant and steady demand for the products of your skill, and are as alive as you ought to be to the influence of free trade on your fortunes—Go and vote for Mr. Babbage. If, in fine, you are a lover of science for its own sake alone, and would desire to see science honoured in those who most adorn it, meet us to-day on Islington-green and vote for Mr. Babbage.'

[12] The *Morning Chronicle*, 10 Dec. 1832. For a discussion of the effects of the stamp duty on newspapers see, Patricia Hollis, *The Pauper Press*, O.U.P., 1970.

[13] *The Times*, 28n3a, 1832.

In the group of philosophic radicals some tended to believe in a wide development of positive social policy, others in a minimum of social policy whatever its form. However education was one field in which Babbage clearly considered that state intervention was necessary. In line with his general objection to all forms of monopoly he was in favour of throwing open the trade with China and India. This aroused the opposition of the East India Company and cost Babbage votes, even leading him to support the secret ballot. But he did so reluctantly and equivocally. He also supported triennial parliaments, another standard radical demand.

At a meeting on 22 November 1832 at Canonbury Tavern, with a 'numerous and highly respectable audience, Mr Hodgskin[14] (the well known writer on political economy) was anxious to put one or two questions to the candidate ... "Did Mr. Babbage think it right or expedient," he asked, "that the dissenter or Irish Catholic should be taxed for the benefit of the parsons of the Established Church? Mr. Babbage told them he held Negro slavery in abhorrence; pray was he prepared to emancipate the English labourer from the tyrannical control of parson justice?" '[15] A trifle cautiously Babbage agreed that tithes should be abolished and other means be found for paying the clergy. Hodgskin's second question was easier to answer. During the latter part of the eighteenth century and the beginning of the nineteenth the abdication by many country gentlemen of their traditional functions of social control in favour of the profits of *laissez-faire* had placed local power in the hands of the clergy as never before or since. Many benefices had been enriched by the parliamentary enclosures, the clergymen moving out of their cob cottages into those large vicarages which have remained such a drain on their successors' resources. Clergymen became gentry, constituting as much as a quarter of the nominal magistracy, and were in effective control of large areas. Babbage was unequivocal: 'I think the practice of investing clergy with the commission of the peace highly objectionable (Cheers).'[16]

Inevitably Hodgskin then raised the question of the £12,000 Babbage had received to construct the Calculating Engine.

Mr. Babbage was glad Mr. Hodgskin's question afforded him the opportunity of publically stating that he had never received a single farthing of public money for his invention (Loud cries of "Hear"). That the machine belonged to the Government, who purchased it as a national benefit [in calculating longitudes and natural logarithms etc.], and that all he had to do with it was superintend its construction.[17]

At a meeting on Tuesday 27 November at the Crown Tavern, Clerkenwell Green, Babbage said that he '... was aware he was addressing by far the largest class in this or any other country, the manufacturing class, with whose pursuits

---

[14] Thomas Hodgskin, a former naval officer, was one of the first socialist political economists.
[15] *The Times*, 23 n1 d, 1832.   [16] Ibid.   [17] Ibid.

and with many of whose leading members he had the good fortune to have acquired an intimate acquaintance'.[18] There is no trace of class analysis here. For Babbage 'the manufacturing class' in England included factory workers and factory owners, all operating in a free market no matter whether they were selling goods or their ability to work.

With respect to the lavish expenditure of Government [he went on] it must be reformed, for it pressed hard on the energies of the country. Sinecures must be the first to be abolished, and he considered that the nation might obtain men efficient for government of the country at less remuneration than at present. The next consideration would be the repeal of those taxes which an economic government could best spare. The first would be those taxes which impeded knowledge ... The next ... the taxes which bore on the prudence and forethought of the nation. And lastly ... taxes which prevented addition to the wealth of the country by pressing on its industry.

At the beginning of the meeting the chairman had reminded the audience that Isaac Newton had been an efficient member of the convention parliament and that Babbage had been unanimously elected to the Lucasian Professorship, Newton's old chair; and had emphasized the advantages of having a man of science in parliament. Babbage took up the theme:

One important tendency was, he claimed, derived from [his] long pursued investigation of the laws of nature as displayed in the physical world—namely a habit of regarding facts solely as facts, and of reasoning on those facts solely with a view to the elucidation of truth. He affirmed that this wholesome habit, so he persuaded himself, he carried with him into political investigations, in which, as in natural philosophy, it was his rule first to be careful in the obtaining and sifting of facts; and in the next place of reasoning on them fearless of the legitimate consequences of his reasoning (Hear, hear!).

On the subject of pledges Babbage said that it was his aim, before he adopted an opinion, whether in morals or in politics, first to collect facts, and then apply all the powers of his reason in order to arrive at the correct conclusion. If he was shown by facts and by clear reasoning that his opinions were wrong, then it would be his duty as well as his determination to alter those opinions.

Babbage has left us one short sketch of his campaign in action.[19]

One day, as I was returning in an omnibus from the City, an opportunity presented itself by which I acquired a few votes. A gentleman at the extreme end of the omnibus being about to leave it, asked the conductor to give him change for a sovereign. Those around expressed their opinion that he would acquire bad silver by the exchange. On hearing this remonstrance, I thought it a good opportunity to make a little political capital, which might perhaps be improved by a slight delay. So I did not volunteer my services until a neighbour of the capitalist who possessed the sovereign had offered him the loan of a sixpence. It was quite clear that the borrower would ask for the address of the lender, and tolerably certain that it would be in some distant locality. So, in fact, it

---

[18] *The Times* 28n3a, 1832.   [19] *Passages*, 272–3.

turned out: Richmond being the abode of the benevolent one. Other liberal individuals offered their services, but they only possessed half-sovereigns and half-crowns.

In the mean time I had taken from my well-loaded breast-pocket one of my own charming addresses to my highly-cultivated and independent constituents, and having also a bright sixpence in my hand, I immediately offered the latter as a loan, and the former as my address for repayment. I remarked at the same time that my committee-room on Holborn Hill, at which I was about to alight, would be open continually for the next five weeks. This offer was immediately accepted, and further extensive demands were instantly made upon my pocket for other copies of my address.

My immediate neighbour, having read its fascinating contents, applied to me for more copies, saying that he highly agreed with my sound and patriotic views, would at once promise me six votes, and added that he would also immediately commence a canvass in his own district. On arriving at my committee-room I had already acquired other supporters. Indeed, I am pretty sure I carried the whole of my fellow-passengers with me: for I left the omnibus amidst the hearty cheers of my newly-acquired friends.

The final result was: Grant 4,278, Spankie 2,848, Babbage 2,311, Wakley 2,151, Temple 787. The first two were elected: but for Babbage and Wakley it was a creditable result.

A dinner was arranged by Babbage's friends. Would John Herschel care to dine with Sir Hugh Middleton? Herschel demurred: he had not had the pleasure of meeting Sir Hugh. Learning that Sir Hugh Middleton's was a tavern near Sadler's Wells, Herschel was happy to come; so were Sir N. H. Nicolas, Mr Lubbock, Mr Brunel and others.

In 1833 Babbage published his pamphlet *A Word to the Wise*, a powerful advocacy of life peerages. Once again the French influence on his ideas is clear. Typically he based his case for the need to modify the political system on advances in industry:

During the last half-century the various classes of society in England have made very great, but very unequal, advances in knowledge and in power. Large branches of manufactures have been created which have given wealth to a body of capitalists possessed not only of great technical skill, but also of general intelligence of an high order; and a population has been called into existence, which depends upon those capitalists for its immediate support. Competition acting upon persons engaged in professions, as well as on those employed in manufactures and commerce, has produced a demand for information upon all subjects, which has given a stimulus to the universal mind of the country.

Amidst this general advance, the aristocracy alone have moved with slow and unequal pace, until at last it is found that amongst all the educated classes of the community, the aristocracy, as a body, are the least enlightened in point of knowledge, and the most separated from the mass of the people.

The splendid exceptions which occasionally present themselves to this too general truth, whilst they command increased respect for the individuals, serve but to render more apparent the general characteristic of the class.

Babbage's conclusion still has a topical ring:

One of the main uses of an Upper Chamber in a popular government, is to give consistency and uniformity to the more fluctuating opinion of the immediate representatives of the people. It is to the political what the fly-wheel is to the mechanical engine. It ought to represent the average but not the extreme opinions of the people. For this purpose it is necessary that none but persons duly qualified should have seats in the House of Lords. Peers should be elected for life: the Peerage should *not* be hereditary.

However, disliking too drastic change, he did not propose the immediate exclusion of hereditary peers, but merely of Ecclesiastical Peers, and '*an immediate and large creation of* PEERS FOR LIFE.'

In June 1834 a seat became vacant at Finsbury through the appointment of the Rt. Hon. Robert Grant as Governor of Bombay. After some persuasion Babbage stood a second time on the same platform as previously. On this occasion the purity of his position as a philosopher had lost its novelty. Equivocation on the question of the ballot was held against him, although he now suggested quite sensibly that it should be tried on a limited scale as an experiment: such a proposal was not likely even to be considered where there was a fundamental clash of interests. At a meeting held at the Crown Tavern[20] on Monday, 23 June, after Babbage had spoken, 'A scene of indescribable confusion followed in the course of which Mr Roebuck, the member for Bath, and Colonel Torrens, the member for Bolton, addressed the meeting' respectively against and for Mr Babbage. When polling began another liberal, Mr Duncome, drew ahead of Babbage whose supporters then deserted him to ensure that a liberal candidate should be elected. Babbage was heavily defeated, and thus ended his political ambitions.

In retrospect it is remarkable that all the support received for his calculating engines came under the old system and that the outstanding champion of technology was the much pilloried Duke of Wellington. After reform Babbage was to receive no support at all. One cannot but feel glad that he was spared from spending the disillusioning years of the second Melbourne administration in parliament: he would have achieved little.

[20] *The Times*, 24 June 4b, 1834.

7

○●○

Science and Reform: The Royal Society

Reflections on the Decline of Science and some of its Causes[1] was published early in the summer of 1830. All Babbage's books contain a campaigning element but the *Decline of Science* is by far the most polemical. It had three principal objectives: to remove Davies Gilbert[2] from the Presidency of the Royal Society, to secure the reform of the Society, and more generally to promote the reform of science in England.

Reform of science had been the credo of the militant young liberals of the Analytical Society from its early boisterous meetings. Their first objective, the introduction of the Continental notation in calculus, was effectively secured by the early 1820s. The work of Babbage, Herschel, and others was beginning to restore the reputation of English mathematics, and foundations were being laid for the English school of De Morgan and Boole. But the young Analyticals had far more ambitious targets in view: they sought to reform the entire approach to science in England.

Among the leading natural philosophers in the country during the 1820s it was common that English science was in a poor state: before he died Humphry Davy had begun writing a book on the subject. Only too obviously English science had declined drastically when compared with the aspirations of the founders of the Royal Society and the peaks reached in the great era of Boyle and Newton. Science was failing to realize its possibilities in the rapidly industrializing Britain which led the world.

When distinct branches developed and established themselves as separate disciplines, scientific reformers in England began to form new societies in spite of the opposition of Sir Joseph Banks, who feared they would weaken the position of the Royal.[3] Indeed Sir Joseph refused to support Babbage for

[1] On 22 May 1830 *The Times* commented on Babbage's *Decline of Science*: 'We last week invited attention by some general remarks, to Mr Babbage's able and intrepid exposure of the Royal Society. We present today a specimen of his charges in distinct form. The interests of science require an instant reform of such degrading transactions as those which are disclosed.' None of Babbage's charges were ever answered for the simple reason that they were entirely justified. (A pamphlet by Moll, he remarked, was irrelevant.)
[2] Banks had in 1820 proposed Gilbert to succeed him as PRS. Babbage to Whewell, May 1820, Add. Ms. a200[192], Trinity Coll. Camb.
[3] Banks even objected to the foundation of the Cambridge Philosophical Society, offspring of the Analytical Society, Bromhead to Babbage, BL Add. Ms. 37,182, ff 201 & 323.

membership of the Board of Longitude because of his part in forming the Astronomical Society.[4] In fact the Royal remained the central body and the former Analyticals' hopes were concentrated on securing the presidency for a progressive figure. When Banks died Babbage and his friends would have liked Wollaston to be elected but as it became clear that Humphry Davy would succeed they retired until the next time the presidency should become vacant. However, when Humphry Davy became ill and resigned, his old patron Davies Gilbert succeeded him. Davies Gilbert was a popular man with some scientific knowledge. His great achievement had been in recognizing and supporting the young Humphry Davy. When Davy had become president of the Royal Society in 1820 his old friend Gilbert became treasurer. Born Davies Giddy he took the name Gilbert on marrying an heiress, Mary Ann Gilbert in 1804. Elected to Parliament in 1806 he became one of the most assiduous committee members of the age. On 6 November 1827, while Babbage was abroad, Gilbert became temporary president. Previously he had supported reform of the Royal. However, surrounding himself by a coterie and hopelessly lost in the habits of political intrigue, instead of turning for support to the active members he attempted to follow Davy's high-handed methods of running the Society. Lacking Davy's prestige and scientific ability he soon alienated Babbage, Herschel, and the leading scientific men. Babbage and his friends were furious at seeing the opportunity pass for a second time.

Many themes in Babbage's life seem to have been leading towards the reform struggle. It was a time when his hopes, as those of many others, seemed to be within reach of attainment, and the book must be considered in the context of those turbulent years. In the writing there is a note of bitter sarcasm which would probably have been absent had Georgiana lived, but the importance of this discordant note can be exaggerated: the times gave rise to forceful statements. Babbage was always a stickler for correct behaviour and the Society's general malpractice, incompetence, and lack of purpose was the basic cause of his objections. Babbage must have enjoyed writing the book. He had no difficulty in showing that the minutes of the Council had been forged, and that the President had been nominating members of the Council without even the formality of an election, so that he was surrounded by a coterie with little interest in science. The method of appointing scientific advisers to the Admiralty was a disgrace; and one of them, Captain Sabine, who had received a Copley medal, was a charlatan. Funds of the Society had been squandered; medals of the Society had been established on one basis and promptly awarded on another.[5] In short Babbage revealed that the Society was in a mess. The

[4] *Passages*, 474.
[5] Babbage had himself hoped to secure the first Royal Medal, which was awarded in 1826, for his paper on what he was later to call his 'mechanical notation'. Although the Royal Medal was

whole book breathes a new militancy stimulated during his Continental tour. He had returned refreshed and full of vigour. Meetings with men of science had suggested new forms of scientific organization[6] while encounters with liberal nationalists had revived the liberalism of his youth. Two years later Babbage was to contribute a book on political economy[7] and during the reform era he was heavily engaged in political activity, but reform of science was closest to his heart. Wollaston and Davy were gone: leadership devolved on Babbage and Herschel.

After pressure for change had been growing within the Royal Society for years, the explosion was sparked by the rapidly rising social and political reform movement in the early months of 1830. Babbage was never a man to shirk a duty which he felt devolved on him. Indeed he was only the more inclined to undertake a task if failure to do so might convey even the slightest suggestion of moral cowardice. At the time his major difficulties with the Government over the Difference Engine had scarcely begun and he could write with complete justification: 'On one point I shall speak decidedly, [*The Decline of Science*] is not connected in any degree with the calculating machine on which I have been engaged; the causes which have led to it have been long operating, and would have produced the result whether I had ever speculated on that subject, and whatever might have been the fate of my speculations.'[8]

The Decline of Science included some positive proposals—a union of scientific societies, an Order of Merit, restricting membership of the Royal Society, a reformed university curriculum—but it was primarily a work of demolition. To form a clear picture of Babbage's positive views on the future of the sciences it should be read in conjunction with his subsequent book, *On the Economy of Machinery and Manufactures*, particularly the final chapter relating what would now be called pure science to technology.

On the first page of *The Decline of Science* Babbage expresses a basic preoccupation which he had shared with Humphry Davy: 'That a country,

to be awarded 'for investigations, completed and made known to the Royal Society in the year preceding the day of the award' (Resolution of the Council, 26 Jan. 1826), the first two medals went to Dalton for his work of twenty years earlier, and to James Ivory for a paper published three years earlier (*Passages*, 144–146).

[6] *Passages*, 431.

[7] *Economy of Machinery and Manufactures*. *The Decline of Science* was dedicated to 'a Nobleman whose exertions in promoting every object that can advance science reflect lustre upon his rank'. The two probable candidates were the Duke of Somerset and Lord Ashley (later the Earl of Shaftesbury). A finely bound copy in the possession of Mr Alfred Van Sinderen, which may be the dedication copy, is inscribed to Lord Ashley, suggesting that he was in fact the anonymous dedicatee. Ashley had studied astronomy with Sir James South, with whom he formed a lasting friendship. Ashley had also helped Babbage in securing finance for the Difference Engine after the latter's return from his Continental tour. It is interesting to speculate on the influence Babbage may have had in directing Ashley's attention to conditions in the factories.

[8] *Decline of Science*, v.

eminently distinguished for its mechanical and manufacturing ingenuity, should be indifferent to the progress of inquiries which form the highest departments of that knowledge on whose more elementary truths its wealth and rank depend, is a fact which is well deserving the attention of those who shall inquire into the causes that influence the progress of nations.'

To the Analyticals the sorry state of British science had seemed so obvious since their days at Cambridge as hardly to need proving: 'It cannot have escaped the attention of those, whose acquirements enable them to judge, and who have had opportunities of examining the state of science in other countries, that in England, particularly with respect to the more difficult and abstract sciences, we are much below other nations, not merely of equal rank but below several even of inferior power.'

To make the point Babbage quoted Davy and also a particularly forthright statement by John Herschel: 'we take this *only* opportunity distinctly to acknowledge our obligations to that most admirably conducted work [the *Annales de Chimie*]. Unlike the crude and undigested scientific matter which suffices, (we are ashamed to say it) for the monthly and quarterly amusement of our own countrymen, whatever is admitted into *its* pages, has at least been taken pains with, and, with few exceptions, has sterling merit.'9

After further similar observations Herschel continued:

[Our] own countrymen ... may rest assured that not a fact they may discover, nor a good experiment they may make, but is instantly repeated, verified, and commented upon, in Germany, and, we may add too, in Italy. We wish the obligation were mutual. Here, whole branches of continental discovery are unstudied, and indeed almost unknown, even by name. It is in vain to conceal the melancholy truth. We are fast dropping behind. In mathematics we have long since drawn the rein, and given over a hopeless race. In chemistry the case is not much better. Who can tell us anything of the Sulfo-salts? Who will explain to us the laws of isomorphism? Nay, who among us has even verified Thenard's experiments on the oxygenated acids,—Oersted's and Berzelius's on the radicals of the earths,—Balard's and Serullas's on the combinations of Brome,—and a hundred other splendid trains of research in that fascinating science? Nor need we stop here. There are, indeed, few sciences which would not furnish matter for similar remark. The causes are at once obvious and deep-seated.

Closely bound up with reforming science was the development of university education. Babbage writes: 'That the state of knowledge in any country will exert a directive influence on the general system of instruction adopted in it, is a principle too obvious to require investigation. And it is equally certain that the tastes and pursuits of our manhood will bear on them the traces of the earlier impressions of our education. It is therefore not unreasonable to suppose that some portion of the neglect of science in England, may be attributed to the system of education we pursue.'10

9 Ibid., vii.   10 Ibid., 3.

Two of the old Analyticals, Peacock and Herschel, were later to play a major part in turning the Cambridge curriculum more towards the sciences, but it was slow and difficult work. 'A young man passes from our public schools to the universities, ignorant almost of the elements of every branch of useful knowledge; and at these latter establishments, formed originally for instructing those who are intended for the clerical profession, classical and mathematical pursuits are nearly the sole objects proposed to the student's ambition.'

Therefore Babbage proposed reforming the university curriculum to include more science:[11]

If it should be thought preferable, the sciences might be grouped, and the following subjects be taken together:

Modern History.	Political Economy.
Laws of England.	Applications of Science to Arts and
Civil Law.	Manufacture.
Chemistry.	Zoology, including Physiology and
Mineralogy.	Comparative Anatomy.
Geology.	Botany, including Vegetable Physiology
	and Anatomy.

Particularly remarkable is the proposal for combining 'Political Economy' with 'Application of Science to Arts and Manufacture.' It is interesting to speculate what effect the establishment of such a degree course in Cambridge, say in 1840, might have had on the future of the country's industry. The Analyticals' demand for the development of science in Cambridge was to be realized later in the century with legendary success, but in well nigh total isolation from manufacturing.

Naturally the attitude of a young man to the choice of subject for university study was much influenced by the rewards offered later in life. In one of the most controversial sections of the book Babbage drew attention to influential positions held by leading men of science on the Continent, particularly in France, Prussia, and Tuscany.

The contrast with England was striking. In the ministerial lists of the entire nineteenth century men of science are hard to find. The country gentry who ran the nation were positively antipathetic to science. This attitude continued when a professional civil service was developed and remains an intractable problem to the present day. During the Second World War there was a saying that 'Scientists should be on tap, not on top'. Indeed this attitude fed back into the very industries which owed their origins to the new technologies, as may be seen clearly in the composition of boards of directors. In drawing attention to the low social prestige attached to science in England Babbage was making a point of fundamental importance.

[11] Ibid., 5–6.

Babbage's book, demanding that the country attach far greater importance to science, caused a storm amongst natural philosophers, and historians turning to the subject are still inclined to argue the question. The story is that the 'anti-declinists' opposed government support for science, much as the classical economists were said to be opposed to all forms of government intervention.[12] In fact for the latter the question really concerned the nature of intervention. And both 'declinists' and 'anti-declinists' favoured government support for science, though the nature of this support and whom it should benefit remained at issue. The 'declinists', Humphry Davy, Babbage,[13] Herschel, were in effect outside the university system. The 'anti-declinists', such as Whewell and Airy, then still minor figures on the national stage but later of central importance, had vested interests in the universities to protect. Certainly they were willing to come down from their academic or, later, government fastnesses and, with little involvement and less understanding, almost *ex-cathedra* to offer advice; but serious involvement in industrial and technological problems, such as Babbage suggested, was for them wholly out of the question.

Babbage delivered a devastating attack on the Royal Society: 'The cultivators of botany were the first to feel that the range of knowledge embraced by the Royal Society was too comprehensive to admit of sufficient attention to their favourite subject, and they established the Linnean Society. After many years, a new science arose, and the Geological Society was produced. At another and more recent epoch, the friends of astronomy, urged by the wants of their science, united to establish the Astronomical Society. Each of these bodies found, that the attention to their science by the parent establishment was insufficient for their wants, and each in succession experienced from the Royal Society the most determined opposition ... '[14]

Babbage continued:

Perhaps I ought to apologize for the large space I have devoted to the Royal Society. Certainly its present state gives it no claim to that attention; and I do it partly from respect for its former services, and partly from the hope that, if such an Institution can be of use to science in the present day, the attention of its members may be excited to take steps for its restoration ...

If those, whose mismanagement of that Society I condemn, should accuse me of hostility to the Royal Society; my answer is, that *the party* which governs it is not the Royal Society; and that I will only admit the justice of the accusation, when the whole body, becoming acquainted with the system I have exposed, shall, by ratifying it with their approbation, appropriate it to themselves: an event of which I need scarcely add I have not the slightest anticipation.[15]

One by one he enumerated the abuses, beginning with the way in which

---

[12] cf. D. P. O'Brien, *The Classical Economists*, 274–5, Clarendon, 1975.

[13] Babbage's position as Lucasian Professor does not invalidate this point: the duties involved him for a couple of days a year and the emolument was insignificant.

[14] *Decline of Science*, 40–1.   [15] Ibid., pref. xiii & xiv.

members were elected. 'A.B.' gets any three Fellows to sign an appropriate certificate. 'At the end of ten weeks, if A.B. has the good fortune to be perfectly unknown by any literary or scientific achievement, however small, he is quite sure of being elected as a matter of course. If, on the other hand, he has unfortunately written on any subject connected with science, or is supposed to be acquainted with any branch of it, the members begin to inquire what he has done to deserve the honour; and, unless he has powerful friends, he has a fair chance of being black-balled.'[16] The Royal Society was far too large, with 685 members compared with 75 of the corresponding body in France, 38 in Prussia, and 40 in Italy.

Next Babbage turned to Gilbert. The resolution proposing him for the Presidency had affirmed that 'Davies Gilbert Esq is by far the most fit person to be ... President'.[17] This was too much for Babbage to stomach, even though Davies Gilbert was a friend and had supported the project for the Difference Engine:

That Mr. Gilbert is a most amiable and kind-hearted man will be instantly admitted by all who are, in the least degree, acquainted with him: that he is fit for the chair of the Royal Society, will be allowed by few ... Possessed of knowledge and of fortune more than sufficient for it, he might have been the restorer of its lustre ... By the firmness of his own conduct he might have taught the subordinate officers of the Society the duties of their station. Instead of paying compliments to Ministers, who must have smiled at his simplicity, he might have maintained the dignity of his Council by the dignity of knowledge ... But he has chosen a different path; with no motives of interest to allure, or of ambition to betray him, instead of making himself respected as the powerful chief of a united republic,—that of science,—he has grasped at despotic power, and stands the feeble occupant of its desolated kingdom, trembling at the force of opinions he might have directed, and refused even the patronage of their names by those whose energies he might have commanded.

Babbage then turned to a number of detailed questions. His views on the suitability of military persons for scientific appointments would find wide acceptance today:

The habits both of obedience and command ... are little fitted for that perfect freedom which should reign in the councils of science. If a military chief commit an oversight or an error, it is necessary, in order to retain the confidence of those he commands, to conceal or mask it as much as possible. If an experimentalist make a mistake, his only course ... is to acknowledge it in the most full and explicit manner. The very qualifications which contribute to the professional excellence of the soldier, constitute his defects when he enters the paths of science; and it is only in those rare cases where the force of genius is able to control and surmount these habits, that his admission to the force of science can be attended with any advantage to it.... persons not imbued with the feelings of men of science, ... are too apt to view every criticism upon [their observations] as a personal question ... Nothing can be more injurious to

[16] Ibid., 50–1.   [17] Ibid., 53–4.

science than that such an opinion should be tolerated. The most unreserved criticism is necessary for truth.[18]

Babbage's views on the medical profession also have a modern ring:

The honour of belonging to the Royal Society is much sought after by medical men, as contributing to the success of their professional efforts, and two consequences result from it. In the first place, the pages of the Transactions of the Royal Society occasionally contain medical papers of very moderate merit; and, in the second, the preponderance of the medical interest introduces into the Society some of the jealousies of that profession.[19]

To make his case Babbage selected a few examples of malpractice for detailed consideration. The first was a case of falsification of the minutes. The serious question behind this occurrence was whether the members of the Council should be nominated by the President, as was Gilbert's practice, or elected after full discussion at the meeting of the Society previous to the Anniversary. 'Another topic, which concerns most vitally the character and integrity of the Royal Society, I hardly know how to approach', Babbage wrote. 'It has been publicly stated that confidence cannot be placed in the written minutes of the Society; and an instance has been adduced, in which an entry has been asserted to have been made, which could not have been the true statement of what actually passed at the Council.'[20] Captain Beaufort, a friend of Babbage's, had declined to sit on the Council and Gilbert, lacking time to substitute a name before the meeting, had simply replaced Beaufort after the meeting by Sir John Franklin, the polar explorer. Gilbert was seeking to conceal the fact that outstanding men of science were unwilling to serve under his leadership.

To the case of Captain Sabine, a Scientific Adviser to the Admiralty, Babbage devoted a tenth of the book. Once again the underlying question was: should appointments be made by the President or should the Royal Society function democratically? The three posts of Scientific Adviser to the Admiralty were sinecures worth £100 a year. Both John Herschel and Captain Kater had declined the positions as degrading. The Admiralty had placed equipment and excellent facilities at Captain Sabine's disposal enabling him to make observations at many points on the Atlantic littoral, and the results were described by Babbage as follows:

The remarkable agreement with each other, which was found to exist amongst each class of observations, was as unexpected by those most conversant with the respective

18 Ibid., 58–9.
19 Ibid., 187–8. Something of this scientific attitude to the medical profession has remained. For example we find Bernhard Katz, in his inaugural lecture as Prof. of Biophysics at University College, London, apologizing for a medical education: *Different Forms of Signalling Employed by the Nervous System*, H. K. Lewis, 1952.
20 *Decline of Science*, 61–2.

processes, as it was creditable to one who had devoted but a few years to the subject, and who, in the course of those voyages, used some of the instruments for the first time in his life.

This accordance amongst the results was such, that naval officers of the greatest experience, confessed themselves unable to take such lunars; whilst other observers, long versed in the use of the transit instrument, avowed their inability to take such transits. Those who were conversant with pendulums, were at a loss how to make, even under more favourable circumstances, similarly concordant observations ... On whatever subject Captain Sabine touched, the observations he published seemed by their accuracy to leave former observers at a distance...

The Council of the Royal Society spared no pains to stamp the accuracy of these observations with their testimony. They seem to have thrust Captain Sabine's name perpetually to their minutes, and in a manner which must have been almost distressing: they recommended him in a letter to the Admiralty, then in another to the Ordnance; and several of the same persons, in their other capacity, as members of the Board of Longitude, after voting him a *thousand pounds* for these observations, are said to have again recommended him to the Master-General of the Ordnance. That an officer, commencing his scientific career, should be misled by such praises, was both natural and pardonable; but that the Council of the Royal Society should adopt their opinion so heedlessly, and maintain it so pertinaciously, was as cruel to the observer as it was injurious to the interests of science.[21]

Sabine's measurements were of course entirely spurious. For example with one instrument it was later found by Captain Kater 'that the divisions of its level, which Captain Sabine had considered to be equal to *one second each*, were, in fact more nearly equal to *eleven seconds*, each one being 10·9'.[22] Thus on a detailed technical question the Council of the Royal Society had made complete public asses of themselves.

After aiming a couple of shots at the Royal Observatory and the Royal Society's handling of publication of the Greenwich Observations, Babbage turned to the Royal Medals which had been instituted in 1825. Two gold medals to the value of 50 guineas each were to be awarded annually 'in such a manner as shall, by the excitement of competition among men of science, seem best calculated to promote the object for which the Royal Society was instituted'. In January 1826 the Royal Society resolved that the medals should be awarded for discoveries or series of investigations completed and made known to the Society in the year preceding the day of the award. In November of the very same year, Babbage pointed out, the medals were awarded for two investigations completed respectively twenty and three years earlier. The Society was incapable of making simple rules and sticking to them.[23]

After pointing to a number of other abuses and anomalies Babbage drew his attack together. 'The Society has, for years, been managed by a party, or

---

[21] Ibid., 78–80.   [22] Ibid., 90.   [23] Ibid., Ch. IV, section 7.

coterie,[24] instead of functioning democratically. To change the Society in practice meant changing the membership and this was widely recognized:

The indiscriminate admission of every candidate became at last so notorious, even beyond the pale of the Society, that some of the members began to perceive the inconvenience to which it led. This feeling ... induced several of the most active members to wish for some reform in its laws and proceedings; and a Committee was appointed ... to enquire,—First, as to the means and propriety of limiting the numbers of this Society; and then, as to other changes which they might think beneficial. The names of the gentlemen composing this Committee were:

Dr Wollaston,	Mr Herschel,
Dr Young,	Mr Babbage,
Mr Davies Gilbert,	Captain Beaufort,
Mr South,	Captain Kater.[25]

The Committee reported early in the summer of 1827, proposing to admit in future only four candidates a year who should all be active men of science. Thus as the old members died the numbers would be reduced gradually to four hundred, the suggested total membership, all actively concerned with science. Babbage himself thought the process much too slow and wished to divide the membership into two classes according to whether or no they had had at least two papers published in the *Transactions of the Royal Society*.

It was not the first time the Society had considered the problem. On 27 August 1674 a Council meeting at which Christopher Wren and William Petty were present had resolved 'that to make the Society prosper, good experiments must be in the first place provided to make the weekly meetings considerable, and that the expenses for making these experiments must be secured by legal subscriptions for paying the contributors; which done, the Council might then with confidence proceed to the ejection of useless Fellows.'[26] Compared with such a drastic proposal Babbage considered his own suggestion moderate.

On 25 June 1827 when the committee reported the Council minute 'regarding the importance of the subject, and its bearings on the essential interests of the Society, ... and considering also the advanced stage of the session, recommend it to the most serious and early consideration of the Council for the ensuing year.'[27] During the ensuing year Babbage was on the Continent and nothing was done.

Davies Gilbert's unwillingness to act on the committee's recommendations and his determination to stifle discussion had led directly to Babbage writing *The Decline of Science*. If the cause of reform was not adequately represented in the Council the question would have to be brought before the general membership or before the last tribunal, public opinion. A printed statement could enter into greater detail than a speech and was likely to come under the

[24] Ibid., 142. [25] Ibid., 152.
[26] Quoted in Ibid., 153. [27] Ibid., 165.

consideration of a far larger body of members. Thus did Babbage justify raising in public the affairs of the Royal Society. Recalling that his book purported to be a general discussion of the state of science, Babbage also turned his attention in one chapter to the question of making observations. The habit of scientific observation is so much a part of present day practice that it takes an effort to recall that even the idea of rigorous observation required explanation. The chapter is beautifully done. Babbage discusses the accuracy of instruments, the liability of observers to error, raising questions of fatigue and bodily health. Then there is an entertaining discussion of hoaxing, forging, and the gentle art of cooking results.

Reflections on the Decline of Science in England was published by B. Fellowes of Ludgate Street and J. Booth of Duke Street, Portland Place at the beginning of May 1830. The effect was predictably explosive. Herschel and other friends approved, if a trifle nervously.[28] Babbage was to dine with the Duke of Somerset in Park House; Davies Gilbert was of the party; Babbage might find the meeting embarrassing after publication of the book and not care to come; Babbage assured the Duke that on the contrary he would be delighted to come.[29] In the corridors of the Royal Society there were dark murmurings and threats of expulsion. At the meeting of the Society where the book was discussed debate was deferred at Dionysus Lardner's suggestion to a special meeting. Babbage was invited to attend in a peremptory note from P. M. Roget, Secretary of the Society. G. Everest wrote to Babbage: 'I have heard today that you are to be castigated at a special meeting on Thursday, and that in the event of your refusing to retreat the society proceed forthwith to the question of expulsion ... command me in any way you like.'[30] Babbage sent a private note to Davies Gilbert[31] saying that he had not had adequate notice of the intending motion, and stating that the book was his alone. On this matter Babbage was connected with no party and had combined with no individual. He also sent a formal letter pointing out that in the book he had already remarked on the impropriety of discussing the affairs of the Society at its ordinary meetings, and declined to attend: 'My appeal has been made openly to the public and if the President and Council of the Royal Society choose *directly to contradict the facts* on which arguments rest—refute the reasoning which is supported by them—or to complain of any undue severity in the opinions they have led me to entertain a similar channel is the most effectual to convey their refutation. I claim from them the same openness which I have exercised.'[32] Early in June the attempt to arraign Babbage fizzled out. Davies Gilbert was doing all he could to smooth the matter over. To have expelled Charles Babbage by the votes of a coterie of

[28] Herschel to Babbage, 22 May 1830. Babbage/Herschel Corr. RS, f 252. Dalton to Babbage, 15 May 1830. BL Add. Ms 37,185, f 176.

[29] BL Add. Ms. 37,185, ff 170–1.   [30] 'Ibid., f 198.

[31] Ibid., f 200.   [32] Ibid., f 201.

mediocrities would have made the Royal Society the laughing stock of scientific Europe.

Meanwhile Babbage continued with his work on the Difference Engine, made visits to industry, and led a busy social life. The Duke of Buckingham had one of Amici's latest and most improved microscopes at Buckingham House, which Babbage wanted to borrow and inspect. In the autumn he and Wolryche Whitmore visited Liverpool for the opening of the Liverpool-Manchester railway, the unfortunate occasion when Huskisson was killed by Stephenson's Rocket.[33] Next we find him visiting a plate-glass works with I. K. Brunel. The most interesting operation, the casting, took place between 6 and 10 at night.

Lonely without Georgiana, Babbage was considering remarrying and prudently enquired about a talented but slightly eccentric young lady in the vicinity of Malvern. His friend replied, 'I think you had better make further enquiries, for the more safely you proceed the more will you secure yourself from all future vexation—certainly nothing I could learn seems a sufficient obstacle, for you know our creed on the subject of talent and oddity is not that of all the world—so goodbye but mind you do not propose and then enquire afterwards ... Mind you are booked for Zummerzet [Somerset] after Malvern.'[34] Shortly afterwards Babbage stayed at Well House, Malvern. But whoever the lady was nothing came of it.

In November Davies Gilbert relinquished the Presidency of the Royal Society, dislodged finally by Babbage's onslaught. Babbage persuaded a reluctant John Herschel to stand and collected signatures to requisition him as President. The old Analyticals rallied round. From his country estate Edward Ffrench Bromhead wrote enthusiastically supporting Herschel: 'You have given our science a jolt and a rough one, but we had got beyond mild remedies— the cold bath and a shock were absolutely necessary. ... We want in England the Order of the Bath extending to civil and literary merit.' In the *'Decline'* Babbage had advocated an Order of Merit to encourage scientific work. He was also suggesting peerages for life. Babbage and his friends remained insistent on the need for some such awards but their attitude was often misunderstood. Personally Babbage had no great interest in decorations and turned down a knighthood when John Herschel and David Brewster both accepted. Decorations were acceptable and indeed desirable only when they contributed to the dignity or social position of science. If acceptable today, when they have

---

[33] *Passages*, 314–16. The Earl of Bridgewater had opposed the railway as a rival to his canal, the second proper canal to be built in England. Having failed to stop the railway he purchased a large block of shares and was said to show a profit of £180,000: Duke of Somerset to Babbage: BL Add. Ms. 37,185, f 304.

[34] BL Add. Ms. 37,185, f 310: Revd. Fr. Lunn to Babbage, Chez Fellowes, en passant. 22 Nov. 1830.

been greatly devalued, they would have had far greater effect in 1830. Bromhead outlined his ideas:

A Sketch
Dignitaries, Knights, Companions answering to our Grand Crosses etc. Each of these classes would contain Naval, Military, Civil.
CDB   Chancellors of Universities, Eminent Patrons of Arts and Sciences, Ambassadors etc.
CKB   Herschel, Ivory, Walter Scott, Shea, Telford, etc. (Babbage, a literary revolutionary, carefully excluded).
CCB   The lay literati and civilians.
This would show Science to be *acknowledged*, and would rally the public mind by bringing into a circle what merits distinction.

In a postscript he added, 'Public awards should not be offered for *indefinite* objects, such as the greatest improvement in magnetism, etc.—but for something which *can be done*, as the best digest of facts on a given subject—a translation, etc.'[35] However, while advocating a system of decorations, Babbage himself doubted whether much could be done until the higher classes were better acquainted with science and the whole climate of public opinion had changed.

In the meantime, unable to find a scientific man of calibre, Gilbert relied on the deference vote, nominating His Royal Highness the Duke of Sussex, third son of George III, for President. An old political hand, he laid his plans carefully: he would exercise through influence the control he could no longer maintain directly as President. The battle grew hot. It was rumoured that His Royal Highness would consider the conduct of those who signed the requisition for Herschel in no other light than a personal insult to himself, precluding him from the meeting in society or sitting down at table with the individuals concerned.[36]

However, support for Herschel was strong and his supporters felt victory was secure. Then Babbage, who was running the campaign, made a fatal mistake: he told Herschel's supporters in more distant parts of the country they need not bother coming to London for the ballot. It is dangerous for a reformer to underestimate the strength of reaction. The chance was lost and the Duke of Sussex elected President by a small majority. Not until much later in the century was entry restricted so that the Royal Society gradually became a professional body. During the formative decades when the professional civil service was beginning to emerge the Royal Society remained relatively ineffectual. From his friend George Harvey in Plymouth Babbage received a sad letter: 'my place on the Mail was *actually taken* when your letter arrived the Evening before my intended departure next Morning saying *it was not at all*

---

35 Ibid., f 351.
36 William Tooke to Babbage, 2 Dec. 1830: Ibid., f 364.

necessary to come up. From what I hear however, if every Man who signed the requisition had been at his post Herschel would have been our chief.'[37]

Safely elected the Duke acted promptly to restore harmony. He let it be known he was anxious to meet on friendly terms those who had supported Herschel and hold out his hand to them. Yet another committee was appointed to consider reform of the Society. Invited to serve, both Babbage and W. H. Fitton declined, considering the exercise futile in the circumstances. Babbage drafted a trenchant letter to the Duke of Sussex:

A few years since the necessity of some considerable alterations in the RS was strongly felt and a Com[mittee] was appointed with nearly the same objects as the present one. I was called upon to become a member of that Com[mittee]—it devoted considerable attention to the subject and made a report—The Council in adopting that report recommended it to the '*earliest and most serious consideration of the Council of the ensuing year.*' The Council of the following year *refused even to take that report into consideration.* When the recommendations of Dr Wollaston, Dr Young and Mr Herschel were treated with such neglect it would be presumptuous in me to anticipate greater attention to any suggestions of my own.

I am aware that it has been said that the late elections into the council have rendered the recurrence of such an event less probable. Having at one of the meetings previous to the election stated to the RS my motives for declining a seat in the Council I may be permitted to offer my own reasoning on the subject without the imputation of personal feelings.

That the society approve of the past management of its affairs appears to follow from the following reasons:

The same gentleman who on the occasion of the reform Com[mittee] filled the office of Pres[ident] is now Vice Pres[ident]. The same preference for men of rank to men of Science exists in a greater degree than before. On the Council of the past year were the names of two persons whose high official station rendered their attendance almost impossible. On the present Council these names have been retained and those of two other gentlemen have been added whose position in Society or whose inclination has equally prevented their attendance: for at the 19 Councils of the past year and of those already held during the present the Council minute-book does not record one single attendance of those gentlemen.

In these remarks it is my wish not to be misapprehended. Those to whom I have alluded have occupied their talents with more advantage to their country in other pursuits, and whilst I think it would be important for science that noblemen and persons of rank should be conversant with its objects and occasionally mingle with its cultivators I am far from thinking that it is conducive to its true interests to reject men of science from its councils for the purpose of paying empty compliments to persons of rank.[38]

Then after discussion with Fitton he thought better of sending the letter and in a brief note declined a seat on the committee.

This was really the last battle the old friends of the Analytical Society fought

[37] 3 Jan. 1831: Ibid., f 429.   [38] Ibid., f 501.

as a group. Their paths were diverging. Herschel had only with much reluctance consented to stand for the Presidency. He was to show little interest in the foundation of the British Association and soon departed for the Cape of Good Hope to make his famous catalogue of the stars of the southern hemisphere. Edward Ryan was already in India. George Peacock in Cambridge played a leading part in reforming the university curriculum, giving more place to science and contributing to the great developments of Cambridge science at the end of the century. Edward Ffrench Bromhead was becoming increasingly isolated in the country. Babbage himself was turning more to political reform and to the systematic application of science to commerce and industry.

With the defeat of Herschel the back of the national science reform movement was broken,[39] and the demoralization which followed later merged into the general disillusion following the compromise of the Reform Bill. Babbage, the leading radical scientific figure, was permanently removed from the centre of power. The development of professional science in England took place on the basis of an exaggerated separation between pure science and applied technology, and this separation, which is in strong contrast to Babbage's doctrine of the union of theory and practice, became one of the most marked characteristics of English science.

[39] Thus in 1831, when the scientific radicals formed the British Association, the lead was taken by the provincial societies. Of interest as a social phenomenon, its importance to the development of science can easily be overestimated. Although it provided an outlet for exasperated feelings, the British Ass. was mainly a sounding board and it was never of comparable scientific importance to the specialist societies, let alone the Royal.

8

○ ● ○

On the Economy of Machinery and Manufactures

Since the inception of the project to construct a Difference Engine Babbage
had been making the most detailed study of industry: touring the country to
visit factories, inspecting every machine, every industrial process he could
discover. Much of his eighteen months on the Continent had also been devoted
to the study of industry. Moreover long before starting on the Difference
Engine he had been interested in political economy. Arising from these studies
Babbage published in 1832 his book *On The Economy of Machinery and
Manufactures*, with great effect. Adam Smith had never really abandoned the
belief, reasonable enough in his day, that agriculture was the principal source
of Britain's wealth; Ricardo's ideas were focused on corn; Babbage for the first
time authoritatively placed the factory in the centre of the stage. The book is
at once a hymn to the machine, an analysis of the development of machine-
based production in the factory, and a discussion of social relations in industry.
But it was something more than analysis: it was intended as a basis for action.
The book was an intellectual response to a set of major policy questions: how
should the resources of the rapidly advancing sciences be deployed to assist
development of British technology and industry? Even to pose such a question
had a novel ring and owed much to French influence (it may be remarked that
Babbage not only had many friends among the Parisian natural philosophers
but was also well-read in French economic literature). Babbage's prescription
was the coherent development of education, science, and technology, and their
application to industry: the union of scientific theory and industrial practice.

Babbage was a familiar figure to the economists of the time. Although not
himself a member he had many friends in the influential Political Economy
Club. Wolryche Whitmore was a member and took Babbage to some of the
club dinners. Babbage himself was the founder of the London Statistical
Society, second in importance among societies of economists only to the
Political Economy Club itself. He was also an early and influential advocate of
decimal coinage: once again the French influence was evident.

The Economy of Manufactures established Babbage's position as a political
economist and its influence is well attested, particularly on John Stuart Mill
and Karl Marx. Babbage's pioneering discussion of the effect of technical
development on the size of industrial organizations was followed by Mill and
the prediction of the continuing increase in the size of factories, often cited as

one of Marx's successful economic predictions, in fact derives from Babbage's analysis. In his *Principles of Political Economy* Mill quotes extensively from Babbage; but beyond this Mill's book is written on the assumption that his readers will be familiar with Babbage's work: it is as if Babbage's book is the transmission system connecting Mill's *Principles* with industrial practice. Nor should it be forgotten that Mill's *Principles* was the bible of economists in the second half of the nineteenth century, the main channel through which the formulations of the classical economists of the first half of the century reached the founders of the new economics in the second. At the same time as Babbage was completing the first edition of *The Economy of Manufactures*, Harriet Martineau was beginning her series of moral tales,[1] popular exposition of political economy. Babbage's book contains a good deal of simple explanation and he was most anxious that it should be read by intelligent working men even if it was not aimed at Harriet Martineau's mass readership. Babbage wrote with many talents: a natural philosopher and mechanical engineer, his knowledge of factory and workshop practice was encyclopaedic; he was well-versed in relevant business practice; and he was without rival as a mathematician among contemporary British political economists. He was also a master of conceptualization and wrote clearly. Although empirical and unashamedly inductive, it is the sure theoretical grasp which gives the book its underlying strength.

The England in which Babbage lived owed its growing wealth and power to the produce of its factories. On the products of its iron works, on Manchester's looms, depended Britain's military power and the expansion and consolidation of its empire. In the rooms of his own house, in the shops crowding the streets of the great cities, in every fabric or article that he saw, there was embodied for Babbage a history of repeated experiment, patient thought, of failures followed by further trials leading to present excellence. The most insignificant manufactured product concealed processes of fascinating simplicity and unlooked for results. It seemed curious to Babbage that the ruling classes should be so singularly indifferent to the processes on which their wealth was increasingly based. The wealthy sons of generations of manufacturers, who had been educated at one of the ancient universities, seemed only too anxious to forget the all too recent associations of their families with manufacturing. Samuel Smiles was to write, with a good deal of justification:

One of the most remarkable things about Engineering in England is, that its principal achievements have been accomplished, not by natural philosophers nor by mathematicians, but by men of humble station, for the most part self-educated.... The educated classes of the last century regarded with contempt mechanical men and mechanical subjects. Dean Swift spoke of 'that fellow Newton over the way—a glass

[1] Harriet Martineau; *Illustrations of Political Economy*, London, 1834

grinder and maker of spectacles.' Smeaton was taken to task by his friends of the Royal Society, for having undertaken 'navvy work' of making a road across the valley of the Trent, between Markham and Newark.[2]

To this supercilious attitude Babbage was the exception. Although educated as a Cambridge mathematician, rising in society and on familiar terms with the great, he involved himself increasingly not merely with such general principles of political economy with which a gentleman might comfortably become passingly acquainted, but with the practical details of manufacturing processes and commercial procedures. By no means satisfied with interpreting industrial phenomena his object was, by the systematic application of mathematical techniques and scientific method, to change and improve commercial and industrial practice. There is not a trace of apologetics in his work: he was completely convinced of the superiority of the capitalist system. His condemnation of improper combinations of workers and of masters is even-handed. Indeed his choice of publishing as an example of improper combinations of masters against the public was to cause Babbage difficulty in publishing the book itself. And in 1832, when the first editions of *The Economy of Manufactures* appeared, he still had confidence in the future of Britain's industrial strength.

Babbage's study of machinery and manufacturing processes originally started in a manner so extraordinary that it has passed almost without comment, as if no one could believe what he was really doing: he settled down to study *all* the manufacturing techniques and processes, more particularly all the mechanical devices and inventions he could find, searching for ideas and techniques which could be of use in the Difference Engine. The manner in which this research led to the elegant devices embodied in the Calculating Engines is itself a fascinating study. Part of the work had been included in the *Encyclopaedia Metropolitana*[3] some years earlier. Babbage had also intended to present in Cambridge in his capacity as Lucasian Professor a series of four lectures on The Political Economy of Manufacture. However, although he was loquacious and a brilliant raconteur, he found lecturing a burden and the lectures were abandoned while he prepared the whole study for presentation as a book.

It is remarkable that while he was working on the Difference Engine and living a busy social life, while he was engaged in the 'Decline of Science' controversy and his other political activities, Babbage should have found time to complete his research. The investigation of Continental industry which he had commenced during his early trips to France had been continued far more

[2] Samuel Smiles; *Lives of the Engineers*, I, XVI.
[3] C.B., Essay on the General Principles which regulate the Application of Machinery, *Encyclopaedia Metropolitana*, London, 1829.

extensively during his long tour through Italy, Austria, and Germany in 1827–8. Returned to England he continued his studies of British industry. This thoroughness, the great capacity for hard work and boundless self-confidence, is characteristic also of some of Babbage's friends. We find it most clearly in the Brunels. If the energy led on occasion to spectacular failure, without it the great engineering achievements of the first half of the nineteenth century would have been inconceivable. It is typical of the era that the development of the Difference Engine should have stretched out over the years, becoming a far greater undertaking than Babbage had originally envisaged. So also did his study of manufacturing. As he combed industry in search of material, minute detail could attract his interest. He kept up a wide correspondence seeking the information he required. But by far his most important source of information was personal inspection and discussion. Only a tiny fraction of all this research is actually incorporated in the book.

We get some glimpses of his trips from the correspondence, but the existing references to journeys are fragmentary and incomplete. He went often to the Midlands where he became a well-known and highly respected figure, particularly after the first edition of his book had been published. When travelling alone Babbage preferred to stay in the commercial inns. There he could benefit from discussions with commercial travellers, collecting many details of commerce. Also such inns were cheap. He was amused to note that his knowledge was such that he could soon pass as a traveller in almost any commodity he chose to name. His upper class friends on occasion found difficulty locating him as it hardly occurred to them to look in such plebeian establishments.

On one occasion we find Anthony R. Strutt of the great textile manufacturing family writing to Babbage from Machiney, two miles south of Belper: 'I hope you will endeavour to come to me on Monday to stay as long as convenient. I keep a bachelor's cottage at Machiney which is six miles north of Derby on the road to Belper. We have considerable works close by—any of the coaches from Nottingham to Manchester pass within 100 yards—you would be set down at *the toll gate at Milford Bridge*. All I can ask is that you come *without ceremony* and *expect none*, but take me as I am—and expect something to eat and a bed.'[4] On his return from the Midlands Babbage was able to forward to the Treasury a petition from the bobbin-net lace manufacturers requesting a reduction in the duty on starch.

He became a familiar figure in court and society, with a circle of aristocratic friends including the Duke of Wellington and Lord Ashley as well as Whig society. But he also knew many industrialists and was thoroughly at home in the company of intelligent working men. The members of trade unions and co-

[4] BL Add. Ms. 37,188, f 76. Anthony Strutt was the member of the family responsible for engineering.

operatives saw that the future lay in the development of industry and the adoption of laws and customs to suit the developing capitalist economy. Their intelligent members were delighted with this practitioner of science who applied his knowledge and sympathetic understanding to the actual problems of the industries which were the centres of their lives.

Babbage tells a story, which typifies his universal curiosity, of a day's visit to Bradford, probably some years later, when he was staying at Leeds. Passing a factory making door-mats, or some such product, Babbage enquired whether a stranger might inspect the works and was duly invited to do so. As they toured the factory Babbage questioned the man appointed to show him round as well as several other workmen. When Babbage's questions passed the limit of their knowledge the local sage, one Sam Brown, was duly sought and he answered them all. It soon transpired that Sam Brown had read *The Economy of Manufactures*, was glad to meet its author, and would be happy to take him round the Bradford co-operative shop if he could spare half an hour. They chanced to meet the secretary of the local co-operative who offered to show Babbage the rules and take him round the local societies. However time pressed, and Babbage had to return to Leeds with the medical friend who had brought him for the day in a carriage. Driving back to Leeds they passed a large iron-works, its tall chimneys sending columns of fire into the night. Babbage knew that in one of the works a tunnel, initially carved from the rock for a canal, was now used as a reservoir of pressurized air for the blast furnaces. A hundred horse-power steam engine continually pumped air into the cavern. The doctor knew the manager of the works and Babbage was able to enter the cavern through a small iron door. He surveyed the inside of the cavern, savouring the power of fire and heat.

While Babbage was continuing his studies of British industry an interesting counterpoint is provided by the travels and work of Wright, the workman who had accompanied him through Europe in 1827 and 1828. Wright's letters date from 1834 and 1835, after the cessation of work on the Difference Engine, and when Babbage's succeeding project to design a more versatile calculating engine was under way. During their tour of the Continent Babbage had made a practice of ensuring that Wright also saw round the workshops and factories which they visited, but only after Babbage had made certain there was no objection and secured the owner's permission. Wright's second letter gives a fascinating picture of a journeyman educating himself in the new industrial age. Such things were rarely put on paper, but a competent workman could write to Babbage and be sure of a comprehending and sympathetic reception. We get a picture of the enormous strength of the British industrial system, with its combination of apprenticeship and night-school, and some indication of how it could manage as well as it did with so rudimentary a formal system of technical education:

No. 3 Shadwick St, Liverpool
2 April 1835

[I] tell you wat I have been doing since I left London—We went direct from hence to Hull and from there to Leeds by Steam. At Leeds we found a great deal of Machinery that wos quite new to us. It took us two weeks to collect the information we required for the information we ware seeking is much more difficult to aquire than that which you generally require. We not only wanted the action and comparative quantity of work done by a machine but also the manner of making each part of it, the piece work and other prices etc. To get this we were obliged to mix with the workmen of different branches which required much time and expense. From Leeds we walked the whole distance to Glasgow and althoe art did not supply us with much of interest in this long line of country, Nature fully supplied us with useful and interesting employment to inspect her wonderful works around the North of Yorkshire, Westmoreland, Cumberland and Scotland and as we had some books with us on Geology and Mineralogy we employed ourselves in learning as much of the outlines of these beautiful sciences as we were able. We went down one of the deepest coal-pits in Yorkshire the shaft of which is 486 feet and from the bottom to were the Miners were at work is nearly half a mile. We also went into the Copper mines in Cumberland, the Iron Works and indeed we saw everything worth our notice in this part of the country. In Glasgow we could find but little machinery that we had no already seen. Their sistem of working and their prices were the principal worth collecting here. Our outlay had exceeded our provision. Therefore we were obliged to branch off in search of a fresh supply and as it depended on chance were we should next work I have not been able to here of my companions since, but when we meet again we shall have the advantage of copying each others notes.

There was much talk about an Engine Factory about 30 miles distance. it was said that their engines would work much better and longer than any made elsewhere from which I was induced to go and learn the cause of this superiority. I worked here (Greenock) for two months. I next took a serquit of 200 miles through the highlands and from there to Londonderry in Ireland and round to Liverpool and am now working in one of the principal factories. Altho I have learned much in this long journey I was not able to get the principal information that I required. One thing among this is locomotive work and this I must collect between here and Manchester and this is the cause of the two months delay as mentioned above [before he can be in London]. Since I left London my wages have been 21/-, 27/-, 34/-, 36/-.

I now intend shortly to settle myself for it is a weary house that we find among strangers were we are only welcome as we are able to pay and the habits and conversation of the Factory are indeed disgusting to a thinking mind. I hope you will not delay your job on my account for I must be in London shortly. I have something to learn here yet (in ornamental turning). I mention this for fear you will inconvenience yourself on my account.

I send this without any fear of offending as I believe if I have taken too great a liberty you will not be offended but tell me of my error and hope to remain

Your humble servant
R WRIGHT.[5]

When Babbage came to discuss the technology of industry he classified

[5] BL Add. Ms. 37,189, f 69.

processes both according to their physical nature and the advantages deriving from them: accumulation and storage of power; regulation and control of power; controlling the velocity of machinery; making accumulated power act over long periods of time; exerting forces far greater than human strength and executing operations too delicate for human touch. Then he discussed counting, measuring, and recording operations, copying of many kinds, and economy of materials. He could not resist the temptation to digress in order to illustrate physical processes or to describe some particularly spectacular effect. These digressions not only indicate the great breadth of Babbage's reading and knowledge but give the book much of its charm.

In the nineteenth century British manufactured goods were striking in their ubiquity, often preceding the most adventurous explorers. An expedition might penetrate to some unknown petty kingdom only to find its food served on a much prized wash-basin of English manufacture. So powerful was the effect of machinery that Indian cotton came to be conveyed to Lancashire to be turned into calico and returned to India where it replaced the local cloth. Today such procedures are commonplace but in an age when the only ocean-going ships were still powered by the winds shipment of cheap cotton over vast distances so that it could be processed by machinery was an extraordinary circumstance of great portent. Where Marx was later to see the result in the bones of cotton weavers bleaching the plains of India[6], Babbage saw that 'distant kingdoms have participated in its [factory production's] advantages'. In the long run, he held, development of technology and industry would benefit the whole world.

If the great new power of steam was to be used to drive industry the power must first be controlled and directed. The best-known control device was the governor but there were many other ingenious contrivances, some special to a particular county or district. In Cornwall, for example, the rate of operation of steam engines was controlled by the time taken to fill a hollow vessel when placed in water, the rate of admission of water to the vessel being determined by an adjustable valve under the control of the engine-man. Another method was to regulate automatically the rate of input of coal. The output of steam engines was kept steady by the use of huge flywheels; but a flywheel could also be used to store power which could then be concentrated in a short time and during that period the rate of power-output could be far greater than that of the steam-engine. By the use of such examples Babbage sought to describe industrial processes and to present the principles which lay behind them. Storage of power could be used to extend the duration of action. For example winding a clock can serve to keep it in action for a week, a month, or even a year. He attached much importance to methods of storing energy and suggested the use of compressed gases.

[6] Report of the Governor General in 1834–5; quoted by Karl Marx, *Capital*, 1, 432; Allen and Unwin, 1946.

The worst effects of Blake's dark, satanic mills did not escape Babbage's attention. But his view was firmly fixed on the positive advantages of machine and factory. Over and beyond the industry of the day Babbage looked far ahead: he was beginning to see the possibility of what has been called the second industrial revolution, with advanced machinery and automatic control. Few now doubt that machinery with modern control systems can make life more pleasant though many question whether it actually does so. Babbage was aware of the boring nature of much industrial work, seeing for example that automatic registering devices could save people from tedious counting of a series of repetitions of a fact. He also writes: 'One great advantage which we may derive from machinery is from the check which it affords against the inattention, the idleness, or the dishonesty of human agents.'[7] Later he was to devise automatic machinery for recording clocking-on and -off times of workmen, a device then known as a 'tell-tale'.

Whenever Babbage turned his attention to a new subject he scattered ideas and proposals broadcast in magnificent profusion. Among the proposals he advanced were the metering of London's water supply to reduce waste, a simple device to record the direction of shocks for use in areas liable to earthquakes, and the hydroplane as a vehicle for fast water transport. He noted that a hydroplane would require a compact and powerful engine and he was taking it for granted that one would become available in due course: such was his confidence in the march of technology. Properly applied machinery could save both materials and human time. Examples he gives of conservation of materials include: saving wood in cutting planks by using improved saws; cutting off veneer in a continuous sheet using a machine designed by Brunel; and saving ink by use of an automatic roller-inking mechanism in printing.

In the book Babbage devotes much attention to processes of copying. Later he was to consider robots which are capable of performing varied individual operations on receiving appropriate instructions, but here he was concerned with identical pieces made with the same tools. The oldest of machine tools apart from the string-driven bow-drill is the lathe. Once the lathe was available the circular form was the easiest to make, far easier than a flat plane. However ornamental turning had long been developed for making elaborate non-circular forms and by Babbage's time several other machines existed for making complex patterns and shapes of various kinds. He was able to list entire classes of copying techniques of which he had seen numerous examples in use: printing from ink held in cavities, for instance, including the copper-plate printing which he finally adopted as the planned output technique for the Difference Engine. Such printing could be on paper, as in the splendid presses of Bodoni which he had seen in Parma, and on the Bank of England's notes, or on Manchester's calicoes and Glasgow's cotton handkerchiefs. Of the contrasting

[7] *Economy of Manufactures*, 54.

process of printing from raised surfaces he gives several examples including block printing on paper, calico or oil-cloth, stereotype, printing on china, and lithography. One interesting use of lithography was for reprinting the Paris newspapers in Brussels. A single copy of the paper was carried rapidly to Brussels and while the ink was still fresh the newspaper was placed on a lithographic stone By application of high pressure a sufficient quantity of ink could be transferred to the stone. The other side of the newspaper was copied on another stone and impressions formed as required in the usual way. Babbage then made an interesting observation: 'It is much to be wished that such a method were applicable to the reprinting of fac-similes of old and scarce books.'[8] His wish was to be realized in photo-lithography. Many other methods of copying were noted including casting, moulding, punching, and indeed copying with altered dimensions.

Despite all his abundant energy, it was obvious that an exhaustive study of the endless different processes being developed so rapidly in the industry of the time was beyond even Babbage's capability. Therefore he urged his readers to take up and continue the study. To help enthusiasts he printed a standard set of questions for use when visiting works and factories. The form provided both for general enquiries and for subdivision into separate processes.

Every book Babbage wrote was to some extent polemical and *The Economy of Manufactures* is no exception. In particular he saw it as a part of the election campaign for the new reformed parliament. But its general object was to secure the reorganization of British industry and commerce on a scientific basis. That required a study of the political economy of manufacturing as well as engineering and production technology.

Thus, while the first part was concerned with the physical processes of manufacture, the second and by far the larger part of the book was concerned with the organization of industry and some aspects of commerce, with what is now called scientific management, and some general considerations concerning the application of science to industry. In the phrase of the time Babbage was considering 'the Domestic and Political Economy of Manufactures'. There is little mathematics in either section. No doubt one reason for avoiding it was the popular nature of the book but another may have been the inadequacy, obvious enough to him, of the theoretical models of the classical economists. He makes a characteristic suggestion: 'The importance of collecting data for the purpose of enabling the manufacturer to ascertain how many additional customers he will acquire by a given reduction in the price of the article he makes cannot be too strongly pressed upon the attention of those who employ themselves in statistical research.'[9] Babbage's whole approach to economic theory and practice was always based on the collection of carefully chosen statistical information.

[8] Ibid., 78–9.   [9] Ibid., 120.

The great factories which Babbage visited on his tours were essentially different in nature from the small handcraft workshops which he had known in his youth in Devonshire. As he put it, seeking to persuade the manufacturers to take a greater interest in theory, 'A considerable difference exists between the terms *making* and *manufacturing* ... If ... the *maker* of an article wish to become a *manufacturer*, in the more extended sense of the term, he must attend to other principles besides those mechanical ones on which the successful execution of his work depends.'[10] He then proceeded to discuss some aspects of the subject of interest to manufacturers.

On the division of labour into many distinct activities, each carried out by one or more workers, the very existence of factories depended. Adam Smith had focused attention on the subject but the division of labour had been greatly extended by the 1830s. Babbage noted that in watchmaking there were no less than one hundred and two distinct branches, to any one of which a boy might be apprenticed. The watch-finisher, whose job was to assemble the separate parts, was the only man of the hundred and two who could work in any department besides his own without further training. After noting the advantages which others had seen as arising from the division of labour, Babbage advanced a new principle: 'That the master manufacturer, by dividing the work to be executed into different processes, each requiring different degrees of skill or of force, can purchase exactly that precise quantity of both which is necessary for each process; whereas, if the whole work were executed by one workman, that person must possess sufficient skill to perform the most difficult and sufficient strength to execute the most laborious, of the operations into which the art is divided.'[11]

During the period following the first reform act the educated classes became more interested in visiting the great factories with their assemblies of machines and concentrations of labour. Indeed it became so fashionable that owners were forced to turn away even eminent foreign visitors, not so much for secrecy but rather because the visitors were becoming a nuisance, interrupting the flow of work. Although he had been most circumspect Babbage himself felt a little guilty on this score after visiting *The Times* to watch it being printed. For him there was no doubt that the causes which had led to the growing size of factories and organizations would continue to operate. The results can be seen today in the giant multinational corporations and, even if modern factories may have fewer workers as automated machinery is introduced, there is little sign of reduction in the concentration of capital. The technical division of labour processes into a sequence of distinct operations, both within and between

[10] Ibid., 120.
[11] Ibid., 175–6. After enunciating the principle, Babbage concluded in the preface to the first edition that he had been anticipated by Melchiorre Gioja (see *Opera Principali*, 1838–40); another indication of the breadth of Babbage's reading in political economy.

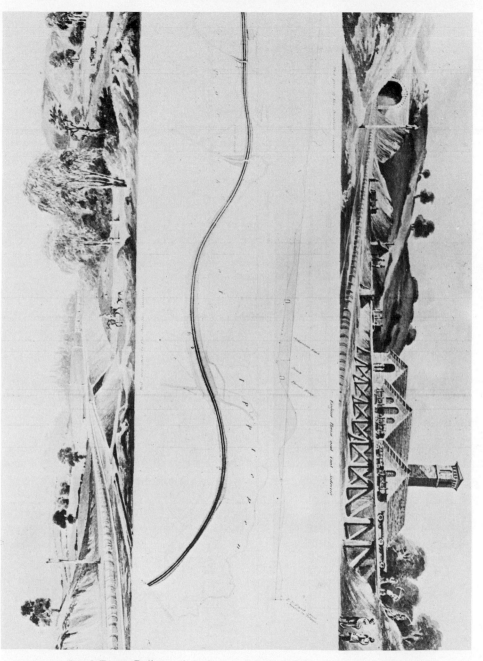

11. South Devon Railway where it passed through Babbage's farm at Dainton.

15. J. M. Jacquard.

machines, which is so characteristic of the modern factory, could already be seen clearly in Babbage's Difference Engine.

Machinery was introduced into the early factories to make many things of the same kind. One such commodity which he discusses in the book was money, both banknotes and coin of the realm: and it was natural for the goldsmith and banker's son to be interested in the technical aspects of money. Together with Marc Brunel and others Babbage was later consulted about new machinery for the Mint. In 1831 he had proposed to the Government a plan for introducing a decimal coinage in three steps, one of the many proposals advanced during the hopeful period of the reform struggle which were destined to come to nothing:

1. Gradually withdraw half-crowns;
2. Coin a piece of silver of value two shillings and call it, say, a Duke of Clarence;
3. A year or two after the Clarences had become familiar introduce a new copper coinage of 100 farthings equal to one Clarence, and in a short time the Public Account could be kept decimally. A new piece of ten farthings would be required called, say, an Earl.[12]

The proposed two shilling piece, the florin, was later coined but further steps were not taken. In 1853 a select committee of the House of Commons was appointed to consider the question of decimal coinage. Although Babbage was called as an expert witness he did not appear.

At the beginning of the second section of the book, recalling that the owner of a large factory was forced to consider many questions beyond the technical problems of manufacture, Babbage leads up to a consideration of more general questions. His own discovery of the subject of political economy had not been through direct practical experience. Like many of his generation he had been interested in the theories of political economy as an undergraduate. And his discussion of crises shows him well aware of the complexity of the subject: 'One of the natural and almost inevitable consequences of competition is the production of a supply much larger than the demand requires. This result usually arises periodically; and it is equally important, both to the masters and to the workmen, to prevent its occurrence or to foresee its arrival.' Although there had been the South Sea Bubble long before, and a financial crisis as recently as 1825/6, and indeed there was to be a serious crisis in 1836/7, the great universal crises which were to place in jeopardy the very existence of the capitalist system could hardly yet be glimpsed on the horizon. Short term palliatives seemed apposite and Babbage proceeded to discuss a number of them. However, 'the subject is difficult', Babbage admitted, 'and, unlike some

[12] Draft dated 21 Jan. 1831; evidently sent by hand to a member of the Government, probably to the Treasury. BL Add. Ms. 37,185, f 4.

of the questions already treated, requires a combined view of the relative influence of many concurring causes.'[13] To the present day there has been little agreement among economists as to the causes of crises, let alone the ability to prevent their occurrence or accurately to predict their arrival.

Like nearly all the political economists of the time Babbage accepted the labour theory of value. Harriet Martineau stated it in its simplest form: 'When equal quantities of any two articles require an equal amount of labour to produce them, they exchange exactly against one another. If one requires more labour than the other, a smaller quantity of the one exchanges against a larger quantity of the other.'[14] She admits only of short term fluctuation of price from value. However accepting the theory is by no means the same as deeming it a useful tool for analysis. When he did turn to consider the subject in the 1840s Babbage chose the path later followed by the marginal analysts. This combination of powerful mathematical tools and theory firmly grounded in extensive statistical information is very rare. He pleads with political economists to use statistical information and with manufacturers not to be afraid of supplying the raw data:

Political economists have been reproached with too small a use of facts, and too large an employment of theory. If facts are wanting, let it be remembered that the closet philosopher is unfortunately too little acquainted with the admirable arrangements of the factory; and that no class of persons can supply so readily ... the data on which all the reasonings of political economists are founded, as the merchant and manufacturer; and, unquestionably, to no class are the deductions to which they give rise so important. Nor let it be feared that erroneous deductions may be made from such recorded facts: the errors which arise from the absence of facts are far more numerous and more durable than those which result from unsound reasoning neglecting true data.[15]

Babbage's interest in economic data was by no means limited to information which the merchant and manufacturer could supply. He was—and once again one can see the influence of his father's banking interests—concerned with securing statistical information about the circulation of money. Always fascinated by the mechanism of clearing houses he also saw them as a valuable source of information. This would include the velocity of circulation, considered essential in modern economic theory. In 1856 he read a paper to the Statistical Society: 'An analysis of the Statistics of the Clearing House during the year 1839'. Later he published the paper as a pamphlet and as usual distributed copies widely.

One of Babbage's most cherished economic ideas was that it cost an appreciable amount to ascertain the quality of an article in order to determine the correct price. Thus for example: 'The goodness of loaf sugar ... can be

13 *Economy of Manufactures*, 231.
14 Martineau, *Illustrations of Political Economy*, 9, 86 London, 1834.
15 *Economy of Manufactures*, 156.

discerned almost at a glance; and the consequence is, that the price is so uniform, and the profit upon it so small, that no grocer is at all anxious to sell it; whilst, on the other hand, tea, of which it is exceedingly difficult to judge, and which can be adulterated by mixture so as to deceive the skill even of a practiced eye, has a great variety of different prices, and is that article which every grocer is most anxious to sell to his customers.'[16] Babbage had arrived at the principle, which he called the 'verification of price', when investigating the high charges levied on the game sent to Georgiana and him by her brother Wolryche Whitmore:

In 1815 I became possessed of a house in London, and commenced my residence in Devonshire Street, Portland Place, in which I resided until 1827. A kind relative of mine sent up a constant supply of game. But although the game cost nothing, the expense charged for its carriage was so great that it really was more expensive than butcher's meat. I endeavoured to get redress for the constant overcharges, but as the game was transferred from one coach to another I found it practically impossible to discover where the overcharge arose, and thus to remedy the evil. These efforts, however, led me to the fact that *verification*, which in this instance constituted a considerable part of the *price of the article, must form a portion of its cost in every case.*[17]

Babbage was led to advocate that the post-office's duties should be extended to include carriage of books and parcels, and both he and his friend Colby advocated a uniform postal rate, later implemented in the penny post. He could not resist mentioning one of his ingenious mechanical devices, which he later described more concisely in his autobiography: 'I then devised means for transmitting letters enclosed in small cylinders, along wires from posts, and from towers, or from church steeples. I made a little model of such an apparatus, and thus transmitted notes from my front drawing-room, through the house, into my workshop.'[18]

In *The Economy of Manufactures* he also noted that steam power was providing a cheap and rapid means of transport which could be used for the mail service, and it was the railway which later permitted the introduction of cheap post. The penny post itself involved more features than merely a uniform rate: the charge was prepaid by means of a stamp and was fixed at the low rate of a penny. Rowland Hill's achievement was not so much his technical plan, which was good, but rather in getting it adopted despite official friction. The important thing which Hill had learned from Babbage, or possibly from Babbage and Colby, was the very idea of an operations research approach to the postal services.

The Economy of Manufactures was published during the most widespread class struggles in England of the whole of the eighteenth and nineteenth centuries. This is reflected in Babbage's insistence that working men should realize that

---

[16] Ibid., 134–5.   [17] *Passages*, 448.
[18] Ibid., 448–9. See also *Economy of Manufactures*, 273–7.

their interests and those of the masters were essentially in harmony. A manufacturer opening a factory might, if he followed Babbage's prescription, instal the most modern and suitable machinery contrived with the utmost ingenuity; he might establish the organization of his factory to take maximum advantage of the possible division of labour, and even follow Babbage in establishing a rigorous system of cost-accounting; and yet there could remain a major obstacle to the satisfactory functioning of his establishment. The difficulty arose because 'A most erroneous and unfortunate opinion prevails amongst workmen in many manufacturing countries, that their own interest and that of their employers are at variance. The consequences are,—that valuable machinery is sometimes neglected, and even privately injured,—and that the talents and observations of the workmen are not directed to the improvement of the processes in which they are employed.' However, he admitted that, 'Convinced as I am, from my own observation, that the prosperity and success of the master manufacturer is essential to the welfare of the workman, I am compelled to admit that this connexion is, in many cases, too remote to be understood by the latter.' He was particularly anxious that the book should be read by intelligent working men lest 'the whole class ... , in some instances, be led by designing persons to pursue a course, which, although plausible in appearance, is in reality at variance with their own best interests'.[19]

The Luddites who smashed the lace frames in Nottinghamshire were the most famous of the groups that opposed the introduction of new machinery. But, as Babbage pointed out, one effect of their action was to force lace factories to move to Devonshire, leaving even heavier unemployment behind. He quoted with approval from the report of the Committee on the Fluctuation of Manufacturers Employment: 'Any violence used by workmen against the property of their masters, and any unreasonable combination on their part, is almost sure thus to be injurious to themselves.'[20] And indeed machine wrecking was a futile gesture by men driven to desperation. The future lay with the development and improvement of manufacturing, the introduction of new technologies and the systematic application of science: the message of Babbage's whole book.

Babbage never liked to leave a major problem without suggesting a solution and he went on to propose 'A New System of Manufacturing'. In the first edition, he described a system of payment by results which was widely used in the Cornish mines. In later editions he generalized from the Cornish system, and from similar systems used in other traditional occupations involving the co-operation of groups of men, and proposed a system of profit sharing. It was designed to give every working man an interest in the prosperity of the whole enterprise, by making wage packets immediately dependent on profits. Thus every workman would have an interest in eliminating waste and mismanage-

---

[19] *Economy of Manufactures*, 250–1.   [20] Ibid., 229.

ment. He also proposed to give each workman a direct reward for innovations and improvements which he suggested or introduced. Babbage completed his proposal by suggesting that a reserve fund be built up during times of prosperity. In circumstances of glut in the market part of the men's time could be employed on innovations and improving their tools, thus facilitating future production. In the simple industry of the time, when workmen were responsible for a large proportion of their own tools, the latter proposal was quite feasible. Designed to damp down class conflict Babbage's 'New System' was in general technically possible: the proposals are not dissimilar to those advocated by the Liberal party today. What was lacking was employers with the intelligence to introduce such a system; there was also the tenacity with which working men held to that 'erroneous and unfortunate opinion' that their own interest and that of their employers were at variance. John Stuart Mill drew particular attention to the importance of Babbage's 'New System' of profit sharing: 'Mr Babbage observes that the payment to the crews of whaling ships is governed by a similar principle; and that "the profits arising from fishing with nets on the south coast of England are thus divided:" Mr Babbage has a great merit of having pointed out the practicability, and the advantage, of extending the principle [of profit sharing] to manufacturing generally.'[21]

Babbage could write with justification that he approached industrial questions scientifically and had no direct pecuniary interest in supporting either side in disagreements between working men and their masters. Although he is quite clear that 'The effects arising from combinations amongst the workmen, are almost always injurious to the parties themselves' he approached the problems of working-men sympathetically and with a wide knowledge of industrial practice, giving some interesting examples of customary arrangements on the shop floor. Indeed there were strict limits to what could be achieved by a trade union at that time. He was bitterly opposed to the 'truck system' (by which workers were compelled to spend their wages at a company shop), and to combinations of masters in restraint of trade. He cites cases of valuable inventions not taken up because of organized opposition. One such was Brunel's veneer-cutting machine. '"the trade" set themselves against it, and after a heavy expense, it was given up.'[22] Generally Babbage's approach is that of a very intelligent master with no sectional axe to grind.

Interested in the problems of publishing at a time when he was writing the book, Babbage chose a combination in the book trade as his prime example of 'Combinations of Masters against the Public':[23]

Some time ago a small number of the large London booksellers [who were also publishers at that time] entered into such a combination. One of their objectives was to prevent any bookseller from selling books for less than ten per cent under published

[21] J. S. Mill, *Principles of Political Economy*, ed. Sir W. J. Ashley, 765, Longman, 1909.
[22] *Economy of Manufactures*, 312.   [23] Ibid., 327.

prices; and in order to enforce this principle, they refuse to sell books, except at the publishing price to any bookseller who declines signing an agreement to that effect. By degrees, many were prevailed upon to join this combination; and the effect of the exclusion it inflicted left the small capitalist no option between signing or having his business destroyed. Ultimately, nearly the whole trade, comprising about two thousand four hundred persons, have been compelled to sign the agreement.

Babbage wrote in the preface to the second edition, when his comments were already the subject of acute controversy: 'I entered upon this enquiry without the slightest feeling of hostility to the trade, nor have I any wish unfavourable to it'. None the less his criticism had been trenchant: 'I think a complete reform in its system would add to its usefulness and its respectability', he wrote, adding that the best body for dealing with the problem would be a society of authors. Before publishing the first edition, Babbage had drawn the attention of Mr Fellowes of Ludgate Street, who had published other books for him in the past, to the chapter in which the book trade was discussed. 'Mr Fellowes, "differing from me *entirely* respecting the conclusion I had arrived at", then declined publication of the volume. If I had chosen to apply to some of the other booksellers whose names appear in the Committee of "The Trade", it is probable that they would also have declined the office of publishing for me; and, had my object been to make a case against the trade, such a course would have assisted me.'[24] However Babbage had no such intention and, having himself arranged the printing, secured a new and independent-minded publisher, Charles Knight of Piccadilly. In spite of hindrance by the trade the book was a sell-out, going through three editions in the year of publication. Babbage attributed much of its success to an awakening interest in the way of life of the voters newly enfranchised by the Reform Act.

Freedom of the manufacturer to conduct his business as he wished; freedom of the worker to sell his labour on the market at the best price; freedom of travel; freedom from improper combinations whether of masters or men; above all freedom of trade—these were the fashionable doctrines. On the question of free trade the acid test was the writer's attitude towards export of machinery. There was serious concern that foreign manufacturers using improved British machinery would be able to compete too effectively with British manufacturers. So seriously was the threat of competition taken that a few years earlier it had been illegal for skilled workmen to take themselves abroad in quest of higher wages. Babbage not only held such restrictions to be futile, knowing perfectly well that industrial development abroad could not be prevented, but held that demand for British machinery should be encouraged because it stimulated the crucial industrial sector of machine building. The solution lay not in restrictive practices but in the continuing development of the skills and technology of British industry. At that date he could still quote

[24] Ibid., preface to 2nd ed., vii.

with confidence from the 'Report of the House of Commons Committee on the Export of Tools and Machinery': 'It is admitted by every one, that our skill is unrivalled; the industry and power of our people unequalled; their ingenuity, as displayed in the continued improvement of machinery, and production of commodities, without parallel; and apparently without limit.'[25]

It is a remarkably persistent myth that free-traders were opposed to all government intervention. True *laissez-faire* might be found in the pages of *The Economist* or in the utterances of the Manchester School but it was mainly to be heard in Parliament issuing from the mouths of politicians defending sectional interests. When they approved of the objective the classical economists saw room for a good deal of government intervention, although they opposed such corrupt government practice as remained from the mercantile era: and corrupt it certainly was. The scope of government action envisaged varied from writer to writer but typically it included public-health regulation, prevention of adulteration of foods, building regulations to combat the terrible developing slums, factory legislation, the care of paupers, and education. Nassau Senior was prepared to endorse publicly provided medical treatment. Taxes of course there must be, and Babbage discusses at some length which taxes will least hamper the development of industry and trade. The pragmatic approach of even so enthusiastic a free-trader as Babbage is nicely illustrated by an example he gives of the cost of verifying the quality and price of commodities:

it has been found so difficult to detect the adulteration of flour, and to measure its good qualities, that contrary to the maxim that *Government can generally purchase any article at a cheaper rate than that at which they can manufacture it*, it has been considered more economical to build extensive flour-mills (such as those at Deptford), and to grind their own corn, than to verify each sack purchased, and to employ persons in continually devising methods of detecting the new methods of adulteration which might be resorted to.[26]

With the book published Babbage's position as an important political economist was established. The foundation of the London Statistical Society, for which he was primarily responsible, brought him into closer contact with leading economists. These included Malthus (who chaired the first small meeting in Richard Jones's rooms in Cambridge of what became the Statistical Section of the British Association), Thomas Tooke, Nassau Senior, and Samuel Jones Loyd (later Lord Overstone). Babbage's friend Drinkwater (later Bethune) played a large part in founding the society. So also did Richard Jones, an old friend of Babbage and Herschel. Like Babbage, Jones was rare among the political economists in insisting that economic theory be firmly grounded in extensive data and subjected to statistical analysis.

At the end of 1845 Babbage was to undertake a mathematical study of the

[25] Ibid., 369–70.   [26] Ibid., 135.

market-place, considering for example the relation between supply and price given that demand is constant. The mathematical tool which he used for analysing 'the law which regulates supply and demand at any given instant of time' was naturally the calculus. A number of men were at this time working on what later became marginal theory. As early as 1838 Cournot[27] had developed the theory considerably further than Babbage. Nevertheless it is interesting that there were men in London who fully understood that a mathematical approach to economic theory was required, decades before the 'marginal revolution' of the 1870s. Babbage was continually urging such an approach in the discussions of the Statistical Society which he attended until shortly before his death. This group of men interested in mathematical theory has been little noticed in the history of economic thought because they were primarily men of science and only occasionally political economists. Among them Babbage was the leading scientific figure, Dionysus Lardner (who may lay claim to having been the world's first science journalist) was the publicist. Lardner's *Railway Economy* became better known in this connection because Jevons cited the book as stimulating his interest in mathematical economics.

The central message of *The Economy of Machinery and Manufactures* was that the future of industry required a consistently scientific approach to all aspects of its problems, both technical and commercial. The book has a frontispiece of Roger Bacon, one of the earliest pioneers of science in England, and the dedication reads: 'To the UNIVERSITY of CAMBRIDGE this volume is inscribed as a tribute of respect and gratitude by The Author.' This implies that he was thinking of no narrow subordination of science to practical ends. On the contrary he was equally interested in the most abstract branches of science. He was so devoted to pure mathematics that he spent much time studying what must then have seemed a quite useless subject, the theory of games. In this book Babbage proclaims 'the union of theory and practice.' But as a political economist he finally concluded he had made no original contribution to theory. In a sense he was mistaken: there was nothing quite like his approach for a century. The combination was remarkable: systematic development of industrial technology; operations research; rational cost-accounting; profit-sharing incentive schemes; economic theory using powerful mathematical techniques and grounded in extensive statistical information. Later he said that when the Analytical Engine should exist it would necessarily determine the course of advance of all the sciences and one of the sciences to which he attached the greatest importance throughout his life was political economy.

What effect did Babbage's *Economy of Machinery and Manufactures* have? Generally his book has received little attention as it is not greatly concerned with such traditional problems of economics as the nature of 'value'. Actually the effect was considerable, his discussion of factories and manufactures

[27] Augustin Cournot, *Recherches sur les principes mathématiques de la théorie des richesses*, 1838.

entering the main currents of economic thought. Here it must suffice to look briefly at its influence on two major figures; John Stuart Mill and Karl Marx.

In his *Principles* Mill quotes Babbage directly on the division of labour and large scale production, and the basic importance of Babbage's work to Mill has been acknowledged by Schumpeter: 'Mill often failed to present references to factual material perhaps because he assumed that his readers could easily supply the deficiency from universally accessible sources—such as the work of Babbage.' Schumpeter goes on to pay tribute to Babbage's *Economy of Manufactures*:

This work which was widely used (also by Marx), is a remarkable performance by a remarkable man. Babbage ... was an economist of note. His chief merit was that he combined a command of simple but sound economic theory with a thorough first-hand knowledge of industrial technology and of the business procedure relevant thereto. This almost unique combination of acquirements enabled him to provide not only a large quantity of well-known facts but also, unlike other writers who did the same thing, interpretations. He excelled, amongst other things, in conceptualization, his definitions of a machine and his conception of invention are deservedly famous ... By contrast to his sound well-balanced treatment, I mention A. Ure's (*Philosophy of Manufacture*)[28], who also presented interesting facts, but was not Babbage's equal as an analyst.[29]

Babbage's influence on Marx can be seen clearly in the chapters of *Capital* on the 'Division of Labour and Manufacture' and on 'Machinery and Modern Industry'. Marx quotes from Babbage: 'When [from the particular nature of the produce of each manufactory], the number of processes into which it is most advantageous to divide it is ascertained, as well as the number of individuals to be employed, then all other manufactories which do not employ a direct multiple of this number will produce the article at a greater cost ... here arises one of the main causes of the great size of manufacturing establishments.'[30] It followed as a corollary that as technology advanced, and also as more auxiliary commercial and industrial functions (such as development laboratories) were required, the size both of manufactories (factories) and commercial organizations would continue to grow. In the chapter on 'Machinery and Modern Industry' Babbage is quoted both on the general nature of machines and on the cost of developing new machinery for industry.

Marx quotes Babbage in the early work *La Misère de la Philosophie* but he had read *The Economy of Manufactures* even earlier, between February and June 1845[31] on the eve of his visit to Manchester when Marx and Engels sat in Cheetham Hill Library sweating through the English political economists.

[28] Andrew Ure, *Philosophy of Manufactures*, London, 1835.

[29] J. A. Schumpeter, *History of Economic Analysis*, 541. Allen and Unwin, 1955.

[30] Karl Marx, *Capital*, 1, 338; Allen and Unwin, 1946. Marx worked initially from the French translation by Biot of the third edition of Babbage's *Economy of Machinery and Manufactures*.

[31] Karl Marx and Friedrich Engels, *Gesamptausgabe*, I, 6, 601.

Marx read Babbage very carefully, recording no less than 73 excerpts. At this date he was not yet fully committed to the labour theory of value. He notes that Babbage belongs in that respect to the school of Ricardo. This is certainly true in Babbage's published work although he later toyed with marginal theory in unpublished notes. It has often been held that the two are incompatible, but the whole question of the relation between the labour theory of value and mathematical economics is complex and the subject of hot debate.[32]

The spring or early summer of 1845 may have been the first time Marx became acquainted with Babbage's book but Engels may have come across it earlier. Particular interest centres on Babbage's exposition of the doctrine of the union of theory and practice. What influence did Babbage have on the adoption by Marx and Engels of this celebrated idea? Babbage was concerned with the union of scientific theory and commercial and industrial practice; Marx, old Hegelian, naturally gave the idea a wider interpretation. It is curious that Marxist scholars have generally given so little attention to Babbage's influence. Possibly they have been reluctant to acknowledge the influence on Marx of so militant a supporter of capitalism; more probably, finding themselves surrounded by Calculating Engines, their eyes glazed over as they turned to more fruitful sources of inspiration.

The Economy of Manufactures was an immediate success, selling 3,000 copies on publication. It was at once translated into French by Édouard Biot, son of Babbage's old friend, and into German by Dr Friedenburg, both translations being published in 1833. It was also translated into Italian and throughout the civilized world the book had much effect, becoming the *locus classicus* of the discussion of machinery and manufacturing. It was followed by other books on manufacturing, and with the growth of industry the decisive importance of the factory became a starting point of discussion. In Britain Babbage's plea for the general application of science and technology to industry, for the union of theory and practice, was systematically ignored; with disastrous consequences for the nation's industry.

[32] cf. Micio Morishima, *Marx's Economics, A dual theory of value and growth*, C.U.P., 1973.

9

○ ● ○

The Great Engine

Babbage's Continental tour of 1827 and 1828 had come at a bad time from the point of view of the Engine. The scale of work was beginning to expand as early design progressed to fabrication, and more sound arrangements were urgently needed. Joseph Clement was a highly skilled mechanic but had not previously been responsible for so large a project. Before leaving England, Babbage was already worried that Clement was less than straightforward, suspecting him of developing and constructing at his employer's cost special lathes and other tools valuable in constructing the Difference Engine but intended to remain in Clement's workshop. The £1,500 advanced by the Government had been spent and Babbage was financing construction from his own resources. At his request Wolryche Whitmore reminded Lord Goderich of the interview in July 1823, but Goderich would not acknowledge any commitment to advance more than the £1,500. Babbage regarded that as only an initial advance and believed that Goderich had made this quite clear. The unfortunate consequences of having no official minute of Babbage's meeting with Lord Goderich were being felt.

On returning to England late in 1828 Babbage wasted little time. After another interview with Lord Goderich, who admitted that the understanding of 1823 was not very definite, Babbage addressed a comprehensive statement of the position to the Duke of Wellington, who was now Prime Minister. This led to effective action. The treasury was soon writing to the Royal Society enquiring: 'Whether the progress of the Machine confirms them in their former opinion, that it will ultimately prove adequate to the important object it was intended to attain.' A committee was appointed which promptly reported that the Royal Society '. . . had not the slightest hesitation in pronouncing their decided opinion in the affirmative', and expressed their hope that 'Whilst Mr Babbage's mind is intensely occupied in an undertaking likely to do so much honour to his country, he may be relieved, as much as possible, from all other sources of anxiety.'[1] Nor was it left at that. In May 1829 a group of Babbage's friends, the Duke of Somerset, Lord Ashley, Sir John Franklin, Wolryche Whitmore, Francis Baily, and John Herschel drew up and signed a memorandum to the Duke of Wellington spelling out the financial consequences in some detail. In due course John Herschel and Wolryche Whitmore had an

[1] *Passages*, 73.

interview with Wellington, who then characteristically decided to see for himself what was happening.

In November 1829 Wellington, Mr Goulburn, the Chancellor of the Exchequer, and Lord Ashley saw the first model of the Engine which Babbage had made in 1822 as well as the drawings and parts being made for the great Engine.[2] When the common run of inventors plagued the Duke of Wellington for funds he usually dismissed them out of hand and with good reason. Most inventions turn out to be worthless, and at that time those which were useful could generally be exploited by private industry. Babbage himself was contemptuous of patents, declining to patent any of his numerous inventions. But the Difference Engine was a great deal more than mere invention: it was a major engineering project. Babbage and his friends had formidable technical forces at their disposal, and when it came to the disposition of forces Wellington was in his natural element. The project received proper backing—£1,500 on 29 April, £3,000 on 3 Dec. 1829, and a further £3,000 on 24 Feb. 1830—and after a pause of nine months, work on the Engine resumed. Support continued under the Reform Ministry of Lord Grey, although there were continuing difficulties in administering finance and Babbage had to advance substantial sums from his private funds.

While these discussions were proceeding Babbage took long overdue steps to organize the work more satisfactorily.[3] He declined paying any more of Clement's bills until a proper system was established for verifying that they accurately represented work carried out. Two engineers, one chosen by each party, were jointly to inspect the work and accounts. If they could not agree a third engineer would be appointed to act as arbitrator. This procedure was standard practice, providing an alternative to recourse to the law courts. Clement chose Henry Maudslay; and Babbage, with treasury approval, chose Bryan Donkin. The accounts were duly inspected and approved, but in reality so novel was the work upon which he was engaged that Clement was fairly free to charge what he wanted and inspection subjected him only to the loosest control. Clement was on a good wicket: he was making a great deal of money and determined to continue doing so while he could.

Babbage remained very worried about him although he had no doubt that he was an excellent craftsman. Babbage was also concerned about the location of the Engine and drawings. He addressed a set of questions to Bryan Donkin for discussion with Clement. Reading it, one's first thought is that such questions should have been settled much earlier. However the project had developed far beyond Babbage's original plans, and organization had been disrupted by the personal crises of 1827 and the Continental tour; moreover he was continually hampered by difficulty in communicating with the Government. Neither

[2] BL Add. Ms. 37,184, ff 415, 421, 451, & 455.
[3] Ibid., f 465; 37,185 ff 57, 63, 69, 86, & 105.

contemporary administration nor the law were adequate to the problem. These were his questions:[4]

1. To whom do the tools belong?
2. To whom do the patterns belong?
3. To whom do the drawings belong?
4. At whose risk are the machine tools and drawings?
5. Ought they to be insured and by whom?
6. How are they to be secured against any creditors of Mr Clement?
7. How is the actual amount of work to be ascertained in order that it may be known what has been added at the next settlement?
8. Ought not Mr C[lement] to engage not to make another such machine without my *written permission*?

The questions were clear and so were the answers:[5]

1. To Mr Clement;
2. To Mr Babbage;
3. To Mr Babbage;
4. The Machine, drawings and patterns are at Mr Babbage's risk, the tools at Mr Clement's;
5. They ought, by their respective owners;
6. Messrs Rennie and Donkin [both of whom were initially appointed by the Government to examine the accounts] consider Mr Clement a careful and prudent man and Mr Babbage need not feel apprehensive;
7. [The result has been described above: Messrs Maudslay and Donkin were appointed to settle the matter.]
8. Mr Clement declines doing so.

One point calls for particular comment. At that date all tools made while working on a job belonged to the workman himself. This was only natural for the hand tools essential for, say, a carpenter, a leather worker, or a bench-worker in a metal shop. A man looked after his own set of tools and took them with him when he moved from job to job. But the system was hopelessly inadequate for lathes and other machine-tools. There was no point in Babbage's arguing whether or not the tools had been constructed during working hours. Even if that could be proved, all the novel and ingenious machine-tools, even though they had been developed at Babbage's instigation, belonged to Clement.

To clarify his position, Babbage was determined that the Government should declare the Engine to be Government property, as he had envisaged it would be from the beginning. Indeed the whole justification for the project was that the Engine would be of national importance, and only if the Government accepted full responsibility would there be any hope that the organization of the project would be satisfactory. Babbage also feared that if the Engine were not Government property he might become personally liable for the funds

[4] BL Add. Ms. 37,184 f 252.   [5] Ibid., f 266.

which the Government had advanced to defray the cost of construction. On 24 February, in an interview with Lord Ashley,[6] the Chancellor of the Exchequer agreed to declare the Engine Government property, and money began to flow. The Treasury disbursed money to Babbage who then paid Clement. Unfortunately the transactions appeared in the Government estimates as money paid to Babbage, giving rise to a rumour that Babbage was himself being paid for the Engine. Later this story was used against him when he stood in Finsbury as candidate for the reformed parliament.

A secure place was soon required to erect the great Calculating Engine. Clement had no suitable building and Babbage was anxious to have the completed Engine near his own house. Clement's workshop, near the Elephant and Castle and Babbage's birthplace, was four miles from Dorset Street. Thus Babbage could only visit the workshops once a day and was forced to work continuously with Clement in stretches of four to six hours at a time. The strain was affecting the health of both of them. It was clearly desirable to erect the Engine near the workshops in which it was being made and Babbage's solution was to move tools, drawings, and parts to a site near Dorset Street: one could not expect a gentleman to move his own residence. Lord Goulburn declared his opinion that it was neither practicable nor advisable to remove the Calculating Engine from the neighbourhood of Clement's present workshops;[7] and was proved right: the project was to founder on difficulties associated with the move, in conjunction with Clement's bloody-mindedness.

In the meantime Babbage's friends rallied round. Marc Isambard Brunel was particularly helpful. A move was technically possible and various sites were considered. One of the first possibilities investigated was that of erecting the Engine in the British Museum. Babbage even thought of designing a building himself, but Marc Brunel brought him back to earth: 'The question of building is now the point on which Lord A[lthorp] expects a solution. Eminent as you are in solutions you will not satisfy his L[o]r[dship] so well as men who are eminent in the purchase of brick and other materials: *chacun son metier* et le trousseau est bien gardé-
The most respectable surveyor and builder in your neighbourhood is Mr [Charles] Jearrad in Oxford St.'[8]

Mr Jearrad was commissioned but could find no suitable vacant site in the district. A Mr Rice owned a cow-yard which backed onto Babbage's garden, but it would cost too much to persuade Mr Rice to move. Jearrad then suggested that Babbage's stables would make a suitable site.[9] The cow-yard remained and its stench in hot weather later caused strenuous complaints from Babbage's workmen, and one wonders that Babbage himself could have tolerated the smell even though his house was a little further away. His coach-house and six-

[6] Ibid., f 69.   [7] Edmund Walpole to Babbage, 6 Aug. 1830; Ibid., f 268.
[8] BL Add. Ms. 37,185, f 382.   [9] Ibid., ff 436-8; 37,186, ff 63 & 90.

stall stables were pulled down and a strong fire-proof building erected. The new workshops had two floors, each fifty feet long, strongly built to support lathes and heavy machinery. Much of the roof was glass to obtain good lighting. A second workshop was built and next to it a fire-proof, dust-free room to house the Calculating Engine. Babbage owned a small house where the carriageway from his old stables passed out into East Street. This house was allocated to Clement. The Government was to pay £125 per annum rent to Babbage for the site of the stables and coach-house and use of the house on East Street. There were also a few contingent alterations to Babbage's property. A coach gateway was made between his garden and the entrance in East Street. And as the roof of his garden shed had to be destroyed he desired that the new roof of the shed be raised six feet to form a small tool-room above. Babbage planned sufficient space actually to print *in situ* the tables produced by the Engine. The tables were to be engraved directly on copper plates according to the plans at that time. Thus space would be needed for storing the plates, and paper, both printed and unprinted, as well as for the operations of printing.

At the end of the summer of 1831, when the building was getting under way, the inaugural meeting of the British Association took place in York.[10] David Brewster and James Forbes wrote to Babbage that they 'were out of all bounds of indignation at your non appearance and . . . never thought the meeting could have answered so well.'[11] Vernon Harcourt, principal architect of the British Association, had also begged Babbage to come but Babbage was desperately anxious to see the new buildings completed before winter set in. Much as he wished to attend the York meeting he could not risk the possible further delays in work on the Engine.

Babbage continued to worry about Clement, but it was exceedingly difficult to alter the arrangements or employ another workman at this stage. It was probably in February 1831 that Babbage received a letter from C. G. Jarvis who was employed as a draftsman by Clement, and was at least as competent as Clement himself. He was far better educated too, and was later to work directly for Babbage on the superb series of drawings for the Analytical Engines. Jarvis was not getting on with Clement, whom he despised, and was thinking of leaving. He wrote to Babbage: 'It should be borne in mind that the inventor of a machine and the *maker* of it have two distinct ends to obtain. The object of the first is to make the machine as complete as possible. The object of the second— and we have no right to expect he will be influenced by any other feeling—is

---

[10] Brewster had written to Babbage on 4 Sept. 1831 : 'You really must not allow the machine to keep you from York. I merely go because you are going . . .'—a reminder of Babbage's central importance in the movement that led to the launching of the British Association ; BL Add. Ms. 37,186, f 74.

[11] Ibid., ff 103 & 114.

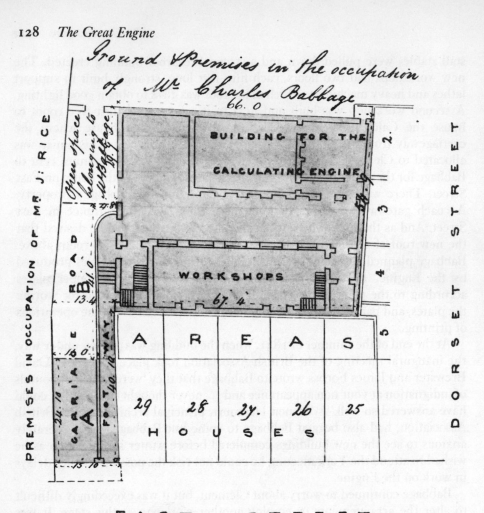

Fig. 1. Ground and premises in the occupation of Charles Babbage.

to gain as much as possible by making the machine; and it is in his interest to make it as complicated as possible.'[12]

In 1832 Babbage instructed Clement to assemble a portion of the Engine. This is the working Difference Engine which can still be seen in the Science Museum. It was an engineering triumph. If the full Engine planned by Babbage had been completed it would have been the wonder of nineteenth-century precision engineering, and even the portion which was assembled is one of the finest achievements of the first half of the century. It still works perfectly and no competent person who has had the privilege of handling it will doubt that the Great Engine would have worked equally satisfactorily. The completed portion of the Engine was moved to Dorset Street and was soon being shown

[12] BL Add. Ms. 37,185, f 419.

to many guests. The Duke of Wellington came and repeatedly brought his friends. The Duke of Somerset of course came and he was able to propose simple problems for the Engine to solve. Babbage invited both the Marquis of Lansdowne and Brougham. Apparently Lansdowne came, but Brougham, although he had been on fairly close terms with Babbage, may have been too busy: he was not only Chancellor but was being gently blackmailed by his former mistress. Naturally the Engine became a showpiece at Babbage's parties. Harriet Martineau gives a brief sketch of Babbage at the time:

As for Mr. Babbage, it seemed to me that few men were more misunderstood,—his sensitiveness about opinions perverting other people's opinions of him quite as much as his of them. For one instance: he was amused, as well as struck, by the small reliance to be placed on opinion, public or private, for and against individuals: and he thought over some method of bringing his observation to a kind of demonstration. Thinking that he was likely to hear most opinions of himself as a then popular author, he collected everything he could gather in print about himself, and pasted the pieces into a large book, with the *pros* and *cons* in parallel columns, from which he obtained a sort of balance, besides some highly curious observations. Soon after he told me this with fun and good humour, I was told repeatedly that he spent all his days in gloating and grumbling over what people said of him having got it all down in a book, which he was perpetually poring over. People who so represented him had little idea what a domestic tenderness is in him,—though to me his singular face seemed to show it, nor how much that was really interesting might be found in him by those who viewed him naturally and kindly. All were eager to go to his glorious soirées; and I always thought he appeared to great advantage as a host. His patience in explaining his machine in those days was really exemplary. I felt it so, the first time I saw the miracle, as it appeared to me; but I thought so much more, a year or two after, when a lady, to whom he had sacrificed some very precious time, on the supposition that she understood as much as she assumed to do, finished by saying, 'Now, Mr. Babbage, there is only one thing that I want to know. If you put the question in wrong, will the answer come out right?' All time and attention devoted to lady examiners of his machine, from that time forward I regarded as sacrifices of genuine good nature.[13]

The parties were but one part of an active social life. Inevitably the heavier aspects of his work loom unduly large in any account of his life, and it must be remembered that the real Charles Babbage was full of fun and delightful company. Charles Lyell gives us one glimpse of the sociable Babbage:

Mill Hill, Hendon, Sunday [January 1832]

I came by coach to this place after breakfast, passing through Hampstead. Dr Fitton is residing here, at Mr Wilberforce's house, a most delightful residence, eleven miles from London. Mrs Fitton and children quite well; six children. Conybeare and Babbage the only visitors; most agreeable but not lying *fallow*. Fitton pronounces me to be rather thin. We have had great fun in laughing at Babbage, who unconsciously jokes and reasons in high mathematics, talks of the 'algebraic equation' of such a one's character in regard to the truth of his stories, &c. I remarked that the paint of Fitton's house

[13] Harriet Martineau, *Autobiography*, 354-5. 3rd ed., London, 1877.

would not stand, on which Babbage said, 'no: painting a house outside is calculating by the index minus one', or some such phrase, which made us stare; so that he said gravely by way of explanation, 'That is to say, I am assuming revenue to be a function.' All this without pedantry, and he bears being well quizzed by it. He says that when the reform is carried he hopes to be secularised Bishop of Winchester. They were speculating on what we should do if we were suddenly put down on Saturn. Babbage said, 'You Mr Leudon (the clergyman there, and schoolmaster, and a scholar), would set about persuading them that some language disused in Saturn for 2,000 years was the only thing worth learning; and you, Conybeare, would try to bamboozle them into a belief that it was to *their interest* to feed you for doing nothing.'

Fitton's carriage brought us from Highwood House to within a mile of Hampstead, and then Babbage and I got out and preferred walking. Although enjoyable, yet staying up till half-past one with three such men, and the continual pelting of new ideas, was anything but a day of rest. We were disputing sometimes on difficult scientific questions, sometimes on other topics, Tom Moore's poetry to wit. I cannot recollect Conybeare's favourite lines, but it is where Moore says to his country that his songs on Ireland are not his, but, like 'the breeze they wake the magic that is all their own.' Babbage thought the Irish melodies superior to all the rest, in which I agree . . . I am not the better for drinking hock or Rhine wine there 'to our friends on the Rhine.' I was as temperate as the rest, but that is not saying much.[14]

A few years later, in a letter to Herschel in South Africa, Lyell made a point which has been little noticed about one serious purpose behind Babbage's soirées: 'I maintain that he [Babbage] has done good, and acquired influence for science by his parties, and the manner in which he has firmly and successfully asserted the rank in society due to science.'[15]

In May 1832 Babbage was pressing Clement to specify the requirements and arrangements for moving the tools, parts of the Engine, and drawings to the premises which the Government had built in place of Babbage's stables and coach-house. Clement's demands were large. In addition to exaggerated expenses for the move he asked £660 a year for the expense of keeping two establishments and running a divided business. The Treasury refused to sanction Mr Clement's 'extraordinary claims':

Mr Clement must recollect that £12,000 of Public Money has been expended on the Engine the great part . . . of which has passed through his hands, from which a very large amount of profit has accrued to him, and that his employment by you has constituted and still constitutes, as my Lords are informed, the principal part of his business, and that he has not been restricted in using for his general business any tools for which the public has paid, and that the further expense of completing the engine will yield him a considerable profit . . . My Lords cannot but express their surprise that Mr Clement should have advanced so unreasonable and inadmissable a claim.[16]

[14] Sir Charles Lyell, *Letters and Journals 1*, Jan. 7-13, 1831, John Murray, 1881.
[15] Lyell to J. Herschel, *Letters and Journals of Sir Charles Lyell, 1*, 1 Jan, 1836; John Murray, 1881.
[16] BL Add. Ms. 37,187, f 134.

Babbage did his best to preserve the peace and find a solution. On 19 September 1832 he wrote to Clement: 'I shall call on Friday and be happy if in any way I can assist in removing the objections of the Lords of the Treasury.'[17]

It was no use. Clement made the reasonable proposal that two engineers should arbitrate on the question of his costs as they had on the work itself. Realizing, however, that his claims were extravagant and would probably be rejected, and moreover that under Babbage's watchful eye there would be little room for continuing the dubious procedures which had been making him so large a profit, Clement changed his mind, withdrew his proposal and refused to move to Dorset Street. Babbage then declined to advance any more of his own money, telling Clement to present his accounts directly to the Treasury in future. Clement sacked his men and work on the Engine stopped. This time it was not to start again. Moreover Clement refused to deliver any of the parts or drawings until he was paid in full. Maudslay had died and been replaced as Clement's nominated assessor by Maudslay's partner, Field. Both Donkin and Field found dealing with Clement distasteful but they examined the work and accounts and the Treasury did finally pay Clement. Even then Clement stalled for as long as possible:[18] if he could not continue to make his extortionate profits, and also take credit for making the Engine, he was determined to be as difficult as possible.

Babbage went on patiently seeking ways to resume work. He was loath to make a final break with Clement because of the difficulty of replacing the specialized machine-tools. Also, although there were certainly other men of Clement's ability they were extremely busy and it was not easy to find a competent workman who could devote sufficient of his time to the Engine. C. G. Jarvis, the draftsman, was no longer working for Clement and could write freely to Babbage:

I cannot help enforcing the necessity of some alteration in the method pursued for the construction of the Calculating Engine, that is, if it is wished or intended to be finished within any reasonable time. The plan I wish to recommend is, that the designs and drawings be all made on your premises and under your immediate inspection, working drawings made from them and directions given in writing to various persons, to construct various parts of the machine, which parts might all be going on at the same time and the entire machine be speedily completed. In this case whenever any difficulty occurred *you* might be at once appealed to, whenever it was found very difficult to produce *nearly* that effect—which is a very common case in machinery—your decision would at once put an end to all perplexity and hesitation and the work would proceed; instead of having to wait days, weeks or *months* for an arrangement which half an hour would complete . . .

Clement was of course bound to oppose any such plan:

To a man who although inactive and unenterprising loves money, it must be

---

[17] Ibid., f 141.   [18] Add. Ms. 37,188, f 437.

agreeable to construct a newly invented machine the cost of the parts of which cannot be taxed; and still more agreeable to be able to charge for time expended upon arranging the parts of that machine without the possibility of the useful employment of that time being disputed, and to doze over the construction year after year ...

It may be suspected I enforce my opinion from interested motives. It is perhaps difficult to divest oneself of selfish feelings and even Messrs Donkin and Field were somewhat under their influence when they prohibited the paying of draftsmen more than a certain sum whatever their worth [they were afraid their own draftsmen might demand higher salaries], and therefore all I shall say on the subject is, that my plan may be followed without my being in any way a gainer by its adoption. In fact the probability is that I shall be out of Britain before the machine is proceeded with in any way.[19]

Babbage suggested that Jarvis might come to work at East Street while continuing under Clement's direction. Jarvis turned the plan down out of hand. What would be the position if he returned?

I must devote all my attention and care to this machine because if anything was made to a drawing which did not answer its intended purpose, I should incur the principal share of the blame, as being necessarily most familiar with the details, whereas all the praise which perfection would secure *would attach to Mr Clement*, who would come over now and then and sanction *my* plans *only* when he could not substitute any of his *own* either *better or worse*, and I should have the indescribable satisfaction of knowing that I was labouring to increase the credit of a man who was envious of my talents and jealous of my influence—whose interest and inclination it would be to use every method in his power however mean and mortifying to guard against the suspicion of my having ability ... No! whatever situation circumstances may force me into I will bear as best I may; but I will never, if I know it, become a party to my own degredation.[20]

On 16 July 1834 Babbage wrote privately to the Hon. James Stuart at the Treasury: 'The drawings and parts of the Engine are at length *in a place of safety*—I am almost worn out with disgust and annoyance at the whole affair.'[21] He also sent a formal letter to the Treasury. On 16 August when Clement's accounts had been examined he received a formal letter from James Stuart conveying 'their Lordships authority to proceed in completing said Engine'.[22] However, Babbage could not continue constructing the Engine without basic reorganization of the work, which would have considerable financial implications. The only solution was to discuss the problem with Lord Melbourne, then Prime Minister, and on 26 September Babbage wrote requesting an interview. Still carried forward by the momentum of Reform Melbourne agreed, but before the meeting could take place the Government fell: the sands of time were running out for the Difference Engine and for many another plan cherished by the reformers.

The Duke of Wellington became Prime Minister again and Babbage again sought an interview. However the Duke this time preferred a statement on

[19] Ibid., f 39.   [20] Ibid., f 58.
[21] Ibid., f 450.   [22] Ibid., f 451.

paper. Babbage's own work was moving rapidly again with Jarvis working directly for him in Dorset Street, but the enterprise was turning in a new direction: Babbage had started developing what were to become the Analytical Engines. Moreover he was developing improved techniques so rapidly that it began to seem *technically preferable* to start again even on the Difference Engine and construct it on simplified and improved lines. Principled to a fault, Babbage felt obliged to place all the options and considerations before the head of government. A personal interview was more than ever necessary and Wellington might very well have given Babbage the only possible advice in political terms: finish constructing the first Difference Engine as begun and then consider new plans.

On 23 December Babbage addressed his statement to the Duke. He suggested four options for dealing with the Difference Engine which the Government might adopt.[23] Firstly, work might be continued with Clement: Babbage considered this effectively impossible, although if the Government insisted he would try to make it work. Secondly, the Difference Engine might be completed by some workman other than Clement. Babbage thought this practicable and had considered possible means. What he probably had in mind was for Jarvis to plan the work which would be executed partly in East Street and partly subcontracted. On this Babbage would consider any Government proposition but would himself advance none. He had worked for thirteen years without thanks or remuneration—indeed at heavy personal sacrifice—and was now determined to have the work on a more satisfactory basis. Almost certainly the dominant consideration was the new Engine on which he was working: Babbage was above all anxious that any arrangement should leave sufficient time and all his personal financial resources free for this new work. In discussion Wellington could probably have cut through the complications directly, but nothing short of an interview was capable of resolving the problem. Thirdly, if the Government wished to appoint some person other than Babbage to complete the Difference Engine he would immediately transfer the Engine to him, but he considered such a proposal of doubtful expediency—as well he might. And the fourth possibility was to abandon the undertaking entirely.

Babbage then went on to discuss his new Engine: it did not replace the Difference Engine in the latter's sphere of making tables, but rather complemented it, thus enhancing its power and range of action. In discussing the significance of the new Engine Babbage made a remarkable statement about what would now be called technological spin-off. The statement deserves to be quoted in its entirety: it not only shows Babbage's perspicacity in these matters but is a reminder of what Britain lost when the Government abandoned work on the Calculating Engines:

Whether I shall ever be able to afford to construct such an Engine from my own

[23] Ibid., f 525.

private resources, injured as they have been by the sacrifices I have made in carrying on the former one is yet uncertain—but it has been suggested to me from several quarters, and has occurred to my own mind that the Government of some other Country forming a different estimate, both of the utility of such undertakings, and of the value of the reputation that attends them, might make some propositions which I should be disposed to entertain—

My right to dispose, as I will, of such inventions cannot be contested: it is more sacred in its nature than any hereditary or acquired property, for they are the absolute creations of my own mind. The consequences of such an arrangement would be, that it would become necessary for me to collect together all that is most excellent in our own workshops—that *Methods* and *Processes* which are equally essential to the perfection of machinery, but which are far less easily transmitted from country to country, would be at once brought into successful practice under the eyes and by the hands of foreign workmen—that I should contrive as I have contrived before, new modes of executing work, and that a school of mechanical engineers might arise, whose influence would give a lasting impulse to the whole of the manufactures of that country, and that the secondary consequences of the acquisition of that Calculating Engine might become far more valuable than the primary object for which it was sought.

Wellington had merely been caretaker for Peel and could only have acted on a specific plan; which Babbage did not feel free to advance. Peel had always been unsympathetic to Babbage and his Engines, but Babbage sent him a copy of the statement prepared for Wellington. In turn Peel's government soon fell, and the country declined into years of Whig financial ineptitude and the doldrums of the second Melbourne administration.

Babbage was becoming disillusioned with the British Government. Probably in 1833 he wrote a letter to the Duke of Somerset in which he says of the assembled portion of the Difference Engine: 'I have ... been compelled to perceive ... that of all countries England is that in which there exist the greatest number of practical engineers who can appreciate the mechanical part whilst at the same time it is of all others that country in which the governing powers are most incompetent to understand the merit either of the mechanical or mathematical.'[24]

Failure to secure an answer from the Government to his request for a decision about the Difference Engine undoubtedly contributed to his growing dissatisfaction. Indeed the failure to complete the Difference Engine, by far the largest Government-sponsored private research project of the time, was itself of considerable importance: had the Engine been completed there might have been more confidence in applied science and other projects could have followed—not only Babbage's personal projects. Economic historians have traced Britain's current economic difficulties to 1870 or sometimes to the 1850s, but Babbage was aware of fundamental problems in the 1830s. His comments are usually couched in terms of his Engines. Typically he wrote to

[24] Add. Ms. 37,187, f 302.

an American in 1835 of the new Engine on which he was then working:[25] '*You* will be able to appreciate the influence of such an engine on the future progress of science.—I live in a country which is incapable of estimating it . . .'

By the end of 1834, although he did not realize it, for Babbage the Difference Engine was history. He was now launched on the course which was to lead to the Analytical Engines.

[25] Babbage to an unidentified American, Burndy Library.

The Ninth Bridgewater Treatise

While he was working on the new engine Babbage wrote his only essay in philosophy, *The Ninth Bridgewater Treatise*, which he also called a 'Fragment'. The essay derived from his work on the engines, and before turning to the Analytical Engines we shall consider briefly this felicitous book, and then in the following chapter some of his other work during the 1830s.

Before the publication of Darwin's *Origin of Species*[1] there developed in England in the first half of the nineteenth century a widespread disturbance of Christian beliefs through the progress of science. Comparative anatomy as well as many literary, philological, and historical studies questioned traditional faith. In particular geology and the study of fossils, by showing the great age of the rocks and the former existence of many unknown species, called in question the truth of the book of Moses. One response by the faithful was to argue through natural theology that, on the contrary, the new evidence of the sciences pointed, through the ever more clear evidence of design which it provided, to the existence of a Heavenly Designer.

The second quarter of the nineteenth century is far better known to the history of religion for the Oxford Movement, but in the developments of natural theology lies the background to the conflicts between science and religion which some decades later were to inflame the discussion of evolution.[2] The Earl of Bridgewater left £8,000 for the publication of arguments in favour of natural religion. This money was placed at the disposal of the President of the Royal Society, who was to appoint and pay people to write, print, and publish works 'On the Power, Wisdom, and Goodness of God, as manifested in the Creation'. Eight men were so appointed, four doctors and four clerics, and the books they wrote were the Bridgewater treatises.[3] More interesting than any of these were two other books concerned with natural theology: one written by the Revd. Baden Powell, Savilian Professor of Geometry in Oxford and a friend of Babbage, entitled *The Connection of Natural and Divine Truth*;[4] the other was Babbage's *Ninth Bridgewater Treatise*. Of the two the latter was much the better known. This essay complements Babbage's two earlier books,

---

[1] C. Darwin, *On the Origin of Species by means of Natural Selection*, London, 1859.
[2] A pioneering study of natural theology is the Cambridge doctoral thesis of John David Yule: 'Impact of Religious Thought in the Second Quarter of the Nineteenth Century', 1976.
[3] *Ninth Bridgewater*, preface xxii.   [4] Ibid., xii.

The Decline of Science and *On the Economy of Manufactures*, and in conjunction with them presents his world picture.

Babbage saw God as a being of science and programmer who defined the entire future of the universe at the time of the Creation as a sort of infinite set of programs. It was a Newtonian universe, determinist and mechanistic, wholly scientific within the limits of Babbage's conception. Few others at the time could really grasp the concept of programs and by the time modern programming came to be developed in the middle of the twentieth century the self-confident Newtonian world had long been replaced by the shifting uncertainties of quantum mechanics. To this extent Babbage's philosophy is unique. It is presented with much grace, and the 'Fragment' is an elegant work.

The first Bridgewater Treatise to be published was by the Revd. William Whewell *On Astronomy and General Physics*. Later a famous Master of Trinity and one of the most important figures in nineteenth century Cambridge, Whewell could be as pugnacious as Babbage himself. In his book Whewell had written: 'We may thus, with great propriety, deny to the mechanical philosophers and mathematicians of recent times any authority with regard to their views of the administration of the universe; we have no reason whatever to expect from their speculations any help, when we ascend to the first cause and supreme ruler of the universe. But we might perhaps go farther, and assert that they are in some respects less likely than men employed in other pursuits, to make any clear advance towards such a subject of speculation.'[5] Coming from his old friend Whewell, who had helped the former Analyticals introduce the modern notation for the calculus into the Cambridge examinations, Babbage considered the statement rank treason. It led to a controversy between these two formidable men which was conducted in the very best academic tradition: openly, trenchantly, and with high good humour.

There are three sources from which it is stated that man can arrive at knowledge of the existence of a Deity: the metaphysical argument, which was unconvincing even for most theologians; from testimony of Revelation, which to so experienced an observer as Babbage seemed full of difficulty; and from examination of the works of the Creator.[6] It was the third that Babbage found convincing. To questions of historical testimony he adopted a sensible critical

---

[5] Quoted in *Ninth Bridgewater*, preface, xii and frontispiece. Whewell is referring particularly to Laplace whose great study of the Newtonian universe was then enjoying extraordinary popularity in Mary Somerville's translation. In spite of their differing views Whewell and Babbage were on close terms. Babbage arranged Whewell's membership of the Royal Society (Babbage to Whewell, May 1820; Trinity College, Cambridge, Add Ms a 200[192]); Babbage asked Whewell to help friends and acquaintances going to Cambridge (Trinity College Add. Mss: Babbage to Whewell, 13 May 1831, a 200[193]; 29 June 1842, c 87[39]; 29 June 1848, c 87[40]); and on at least one occasion when he was visiting Cambridge stayed in Whewell's rooms in Trinity (Whewell to Babbage, May 5, 1820, BL Add. Ms. 37,182, f 258).

[6] See Yule's thesis (note 2 above).

approach which was to become common later in the century. Well aware of the difficulty of copying manuscripts accurately—he had suffered enough from errors in copying mathematical tables, and indeed removal of such errors had been a prime object in building the first Difference Engine—he focussed attention on the great difficulty in interpreting the meaning of words and phrases, and above all subtle analogies, from so remote a period as that of the Biblical past. In 1837, when the *Ninth Bridgewater Treatise* was published, although abundant fossil evidence showed that innumerable sequences of creatures had gradually developed and adapted to suit changing environments, it was still possible to consider evolving natural forms as the result of a long series of special creations. Only following publication of Darwin's theory did such a view became quite untenable for a natural philosopher. However a set of special creations seemed to imply a God 'perpetually interfering, to alter for a time the laws he had previously ordained; thus denying to him the possession of that foresight which is the highest attribute of omnipotence'.[7] Drawing on his experience with the calculating engines Babbage succeeded in devising a novel picture of God whose undeviating law would be consistent both with successive special creations and with miracles. This was Babbage's unique view of the Creation.

His starting point is that of the man of science, whose work can be judged according to the degree of foresight manifest in it. Then God is man of science and programmer writ large:

The estimate we form of the intellectual capacity of our race, is founded on an examination of those productions which have resulted from the loftiest flights of individual genius, or from the accumulated labours of generations of men, by whose long-continued exertions a body of science has been raised up, surpassing in its extent the creative powers of any individual, and demanding for its development a length of time, to which no single life extends.

The estimate we form of the Creator of the visible world rests ultimately on the same foundation . . .

The greater the number of consequences resulting from any law, and the more they are foreseen, the greater the knowledge and intelligence we ascribe to the being by which it was ordained.[8]

Now the calculating engines—and also of course a modern computer— could easily be instructed to proceed according to one law for any number of operations and then proceed according to some other law, the change in operation being programmed *ab initio*. Similarly, reasoned Babbage, the changes in natural law, as evidenced by the creation of new species, were not proof of Heavenly intervention but could also have been programmed by the Creator *ab initio*: that is to say at the time of the Creation. In a similar manner miracles appeared as singularities in the Celestial Program: a miracle was merely a

[7] *Ninth Bridgewater*, 24–5.   [8] Ibid., 30–1.

subroutine called down from the Heavenly store. To make the concept more comprehensible to his contemporaries Babbage explained it in terms of singularities in equations of the fourth degree,[9] but he was really thinking in terms of his beloved engines. He then turned aside to demolish mathematically David Hume's celebrated argument against miracles, remarking in a mandarin fashion that 'Hume appears to have been but very slightly acquainted with the doctrine of probabilities.'[10]

Not only the theology but also the science of the *Ninth Bridgewater Treatise* foreshadows the evolution controversy. In the *Treatise* Babbage first published his theory of surfaces of equal temperature within the earth, discussing the movement of the earth's crust upon a heated fluid interior. The earth itself was in a state of continuing flux and change. He also included a remarkable extract from a letter of John Herschel, a passage which later became famous as one of the steps towards the theory of evolution. Herschel referred to:

that mystery of mysteries, the replacement of extinct species by others. Many will doubtless think your speculations too bold, but it is as well to face the difficulty at once. For my own part, I cannot but think it an inadequate conception of the Creator, to assume it as granted that his combinations are exhausted upon any one of the theatres of their former exercise, though in this, as in all his other works, we are led, by all analogy, to suppose that he operates through a series of intermediate causes, and that in consequence the origination of fresh species, could it ever come under our cognizance, would be found to be a natural in contradistinction to a miraculous process—although we perceive no indications of any process actually in progress which is likely to issue in such a result.[11]

Whewell's reply when it came was in the form of an open 'Letter to Charles Babbage Esq.'. Expressing 'my satisfaction at having you for my volunteer fellow-labourer in such a cause', Whewell continued 'There have been in the recent literature of our country, many proofs how generally acceptable the subject [of natural religion] is, but none in which the sympathy of others with regard to it has given me more pleasure, and none in which it is treated in a more original manner.'[12] Whewell nevertheless maintained his position courteously but firmly.

Babbage's answer came in an 'Advertisement to the second edition' of the Treatise. The two were agreed that the relation of man to his Maker was of far greater importance than mathematics or physics. But their attitudes to science and in particular to scientific evidence as a basis for belief were fundamentally different. In practice for Babbage experimental evidence interpreted with the aid of pure mathematics was decisive; for Whewell it was subordinate.

---

[9] Ibid., 99–101. It may be doubted whether the discussion of equations of the fourth degree would greatly clarify the concept for a modern clerical audience, but Babbage was writing largely for the mathematically educated clerical products of Cambridge.

[10] Ibid., 123   [11] Ibid., Appendix, note I.

[12] Whewell, *Open Letter to Charles Babbage*, 1837.

Whewell, though he would have been reluctant to acknowledge the full implications, had serious reservations about the limits of the application of science: this shows, for example, in his later attitude towards the development of science examinations in Cambridge.[13] Babbage's views, though he would certainly have repudiated such a suggestion, would as science advanced lead to the abandonment of the hypothesis of a Deity.

In Babbage's world picture there seems to be a close relation between the physical particle, moving freely but subject to natural law and affected by physical forces, and the commodity in the capitalist system. Both the Newtonian physical universe and the capitalist system are in some sense designed by the Deity to be the best of all possible systems and therefore, from the theological point of view, the actual systems.

The more subtle metaphysical arguments of the 'Fragment' left little trace but another idea Babbage put forward made a lasting impression. It was a consequence of Newtonian physics that words once spoken, though the loudness of the sound would continuously diminish, would never entirely be lost. Dickens later took the idea up, presenting it in a speech[14] and giving it wide currency. The idea is best expressed in Babbage's own words:

The principle of the equality of action and reaction, when traced through all its consequences, opens views which will appear to many persons most unexpected.

The pulsations of the air, once set in motion by the human voice, cease not to exist with the sounds to which they gave rise. Strong and audible as they may be in the immediate neighbourhood of the speaker, and at the immediate moment of utterance, their quickly attenuated force soon becomes inaudible to human ears. The motions they have impressed on the particles of one portion of our atmosphere, are communicated to constantly increasing numbers, but the total quantity of motion measured in the same direction receives no addition. Each atom loses as much as it gives, and regains again from other atoms a portion of those motions which they in turn give up.

The waves of air thus raised, perambulate the earth and ocean's surface, and in less than twenty hours every atom of its atmosphere takes up the altered movement due to that infinitesimal portion of the primitive motion which has been conveyed to it through countless channels, and which must continue to influence its path throughout its future existence.

Thus considered, what a strange chaos is this wide atmosphere we breathe! Every atom, impressed with good and with ill, retains at once the motions which philosophers and sages have imparted to it, mixed and combined in ten thousand ways with all that

---

[13] D. A. Winstanley, *Early Victorian Cambridge*, C.U.P., 1940. See the discussion leading up to the Royal Commission of 1850 on University Reform. On this important Commission were Dr Graham, George Peacock, Sir John Herschel, Sir John Romilly, and Adam Sedgwick. One notes two old Analyticals (Herschel and Peacock) and the liberal Sedgwick, a good friend of Babbage's. Sedgwick and Herschel had many clashes with the conservative Whewell, then Master of Trinity.

[14] *Dickens's Speeches*, ed. Fielding, 399, O.U.P., 1960. Speech at Birmingham, 27 Sept. 1869.

is worthless and base. The air itself is one vast library, on whose pages are for ever written all that man has ever said or woman whispered. There, in their mutable but unerring characters, mixed with the earliest, as well as with the latest sighs of mortality, stand for ever recorded, vows unredeemed, promises unfulfilled, perpetuating in the united movements of each particle, the testimony of man's changeful will.

But if the air we breathe is the never-failing historian of the sentiments we have uttered, earth, air, and ocean, are the eternal witnesses of the acts we have done.[15]

Babbage then took the opportunity for a devastating attack on the slavers, quoting from some contemporaty reports:

The soul of the negro, whose fettered body surviving the living charnel-house of his infected prison, was thrown into the sea to lighten the ship, that his christian master might escape the limited justice at length assigned by civilized man to crimes whose profit had long gilded their atrocity,—will need, at the last great day of human account, no living witness of his earthly agony. When man and all his race shall have disappeared from the face of our planet, ask every particle of air still floating over the unpeopled earth, and it will record the cruel mandate of the tyrant. Interrogate every wave which breaks unimpeded on ten thousand desolate shores, and it will give evidence of the last gurgle of the waters which closed over the head of his dying victim: confront the murderer with every corporeal atom of his immolated slave, and in its still quivering movements he will read the prophet's denunciation of the prophet king:—
And Nathan said unto David—*Thou art the man*.[16]

Some years later Babbage was intrigued to learn that the idea of the permanent record of our words reaching a superintending providence had been anticipated in Chaucer,[17] though naturally without so clear a physical basis.

To complete the picture, from the theory of the eternal physical record of man's words and deeds Babbage proceeded to create his own intellectual and highly moral concept of future punishment:

Who has not felt the painful memory of departed folly? who has not at times found crowding on his recollection, thoughts, feelings, scenes, by all perhaps but himself forgotten, which force themselves involuntarily on his attention? Who has not reproached himself with the bitterest regret at the follies he has thought, or said, or acted? Time brings no alleviation to these periods of morbid memory: the weaknesses of our youthful days, as well as those of later life, come equally unbidden and unarranged, to mock our attention and claim their condemnation from our severer judgment.

It is remarkable that those whom the world least accuses, accuse themselves the most; and that a foolish speech, which at the time of its utterance was unobserved as such by all who heard it, shall yet remain fixed in the memory of him who pronounced it, with a tenacity which he vainly seeks to communicate to more agreeable subjects of reflection ...

If such be the pain, the penalty of thoughtless folly, who shall describe the punishment of real guilt? Make but the offender better, and he is already severely

---

[15] *Ninth Bridgewater*, Ch. IX.   [16] Ibid.
[17] Chaucer, *Hall of Fame*, BL Add. Ms. 37,196, f 475.

punished. Memory, that treacherous friend but faithful monitor, recalls the existence of the past, to a mind now imbued with finer feelings, with sterner notions of justice than when it enacted the deeds thus punished by their recollection.

If we imagine the soul in an after stage of our existence, to be connected with a bodily organ of hearing so sensitive, as to vibrate with motions of the air, even of infinitesimal force, and if it be still within the precincts of its ancient abode, all the accumulated words pronounced from the creation of mankind, will fall at once upon that ear. Imagine, in addition, a power of directing the attention of that organ entirely to any one class of those vibrations: then will the apparent confusion vanish at once; and the punished offender may hear still vibrating on his ear the very words uttered, perhaps, thousands of centuries before, which at once caused and registered his own condemnation.[18]

At the same time Babbage saw a vision of heavenly reward for the good:

But if, in a future state, we could turn from the contemplation of our own imperfections, and with increased powers apply our minds to the discovery of nature's laws, and to the invention of new methods by which our faculties might be aided in that research, pleasure the most unalloyed would await us at every stage of our progress. Undogged by the dull corporeal load of matter which tyrannizes even over our most intellectual moments, and chains the ardent spirit to its unkindred clay, we should advance in the pursuit, stimulated instead of wearied by our past exertions, and encountering each new difficulty in the inquiry, with the accumulated power derived from the experience of the past, and the irresistible energy resulting from the confidence of ultimate success.[19]

It is a pleasant foible in Babbage how close an analogy there is between his concept of the Deity and his own pursuits on earth, and it was natural for him to extend the parallel to the afterlife as well.

[18] *Ninth Bridgewater*, Ch. XIV.   [19] Ibid.

11

○ ● ○

Rail, Steam, and the British Association

The 1830s were a heroic age for the engineers in Britain: the Brunels and the Stephensons were folk heroes of the time. Railways marched across the countryside, bridges and viaducts leaped across rivers and valleys, tunnels were driven through inconvenient hills, while steamships began to cross the Atlantic. There was great rivalry between the Stephensons in the North and the London-based Brunels.[1] In this rivalry and competition, which developed into the great battle of the gauges, Babbage became deeply involved. There was a fundamental difference in approach between the two groups. The Stephensons were more cautious and pragmatic, less imaginative; the Brunels were brilliantly imaginative, more scientific, and they had far wider interests than their northern rivals. With the Brunels the francophile Babbage formed a natural alliance.

Old George Stephenson,[2] founding father of the public railways, did not learn to read or write until his eighteenth year. He was an engine man in the mines, looking after both Newcomen and Boulton and Watt engines. He also invented the safety lamp independently of Humphry Davy, and received a £1,000 award for the invention. From 1813 onwards Stephenson was working on steam locomotion for use in the mines. The first public railway, between Stockton and Darlington, was opened for public traffic on 27 September 1825. It was natural for Stephenson to use the 4′ 8½″ gauge which was used in the small railway tracks in the mines. He simply did not think of using any other, and he continued to use the narrow gauge when he began working on the Liverpool to Manchester railway, which was really the time when he should have moved to a wider gauge.[3] The narrow gauge spread and established itself: Britain's railway network has been lumbered with it and the narrow gauge is still a handicap. As happens so often in engineering, a standard can be established on a largely accidental basis, but once the standard is established it becomes almost impossible to change.

Babbage and Wolryche Whitmore went north for the opening of the Manchester to Liverpool railway. At that time almost anything in railways was novel and Babbage suggested to Hodgson of the railway company what was

[1] Really between the Stephensons and I. K. Brunel, although the latter drew much strength from his father.
[2] Samuel Smiles, *Lives of the Engineers*, London, 1861.
[3] BL Add. Ms. 37,185, f 326; *Passages*, 317–18.

later to be called a 'cow-catcher' for sweeping obstacles off the line. At the official opening of the railway on 15 September 1830 the famous first railway tragedy occurred :[4] the politician Huskisson was killed by Stephenson's Rocket as it showed its paces. The Duke of Wellington, then Prime Minister, was present and a galaxy of political and society figures. The death cast a shadow over the railways which was not easily lifted. Death while riding was commonplace but this was a reminder of the awesome power of machinery and was to be followed by many railway deaths in fiction.

George Stephenson made sure that his son Robert received a good schooling. In 1822, after six months at Edinburgh University, he became manager of the locomotive factory which his father had established at Newcastle. Tiring of his father's domination, Robert went to Colombia in South America to superintend the mining of gold and silver, useful experience when he came to build railways abroad. Returning in 1827 he assisted his father and supervised the construction of the Rocket. In 1833 he was appointed engineer of the projected London to Birmingham railway, thus building the first great railway line into London. Later he built a tubular railway bridge across the Menai Straits and many railways and bridges in different parts of the world. Isambard Kingdom Brunel and Robert Stephenson kept up a friendly rivalry but they had the greatest respect for each other; and it is pleasant to note that in difficulty each could rely on the other for encouragement and help.

The background of the Brunels was totally different. Marc Isambard Brunel came from a family of well-off Normandy peasants. As a younger son he was destined for the priesthood and learned mathematics, Latin, Greek, and drawing. He haunted the village carpenter's shop and was fascinated by architecture, making detailed drawings of old buildings wherever he went. He disliked the idea of the church and turned instead to the navy, learning instrument making on the way. At that time applied science in France was readying itself for the rapid development which took place after the revolution and formed the background of his theoretical approach to engineering, an approach further developed by his son. Brunel was a royalist and after service on a corvette escaped to the United States. There he worked for six years as architect and engineer, taking American citizenship.

Brunel came to England seeking a contract to design block-making machinery for the Portsmouth naval dockyard.[5] The set of machines he developed was the first to make by machine tools equipment with interchangeable parts. It saved the Admiralty tens of thousands of pounds a year. So impressed was the Tsar on seeing the machinery that he gave Brunel a large ruby set in diamonds and invited him to come to Russia. The drawings of the block-making machinery changed British engineering drawing practice.

[4] Elizabeth Longford, *Wellington, Pillar of State*, 221, Weidenfeld, 1972.
[5] K. R. Gilbert, *The Portsmouth Blockmaking Machinery*, H.M.S.O., 1965.

Moreover the machinery itself was made by Henry Maudslay and the project established his famous workshop which made so much of the most important machinery in the decades ahead and cradled a generation of engineers. Babbage soon established a close rapport with Brunel, and the Difference Engine was the next important piece of equipment with interchangeable parts to be made by machine tools. The parts of Babbage's Engine were made with an altogether higher degree of precision and the Difference Engine gave a stimulus to the development of machine tools similar to the stimulus of Brunel's project twenty years earlier: the line of succession is direct, and as a matter of fact Brunel recommended Clement to Babbage.[6] These machine tools and the associated engineering techniques were decisive in the continuation of the industrial revolution: they equipped the factories of Britain and much of the rest of the world.

When the block-making machinery was finished Brunel built himself a sawmill at Battersea. He also built timber handling and sawing machinery at the Chatham naval dockyard, using a wide gauge railway for carrying logs, an idea which his son later adopted and developed for the Great Western Railway. The elder Brunel also built with the Government's encouragement a factory to make army boots. The peace after Waterloo left him with a mountain of boots unwanted by the army. Then his saw-mill at Battersea burned down. Bankrupt, he went to the King's Bench prison. Wellington, always sensitive to the needs of a good engineer, obtained £5,000 to release Brunel, on the sensible condition that he should stay in England and not go to Russia.

Brunel became famous for his tunnel under the Thames, not for the far more important block-making machinery. In building the tunnel he was helped by his son, Isambard Kingdom Brunel: indeed for some years the younger Brunel was largely responsible for the work. Young Brunel had an excellent education, first at the College of Caen in Normandy, and then at the Lycée Henri-Quatre in Paris, at that time famous for mathematics. Next he was apprenticed to Abraham Louis Breguet, maker of precision instruments, including chronometers and watches. A superb horologist, Breguet sired a school of craftsmen as did Maudslay in England. I. K. Brunel's training was completed by a spell as his father's assistant, during which he actually spent a great deal of his time with Maudslay, Sons and Field in Lambeth: he emerged as a well educated, professionally trained engineer.

But in spite of his training and experience, when work on the Thames tunnel ceased due to flooding in 1828 and lack of funds by the company financing it, his position did not look at all good. He had no completed work to his credit, while his young rivals seemed far better placed. Robert Stephenson was engaged on his father's highly successful railway projects. The young Rennies were completing the splendid new London Bridge and establishing first rate

[6] S. Smiles, *Industrial Biography*, 343, John Murray, 1879.

government contacts. In contrast young Brunel found himself involved in a series of disappointing attempts to obtain work engineering harbours and canals. In January 1830 he applied for the post of engineer to the Newcastle and Carlisle Railway, only to have the humiliation of being overlooked in favour of a mediocre canal engineer.

Then in October 1830 Babbage introduced I. K. Brunel to the solicitor of the projected Bristol to Birmingham railway. Soon young Brunel was writing to Babbage:

Birmingham,   Nov 10th
1830

My Dear Sir,

I might naturally expect that a letter from you would be the best of introductions but I certainly never even hoped to find it the powerful spill it has proved.

In three weeks there [in Birmingham] it has brought me in contact with a most gentlemanly and agreeable man and thro' his introduction I find myself in two days the appointed engineer to the provisional committee of the Birmingham and Bristol railway. I am to commence a rough survey next week—to Gloster where for the present the line is to terminate as the Bristol people do not yet come forward liberally—

You know how important such an opening into the railway world is to me and therefore you can conceive how grateful I am towards you to whom I owe it—

I trust I shall not disgrace your recommendation—as I hope to have the pleasure of seeing you on Sunday I will leave till then any further account of this affair.

Believe me dear sir,
Very gratefully and sincerely yours,
I. Brunel.[7]

Although it served as an invaluable introduction for Brunel into the railway world the Bristol and Birmingham project proved a false start because the required funds were not forthcoming. For some time longer he had to content himself with minor projects. One was the observatory for Sir James South on Campden Hill. Brunel designed a revolving dome with mechanically operated shutters in it. Naturally the mechanism was built by Messrs Maudslay, Sons and Field. On 20 May forty-five people including Babbage, I. K. Brunel, and John Herschel sat down to dinner on the lawn to celebrate the completion of the observatory. However the cost of the work had exceeded the original estimate and, although he had approved the design, South refused to pay the excess. Everyone was on Brunel's side but South published an anonymous article in the *Athenaeum* attacking Brunel. A later quarrel of South's, on the matter of the mounting for the telescope intended for Brunel's dome, was to have more serious consequences in which Babbage reluctantly became involved.[8]

Isambard Kingdom Brunel's chance came in Bristol. He made plans for

[7] BL Add. Ms. 37,185, f 343.   [8] See *The Exposition of 1851*.

improving the docks; designed a suspension bridge across the Avon gorge; and became engineer to the Great Western Railway. He engineered the G.W.R. with a 7 foot gauge thus leading to the great fight over the standard gauge for Britain's railways. The background to the 'battle of the gauges', as it was called, and of the associated competition over Atlantic steam navigation, was the decline of Bristol as a port and the rise of Liverpool. Attempts to restore Bristol's position might have been more successful if the docks and harbour had been improved, but Brunel's plans were pigeonholed after the Bristol riots in October 1831.

The suspension bridge across the Avon which he designed was not actually constructed until after his death. There had been a competition in which twelve designs were submitted. After being whittled down to four they were submitted to the judges, Davies Gilbert and, curiously, John Seaward, builder of marine engines. Davies Gilbert preferred Brunel's design but allowed his mind to be changed by Seaward. Brunel managed to have a talk with the judges and changed Davies Gilbert's mind back again to support for the Brunel design. Brunel was horrified, writing in his diary: 'D. Gilbert came down with *Seaward* to assist him!!!!!! Seaward!!! . . . It appears that my details are found *very bad*, quite inadmissable.' But after Brunel had explained his design to Gilbert ' . . . he returned to his original opinion viz approval of all the details. Oh quel homme! P.R.S.!!!'[9] This is a reference to the Decline of Science in which Babbage had blasted Davies Gilbert out of the Presidency of the Royal Society. Actually he had some quite sensible suggestions to make, including the use of four great chains, two on either side, so that if one were to break it could be replaced without danger.

Brunel's opportunity came when he was appointed engineer to the London to Bristol railway, soon to become the proud Great Western Railway. At each stage of its construction, whether he was building the Thames bridge at Maidenhead, planning the Box tunnel leading down to Bristol, designing the timber viaducts that straddled the Cornish valleys, or planning his last great bridge at Saltash across the Tamar, Brunel relied on both theory and experiment. He could use more powerful theoretical tools than his rivals, but not one of them was a more thorough experimentalist than Brunel. He was quite as practical as Robert Stephenson, and in contrast to the academic dabblers Brunel was master of all the relevant theory of the time.[10]

Dionysus Lardner was the comedy act in the show: he ballooned across the engineering landscape of the time sustained by an inexhaustible supply of hot air. Intensely jealous of those with real scientific and engineering knowledge, such as Brunel and Babbage, Lardner could be relied upon in any engineering

[9] L. T. C. Rolt, *Isambard Kingdom Brunel*, 56, Longman, 1957; quoted from Brunel diaries.

[10] Sir Alfred Pugsley (ed.), *The Works of Isambard Kingdom Brunel, an Engineering Appreciation*, Institute of Civil Engineers, 1976.

controversy to get hold of the wrong end of the stick. His pretensions as either experimentalist or theoretician would not have been taken seriously and he could have joined innumerable eccentric divines as mere sources of entertainment, but the controversies in which he intervened had immediate practical consequences in which powerful commercial interests and pressure groups were directly involved; and they made use of Lardner's colourful if unfounded assertions.

The first time Lardner clashed with Brunel was over the tunnel under Box Hill leading the Great Western Railway down into Bristol. The Box tunnel was to be constructed at a gradient of 1 in 100, and was to be nearly two miles long. The opponents of the G.W.R. were a diverse lot: landowners, coach-proprietors, canal owners; even the provost of Eton, who declared that the proximity of the railway would undermine morals and discipline in his school. They seized on the Box tunnel, declaring that the noise of two trains passing would be intolerable. Lardner added his pseudo-science to the debate by arguing that if the brakes failed in the tunnel a train would by his calculations emerge at 110 miles per hour. Brunel pointed out that Lardner had forgotten about friction and air-resistance. The opposition was duly defeated and the G.W.R. bill passed through parliament, but the point about air-resistance sank in: it dawned on Lardner that if air was resistant, why then water was even more so. Thus when Brunel was later designing a steamship to cross the Atlantic, Lardner did a volte-face, proving to his own satisfaction that the resistance of the water would make such a crossing impossible. When Brunel's *Great Western* actually crossed the Atlantic, Lardner then decided that crossing the Atlantic by steam was not impossible but rather that it would be uneconomic, only to be discomfited once again. Lardner also took up Brunel's specific points about the effects of air resistance and friction on trains, seeking to prove that the wide gauge railway was inefficient. On that occasion he was put down by the combined forces of Babbage and Brunel.

A brilliant popular lecturer, Lardner was undoubtedly a likeable rogue and Babbage kept up a frequent correspondence with him in the late 1820s and early 1830s. In his diary Macready noted on 18 June 1835:[11]'Went on box of Sheil's carriage to Dr. Lardner's, where I saw and was introduced to the Guiccioli—saw Mrs Norton, Mrs L. Stanhope, etc. Was surprised to see Mr Cooper, and Miss Betts, and Miss —— Oh Dr. Lardner! Is this the society for a philosopher.' Lardner was a scientific Falstaff, though even now he is occasionally mistaken for a serious figure.

By 1841 the Great Western Railway was nearing completion and the Box tunnel was opened. Lardner kept quiet, but in his place the Revd. William Buckland, the Oxford geologist, spoke up declaring that, although he had never

[11] William Toynbee, *The Diaries of William Charles Macready*, Chapman and Hall, 1912.

dared venture into the tunnel to investigate, he was sure that the portion of the tunnel which was unlined with brick was unsafe because rock would certainly fall under the impact of sound waves and the effect of vibration. Brunel contemptuously replied that although he had little academic knowledge of geology he had some recent experience of tunnelling through that particular rock and considered it quite safe. In 1957 when Rolt wrote his biography of Brunel the tunnel continued in use unlined: it still does. Buckland is a prize example of the growing divorce between academic science and engineering practice; and Peel often consulted him about scientific questions. In 1845 on Peel's recommendation Buckland was appointed Dean of Westminster.[12]

Science was becoming institutionalized and also more conservative. As in the case of the Analytical Society, once again the central group of academic men of science was based on Trinity College, Cambridge. This time it was known as the 'northern lights', and included Whewell, Airy, and Sheepshanks. Far more conservative than the Analyticals, without anyone with anything like the scientific ability of Babbage or Herschel, this group played a crucial part in the academic establishment and the embryonic scientific bureaucracy. Airy clashed with Babbage, and also with I. K. Brunel over the railway gauges. Whewell clashed repeatedly with Babbage over the position of science, and even had a bitter argument with the peaceable Sir John Herschel over the place of science in Cambridge degrees. Whewell wanted men to take an arts course before they turned to science in earnest.[13] In the end it was another old Analytical, George Peacock, who established science in its proper place in the Cambridge University examinations. Nor should it be thought that the role of Whewell and Airy in the developing split between pure academic science and lowly, inferior applied science was purely accidental. Airy received his appointment as Astronomer Royal and Whewell the mastership of Trinity through the good offices of Robert Peel. Indeed the previous master had held on until the appointment could be made by a conservative prime minister: otherwise the post would have gone to the liberal Adam Sedgwick.[14]

An alternative body was needed to represent the interests of science to government, and also to act as a co-ordinating centre for supporting research,

[12] Buckland had been a geologist of note. Lyell, who went to Exeter College, Oxford, had attended Buckland's lectures. But Peel would use this impractical man for advice on the *applications* of science.

[13] By this time Whewell had become an overt reactionary in matters of scientific education, causing even Robert Peel to protest: 'The Doctor's [that is to say Whewell's] assumption that *a century should pass* before new discoveries in science are admitted into the course of academical instruction, exceeds in absurdity anything which the bitterest enemy of University education could have imputed to its advocates.' Sir Theodore Martin, *Life of the Prince Consort*, 2nd ed., 118, 1876.

[14] D. A. Winstanley, *Early Victorian Cambridge*, C.U.P., 1940.

where appropriate, with finance and encouragement. Following the election of the Duke of Sussex as president, Babbage and his friends mostly wrote off the Royal Society as useless for the forseeable future.[15] Practitioners of science— the term 'scientist' was not coined until the end of the decade—decided to form a new organization which could forward the interests of British science. David Brewster in Scotland took the initiative in proposing early in 1831 'a meeting of British men of science in July or August next'. The plan derived from Babbage's report of the congress of German philosophers, which he had attended in Berlin in 1828, and also from a report of a later congress of the same body. Together these reports had roused much interest and it was on the Deutsche Naturforscher Versammlung that the British Association for the Advancement of Science was based.[16]

The defeat of Herschel for the Presidency of the Royal had taken the heart out of the National Science Reform Movement in England, and the principal effort in the initial organization of the British Association came from Edinburgh and the provincial scientific societies. Babbage himself was prevented from attending the first meeting at York by pressing work in erecting workshops and buildings for the Difference Engine. However his central position was acknowledged when he was appointed one of the three trustees, the only permanent officials of the Association. At the outset most of the other leading lights of the Royal Society of London were less interested. Herschel in particular was satiated with political controversy and disenchanted with scientific organization in general: his stint as secretary of the Royal Society had been a great distraction from scientific work and he had stood for the presidency with reluctance and only because he felt it his inescapable duty to do so. A year or two later, when the Association was well established and obviously destined to be successful, leading academics were only too glad to join. The meeting at York was attended by about 350 members and established the Association on a firm basis. The meetings grew, reaching nearly two and a half thousand at the Newcastle meeting in 1838, and then fell away until by the second meeting at

---

[15] Lubbock and some others stayed on in the hope of better times, but most of Herschel's supporters would have nothing to do with the Society save for continuing their membership. For example W. H. Fitton wrote on 7 Dec. 1830 to Lubbock: 'You may rest assured that a very large body of the Royal Society—(*besides* those who signed the requisition for Mr Herschel) are deeply offended by the manner in which the new President had been forced into his office—and that opportunity only is wanted to produce a demonstration of their feeling. It is obvious in fact that there can be no *hope* of saving the society—or even restoring it again as an useful instrument for the advancement of knowledge—till it is taken out of the hands by whom the late disgraceful intrigue has been conceived and accomplished'. Lubbock papers, f 15, RS.

[16] For example James D. Forbes wrote on 26 Aug. 1831 to Babbage: 'I had a letter from Whewell saying that he was not much disposed to rally under Dr. Brewster's banners at York after his late attacks on Professors. Now this really is too bad; making it a party matter at once, when in fact Dr B has nothing to do with it but having suggested the meeting ... The Naturforscher of England ...'

York, in 1844, attendance was little more than at the original meeting in 1831. During the 1840s many provincial scientific societies were also in decline.[17]

The principal responsibility for arranging the first meeting at York fell on the Yorkshire Philosophical Society and on its leading spirit, William Vernon Harcourt. He took on the responsibility with enthusiasm and energy, writing a long standard letter to Babbage, Herschel, Whewell, and others, and effectively giving the organization its shape.[18] Whereas the principal object of the German association's meetings had been to give scientific men the opportunity of meeting each other and exchanging ideas, Vernon Harcourt wished also to give the British Association a directing role in scientific research, proposing problems to be solved and to some extent actually funding research projects. The Association did in fact fund many projects with a few tens of pounds, occasionally hundreds, and on rare occasions with over a thousand pounds, but it never had the resources to finance research on a large scale. The Association was also able to make occasional representations to government on matters of scientific interest, such as the rate of duty on scientific instruments. For this purpose Babbage, who was personally acquainted with the important ministers, was able to be of service. Such functions were useful in the 1830s when the Royal Society was at a low ebb, but it was none the less the Royal Society that was the important central body and so it has remained. The peripheral nature of the British Association is shown by the fact that it never established a journal. Even as a meeting place its importance can be exaggerated: these men met each other regularly in any case.

The British Association was subdivided into a number of sections, each concerned with a particular branch of science. At the third meeting, held in Cambridge in 1833 and attended by M. Quetelet on behalf of the Belgian government, Babbage proposed the establishment of a statistical section. This he promptly organized, becoming its chairman. He did not leave the matter there. Returning to London he organized in his own house a preliminary meeting to form a national society independent of the British Association. The meeting at Dorset Street was followed on 15 March 1834 by a public meeting chaired by the Marquis of Lansdowne. It was there resolved to form the Statistical Society of London and Lansdowne became its first president: a fitting appointment as the Lansdownes never forgot their relationship to Sir William Petty, founding father of econometrics. Babbage was a trustee of the Statistical Society for the rest of his life.[19]

There was always something comic about the great jamborees of the British

---

[17] See reports in the *Athenaeum*.

[18] BL Add. Ms. 37,186 f 136. William Vernon Harcourt (1789–1871) was a friend of Davy and Wollaston. He was first president of the Yorkshire Philosophical Society.

[19] Michael J. Cullen, *The Statistical Movement in Early Victorian Britain*, Harvester, 1975.

Association, with their holidaymaking, junketing, backslapping, and unending mutual congratulations. *The Times* regularly sneered at the 'British Ass.' and refused to publish reports of the meetings except as paid advertisements. The continual mutual praise was highly entertaining to the onlooker and the sections provided opportunities for all the cranks in the country to air their views on every subject under the sun. Charles Dickens parodied the proceedings delightfully in the Mudfog Papers. It has been suggested that the immortal Pickwick Club owes something to the British Ass.: possibly also to the Royal Society under Sir Joseph Banks. Most of the members of the Association took themselves far too seriously to realize that they were being mocked.

The 1836 meeting of the Association was held in Bristol. There Lardner might have been heard sounding off about the impossibility of crossing the Atlantic by steam powered vessels. The foundation stone was laid for Brunel's suspension bridge across the Avon. One evening light entertainment was provided by Buckland, best known for his collection of geological samples. On that evening in Bristol he was lecturing about fossil footprints. He portrayed a cock or a hen on the edge of a prehistoric muddy pond. The audience was much amused as he lifted one leg after the other in a ballet to show the imprinting of the footmarks. This was pure Mudfog, and Dickens could have done no better. Babbage, David Brewster, and others had for some time wished to hold one of the meetings in a great industrial city, and a debate took place whether the 1837 meeting should be in Manchester or Liverpool. When it was pointed out as an attraction that Manchester had a statistical society, Whewell, inveterate opponent of applied science, remarked that he thought that an excellent reason for not holding the meeting in Manchester.[20] In the event Liverpool was chosen, the natural venue after its rival Bristol. Newcastle–upon–Tyne followed in 1838 and Birmingham in 1839. By then the Association was in the hands of the committee men and rapidly declining.

Babbage was concerned that the British Association was doing little to attract the manufacturers and shop-keepers, and also that there was little to attract the landed gentry. The statistical section had gone some way to remedying these gaps, as its rules confined its enquiries to the collection of data, in contradistinction to mathematical statistics, and such data might well be of commercial interest. In 1837 at the Liverpool meeting he proposed that the meetings of the Association should be accompanied by substantial exhibitions of the products and manufactures of each district the Association visited. Such an exhibition was held at Newcastle in 1838 and on a much larger

---

[20] The comical activities of the early statistical societies went a long way towards justifying Whewell's remarks. Dickens designated a good part of the British Ass's activities as 'Umbugology and Ditchwateristics'. The London Statistical Society was the very home of 'Umbugology' and the little Manchester Society no better. From an irreverent Mudfog point of view, sociological study of the early years of the British Ass. becomes an 'Umbugological' study of 'Umbugology'.

scale the following year in Birmingham. These exhibitions were precursors of the Great Exhibition of 1851.

It was the 1837 Liverpool meeting that inspired the Mudfog Association for the Advancement of Everything, and if anyone would seek to identify the 'image' of the British Association they have only to look at one of the later editions of *Sketches by Boz*. At the beginning of 1837 Dickens had published in *Bentley's Miscellany* an amusing piece:[21] The 'Public Life of Mr Tulrumble, once Mayor of Mudfog'. The description of Mudfog with 'an agreeable scent of pitch, tar, coals and rope-yarn, a roving population in oilskin hats, a pretty steady influx of drunken bargemen, and a great many other maritime advantages' did well enough for Liverpool, and Mr Tulrumble's efforts at civic display parodied the official municipal welcoming of the Association.

The 'Full Report of the First Meeting of the Mudfog Association for the Advancement of Everything' is in fact a parody of the reports of the British Association in the *Athenaeum*. Just as regular features in the *Athenaeum* were deferred to make room for reports of the Association, so Dickens announced that the regular instalment of Oliver Twist was delayed for reports from Mudfog, and England rocked with laughter. Actually some of the reports in the *Athenaeum* are themselves worthy of the Mudfog Papers. For example, in 1834 there appeared this report on the journey to Edinburgh:

As we approached northward from London, the symptoms of preparation for an important meeting forced themselves on our attention. At every stage places were eagerly sought by members, more than sufficient for a dozen coaches, and when we reached York, we found that there were more than twenty persons in that city who had got so far but were unable to advance, because more adventurous, or more prudent, travellers had secured places for the entire distance. At Newcastle the hotels were so crowded that it was scarcely possible to get beds, and the seats on the coaches were the subject of fierce strife.

We entered Scotland over the Cheviot hills. Their appearance attracted the notice of all, and it was soon evident that our fellow-travellers were members of the Association, full of their respective subjects, eager to impart and receive information ... Science destroyed romance—the field of Chevy Chase scarce elicited a remark— the cross marking the spot where Percy fell was observed by one of the geologists to belong to the secondary formation; the mathematician observed that it had swerved from the perpendicular, and the statisticians began a debate on the comparative carnage of ancient and modern warfare.[22]

When they reached Edinburgh they found, sadly, that the money subscribed for a public dinner for the Association had been lodged in a bank which had failed.

[21] C. Dickens, *Sketches by Boz, The Public Life of Mr Tulrumble, once Mayor of Mudfog*, 608–24, Oxford Illustrated Dickens.
[22] The *Athenaeum*, 675, 1834.

The *Athenaeum's* report from Dublin in the following year gave a picture of the gruelling toil involved in participating in one of these great gatherings:

Imagine the Rotundo—a room capable of accommodating from 1,500 to 2,000 individuals,—thronged to excess on some of the hottest evenings of this hot and comatory season; the ladies flirting and fanning; the gentlemen casting one eye upon Science and the other upon Beauty; and the whole scene (saving the reader's presence), mopping and puffing, and ready to drop with exhaustion and fatigue. Then reflect on the scant attention which those in earnest about the business in hand could give to the discourses of the orators. First, they were fatigued with the labour of the Sections; then trotted about the city to see the sights and walk off the repletion of the copious and elegant repast which preceded them;—then came the hot and crowded ordinaries with hundreds seated around the smoking viands; and finally hurried off to encounter the stewing of the evening meeting. But the business of the day was not even then concluded, for the rout and the supper had then to be gone through; and the next morning, with bodies jaded by the labours of the previous day, and minds still clouded with the yesterday's feast, the itinerant *savans* had again to brace themselves for encountering the like routine.[23]

Of the actual discussions virtually nothing appears in the anodyne reports in the *Athenaeum* or the pompous annual official Proceedings. Thus to get a picture of the scene we must turn to the Mudfog Papers. Professors Snore, Doze, and Wheezy; Mr Slug, so celebrated for his statistical researches; Mr Woodensconce, Mr Leadbrain, and Mr Timberead; Professors Mugg and Nogo dissecting the pet pug-dog on a pie-board; that celebrated group of physicians, Dr Kutankumagen, Dr W. R. Fee, Dr Neeshorts, and Mr Knight Bell: these and their colleagues deliberate forever, considering new inventions and discoveries; the cauliflower parachute, the high-speed miniature railway, a method of feeding all able-bodied paupers—in workhouses—on one twentieth of a grain of bread and cheese per day, the forcing machine for bringing joint-stock railway shares prematurely to a premium.

In the following autumn the celebrated band was observed in action again, joined by new colleagues at 'The Second Meeting of the Mudfog Association for the Advancement of Everything'. On this occasion the correspondent went by steam-packet to 'Oldcastle' and the description of the trip formed a large part of the report. At the second meeting there was a supplementary section on 'Umbugology and Ditchwateristics'.[24] A highlight of the meeting was the Automaton Police Office—and the real offenders. Sadly that was the last meeting of the immortal Association. Dickens became bored, and so did many others: no doubt one of the main immediate causes for the decline of the British Association.

The British Ass. was, however, the scene of the great popular debates of the

[23] Ibid., 640, 1835.
[24] *Sketches by Boz; Full report of the Second Meeting of the Mudfog Association for the Advancement of Everything*, 665, Oxford Illustrated Dickens.

nineteenth century. By far the most famous was the debate over Darwin's theory of evolution at which T. H. Huxley triumphed. But even here the importance of the Association for science can be exaggerated: after all, Darwin's work had long been carried out and the British Association was merely a sounding board. Babbage presided over two less well-known but important debates during the meeting in Newcastle in 1838: the great debate on the Atlantic Steam Navigation, and a debate on the railway gauges.

Babbage was deputed by the council of the Association to invite George Stephenson to be president of the Mechanical Section, but Stephenson declined. The second choice, Mr Buddle, also declined. So did the third, Bryan Donkin. Thus Babbage found himself in the chair with George Stephenson and Donkin among his vice-presidents. Dionysus Lardner was to address the meeting on the feasibility of crossing the Atlantic by steam vessels. Babbage has left us a brief description of the debate.[25] Several members of the audience were after Lardner's blood and the atmosphere was tense. Babbage had a quiet word with Lardner before the meeting to persuade him that some of his views were hasty and to admit as much as he could. From the chair Babbage said that stronger views had been ascribed to Lardner than he actually held and that further information had led him to modify them. 'Nothing', said Babbage, 'could be more injurious to truth than to reproach any man who honestly admitted that he had been in error.' The fact was that Brunel's *Great Western*, a wooden-hulled paddle boat, had been launched in 1837 and was proving very successful. Soon the *Great Western* was providing a regular service. Lardner gave a beautifully illustrated lecture; the audience was denied its fun; and one of Babbage's acquaintances said on leaving the meeting: 'You have saved that _____ _____ Lardner.'

The Battle of the Gauges was not to be settled so simply or so logically. The battle lines were clearly drawn: Bristol and the West country were associated with Brunel's wide gauge; Liverpool and powerful Northern interests with the narrow 4′ 8½″ gauge that old George Stephenson had taken from the tramways in the mines. There were discussions of several technical aspects at the Newcastle meeting but the problem could not be settled on technical grounds. George Stephenson later admitted to Babbage that if he could have started again he would have chosen a wider gauge,[26] but by that time the question was academic. The narrow gauge had spread throughout the country and became the established standard.

The Newcastle meeting was also the scene of a major row which led in the end to Babbage ceasing even to attend the annual meetings. Murchison, a secretary of the Association as well as one of the trustees, had asked Babbage if he were willing to be the next president, and Babbage had duly agreed. However when John Herschel returned from the Cape to become the

[25] *Passages*, 326–8.   [26] Ibid., 335.

fashionable scientific figure of the moment, Murchison sounded Herschel to see if he would be president. Knowing nothing of Murchison's previous suggestion to Babbage, Herschel agreed. Babbage was furious: he had been put in an embarrassing position and a source of unpleasantness created between himself and his old friend. Discovering what had happened, Herschel declined the presidency or any other office. Babbage proposed Vernon Harcourt for president and he was duly elected.

This was not sufficient for Babbage, who felt he could no longer work with Murchison whose resignation as secretary he demanded. That failing, Babbage made a public statement and retired from his position as trustee of the Association. He agreed to let matters rest quietly but in the following year the Duke of Newcastle placed Babbage's hand in Murchison's as a gesture of reconciliation. Babbage was not prepared to have any personal relations with Murchison, whom he cordially despised, and the row blew up again. Babbage was jockeyed off the council during the meeting so that he could take no further part in its deliberations. Although this manoeuvre was the occasion of Babbage finally ceasing even to attend the meetings it was not the basic cause. The Association was not playing the positive part in encouraging applied science that Babbage had hoped it would: it was in fact a mere talking shop. Brewster also often felt like resigning from the Association although he did not actually do so. A row with Babbage was more than Murchison had expected. He wrote plaintively to Wolryche Whitmore but only provoked him into expressing his own views on Murchison: 'One thing is clear—the creature has been at its dirty work again—shuffling—flattering—and intriguing—it is *his* nature.'[27]

After the meeting at Newcastle Babbage headed north for a holiday in the Scottish highlands, but he cut the tour short in order to make a study of the English railway system so far completed. As appears in the Mudfog Papers, many of the people travelling to Newcastle for the meeting had gone by steamer. One reason was that the early railways were very uncomfortable, even where they had been completed: the trains were considered worse than the worst of stage coaches.[28] This was the background against which Babbage began to investigate the railways in earnest. His approach was to replace general impressions of shaking and discomfort by quantitative observations, and his initial method could not have been more simple: he travelled over most of the then existing railway network with a watch, observing and recording how frequently the carriage gave a hard shake. Later he was to develop these preliminary investigations into the most extraordinarily thorough series of observations and measurements.

The first stage of the battle of the gauges was approaching its climax. The

[27] Wolryche Whitmore to Babbage, BL Add. Ms. 37,190, f 558.
[28] cf. Mr R. Castle at the deferred General Meeting of the proprietors of the G.W.R. 10 October 1838 at the Merchants Hall, Bristol. *The Times*, Sat. 13 Oct. (3d).

question at issue was whether the Great Western Railway should continue to be built with Brunel's wide gauge. The matter was fought out at the general meetings of the proprietors of the G.W.R. In the background was the rivalry over the development of steam-powered vessels for the Atlantic crossing. Brunel's soaring imagination had seen passengers travelling to Bristol and then being able to board a steamship to cross the Atlantic. During a meeting of the board of the G.W.R. in 1835 a timid member suggested that it was over-ambitious to build a railway the whole distance from London to Bristol at one go. Brunel then made his famous challenge: 'Why not make it longer and have a steamboat go from Bristol to New York and call it the Great Western.' The *Great Western* was launched in 1837. Companies in the rival ports of Liverpool and London hastily laid down rival vessels, but when it was clear that they would not be ready in time the companies chartered two small vessels originally designed for the crossing to Ireland. The tiny *Sirius*, hired by the London group, arrived in New York on 22 April 1838 with only four tons of coal left in her bunkers. It had been a heroic voyage, but the *Great Western* steaming in on the afternoon of the same day with a comfortable 200 tons of coal to spare really established the Atlantic crossing. This success immensely strengthened Brunel's position.

The engineering of the G.W.R. was discussed at the half-yearly general meeting[29] of the proprietors held in the Merchants' Hall in Bristol on Wednesday, August 1 1838. It was attended by more than four hundred gentlemen, with W. V. Simms Esq., chairman of the board, presiding. Previously the board had been bitterly attacked by the northern interests, and as a defensive measure it had invited three eminent engineers to investigate and report to the general meeting. The engineers were Robert Stephenson, Walker, president of the Institute of Civil Engineers, and Nicholas Wood, who had extensive experience of railway construction. The two former had declined— Stephenson would have been in a most invidious position—but Nicholas Wood was to commence experiments in the near future. Brunel made a long and detailed report proving that his system of constructing railways was greatly superior to all others. The method of laying tracks, the width of the gauge, size of tunnels, and other matters were discussed. Mr Simms declared that the directors had heard nothing to shake their confidence in the width of gauge chosen for the Great Western Railway. On a show of hands consideration of Brunel's report was deferred to a special general meeting to be held in October. However Nicholas Wood was an avowed supporter of Brunel and the northern group insisted that a Mr Hawkshaw, described as 'the Engineer of the Bolton and Bury Railway', should also make an investigation.

At this stage Babbage decided to intervene and himself appeared at the deferred meeting in October, also held in the Merchants' Hall in Bristol. On

[29] *The Times*, 22 Aug. 1838 (3c), *Supplement to the Railway Times, 43*, No. 32, 18 Aug. 1838.

this occasion the large hall was packed. Mr Hawkshaw had returned a report recommending a total change of plan : '. . . they should begin again *de novo*, take up the rails and lay them down at 4′ 8½″ gauge.' However Wood had not had time to complete his report and the directors took the position that it would be unfair to present Hawkshawe's report until Wood's had also been submitted. The Revd. Dr L. Carpenter and others defended the directors while a group led by Mr Hayes launched a vehement attack. Babbage made a long and effective speech, which was reported :

He appeared as a Shareholder, not an engineer. In a tour recently made of the manufacturing districts he had heard a great deal of railways, but in all his experience of popular delusions he had never heard so many misstatements made upon any one point as upon the Great Western Railway . . . Every coffee room from Bristol to the highlands teemed with the most unfavourable reports as to the inconvenience of travelling upon the line. To all these rumours he (Mr Babbage) could at that time only say, that such inconvenience did not exist when he travelled upon the line. Having, however, heard these reports in so many shapes, he thought it his duty to examine personally into the matter. He had accordingly given up a tour of pleasure for the purpose of travelling upon as many railways as possible . . . He had examined . . . twelve railways and conversed with their respective Managements. Upon seven of these lines he had himself travelled ; . . . He had noted in his pocketbook, at the time . . . the number of vibratory or irregular movements which had occurred to the carriages . . . as well as other exploratory circumstances . . . taking them in the numerical order of their superiority the Great Western stood second. (Hear, hear). He might add that the comparison was scarcely fair so far as regarded the Great Western ; for while the notes were taken on the other lines as their trains were proceeding at their usual speed of from twelve to twenty miles an hour, those on the Great Western were taken upon the experimental train while it was progressing at a speed of forty miles (Hear, hear).[30]

He added that the 4′ 8½″ gauge made the engines so cramped that they were most inconvenient to repair. It was settled that there should be a further meeting on Thursday, 20 December next in London. Meanwhile track-laying was suspended although other work could proceed.

The *Railway Times*, supporter of the Stephensons, made a bitter attack on the directors, Brunel and Babbage. It said of Brunel : . . . 'all that "the gifted" has ever done for them [the railways] is to throw them into great confusion— to unsettle the minds of men in respect to the merits of all other systems, and to fail utterly in substituting anything better of his own'.[31]

Babbage decided to carry out a systematic study of the G.W.R. Brunel and the directors were in a delicate position : they valued Babbage's help but they could not associate themselves directly with his study after the general meeting of proprietors had appointed Wood and Hawkshawe to report. However they offered him every assistance, as was appropriate to his position as shareholder and eminent man of science, placing a carriage at his disposal and appointing

[30] The *Railway Times*, Sat. 13 Oct. 1838, 558.   [31] Ibid., 595.

a man to join him and make all the necessary arrangements. Thus began Babbage's quite extraordinary series of studies on railways. There was no other person at the time remotely capable of conceiving and carrying out a comparable project in what is now known as operations research. It is reminiscent of the work of Blackett and Bernal in the second world war. They did not carry out·experiments so much as make systematic measurements under actual operating conditions and build up a theoretical picture from their results. They were concerned with military matters, with the effectiveness of anti-aircraft guns for instance, but the principle was similar. Instead of sitting in a carriage on a moving train and simply counting the number of times he was noticeably shaken, Babbage decided to build and install in the carriage means for recording automatically the movement of several parts as it travelled over different stretches of line at chosen speeds.[32]

His son Herschel, who was already working for Brunel on the G.W.R., designed the apparatus under Babbage's careful supervision and for five months Babbage's team of workmen was diverted from the Analytical Engines to making equipment and conducting experiments on the G.W.R. One of the smaller of the G.W.R.'s second class carriages was stripped internally. A long table was mounted on firm supports passing right through the carriage and fixed to the framework underneath. On the table long sheets of paper were pulled while pens automatically traced curves on the paper. Babbage had some difficulty in finding suitable paper. Bryan Donkin recommended wall-paper which was made in long rolls as it would be expensive to have rolls especially made in the small quantities which Babbage required; and apparently wall-paper was used. Castings and fittings for connecting Babbage's apparatus to a pair of wheels on the carriage were made on Brunel's instructions because that part of the apparatus would be permanently useful to him, thus saving Babbage some expense. The apparatus for moving the paper and also the pens and their associated equipment were made by Holtzapffel and Co. of Charing Cross.[33] It was driven through a clutch box and the paper could be moved at velocities of between 3 inches and 60 feet to the mile (Holtzapffel's specification gives between 9 inches and 50 feet to a mile, but this may have been altered). Faraday advised Babbage on the choice of lubricants which would not thicken in the bitter cold to be expected during a series of experiments to be conducted in mid-winter. In practice Babbage found it convenient and satisfactory to feed the paper at between eleven and forty-two feet to every mile travelled by the carriage. The paper was drawn from the main roll by three mahogany rollers, 18″ long by 4½″ diameter with brass caps hinged to iron axes: the best equipment made at that time with such materials was superb and a joy to handle. In case the paper should slip the last of the train of gears had means for

[32] *Passages*, 320–6.
[33] BL Add. Ms. 37,191, ff 63, 70, 98.

automatically inking the paper with twenty dots per revolution. Thus if paper-slippage could not be completely prevented it could be detected.

A powerful spring-driven clock, made by Edward J. Dent of the Strand, was adapted to raise and lower a special pen which made a dot on the paper every half second. Thus from the spacing of these dots the speed of the train could immediately be determined. The main inking pens and the ink feed gave Babbage a good deal of difficulty but ultimately the pens worked well, tracing out their curves, followed closely by a roller faced with blotting-paper. The pens were connected mechanically to different parts of the carriage or to some special piece of apparatus. Thus was formed a multi-channel pen-recorder with mechanical linkage. Up to twelve pens were commonly in use expressing the following measures, or one or two others as might be required:

1. force of traction
2. vertical shake of carriage at its midpoint
3. lateral ditto
4. end ditto
5, 6, and 7. the same shakes at the end of the carriage
8. the curve described upon the earth by the centre of the frame on which the apparatus was mounted
9. The chronometer marked half seconds on the paper.

About two miles of paper was thus covered.

Babbage described in some detail how he measured curve number 8:

Finding this a very important element, I caused a plate of hardened steel to be pressed by a strong spring against the inner edge of the rail. It was supported by a hinge upon a strong piece of timber descending from the platform supporting the carriage itself. The motion of this piece of steel, arising from the varying position of the wheels themselves upon the rail, was conveyed to a pen which transferred to the paper the curve traversed by the centre of the carriage referred to the plane of the rail itself.

The contrivance and management of this portion of my apparatus was certainly the most difficult part of my task, and probably the most dangerous. I had several friendly cautions, but I knew the danger, and having examined its various causes, adopted means of counteracting its effect.

After a few trials we found out how to manage it, and although it often broke four or five times in the course of the day's work, the fracture inevitably occurred at the place intended for it, and my first notice of the fact often arose from the blow the fragment made when suddenly drawn by a strong rope up to the under side of the floor of our experimental carriage.[34]

So impressed was Babbage with the value of the permanent records which he obtained with his apparatus, that he suggested that similar equipment should be installed on railway trains as a matter of routine, so that in the event of accident it should be possible to determine the causes. Interestingly enough

[34] *Passages*, Ch. xxv, Inst. Civil Eng. *Minutes of Proceedings*, 21, 385–7.

it is now standard practice to carry such recording equipment, particularly in aircraft, where it is often called the 'black box'.

While Babbage was preparing his apparatus a complication arose: Wood was unable to spare the time and asked the ubiquitous Dionysus Lardner to conduct the experiments for him. Lardner really could not cope and Babbage invited him to carry out the experiments jointly, but the invitation was declined. Lardner had the idea of measuring vibration by watching the oscillation of mercury in a tube as the train shook. He also conducted some experiments by releasing trucks and measuring how long they took to come to rest. Thus, as in the Atlantic Steamship debate at the British Association meeting in Newcastle, Babbage's problem was to deal with Lardner. The meeting on 20 December was pro-forma and delayed, this time to 9 January 1839. Once again the *Railway Times* was furious, accusing the directors of dark misdeeds. Babbage had just enough time to get his apparatus working and make a few sets of measurements. A letter he wrote to John Herschel at this time gives the picture:

Great Western Railway
London Terminus
Paddington

5th January 1839

Dear Herschel,

I start tomorrow from this place and in about 40 mins shall be at Slough with an Engine and my carriage full of apparatus. I propose going several times backwards and forwards over 4 miles of road which has been examined and measured as to the piles and junctions and shall then go to Maidenhead and spend the whole day on the line.— I can now get my pens to write pretty well and I have got Dr Lardners App in the carriage which I have also taught to write [Presumably Babbage had connected the mercury in the tube by some device to his pens so that the oscillations in the mercury could also be recorded]. I am anxious that you should see the contrivances and the whole scheme and probably this will be the only opportunity of our having sufficient time. As soon as I reach Slough I will send to you—but if you cannot come pray have a note at the station for a few minutes are sometimes of importance to,

Yours ever
C. Babbage.[35]

It was not very convenient for Babbage to carry out his studies when his experimental carriage was attached to a public train, because the measurements were continually interrupted when the train stopped at the stations. Nor was a special engine the solution as it had frequently to run into sidings to avoid an approaching train. Brunel told Babbage that it would be far too dangerous to work at night because ballast-waggons and others carrying equipment and materials for constructing the railway used the track at uncertain times. An incident soon occurred to confirm this advice. One night a train of twenty-five

[35] Babbage/Herschel correspondence, f 293, RS.

empty ballast-waggons, each carrying two men, had driven right through the engine-house, damaging the *North Star*, which was the company's finest engine and had been assigned to Babbage's use on the following day. There was no 'dead man's handle' to stop the train automatically and the driver and stoker, exhausted from overwork, had fallen asleep. The men were lucky to escape with a bad shaking.

Babbage decided to work on Sundays. Unwilling to challenge public opinion the directors declined to give formal permission for violation of the peace of the Lord's day but were glad to turn a blind eye. One Sunday he decided to test the effect of additional weight and directed that three waggons loaded with thirty tons of iron should be attached to the experimental carriage. Babbage was alarmed when told by his aide-de-camp that they were to travel on the northern line. The man assured Babbage that there would be no possible danger as no other engine could be on either line before five o'clock that evening. A messenger arrived to announce that the line was clear.

At that point there was a quarter of an hour's delay because the fireman had neglected to have the engine ready with steam up. Then Babbage's ear, acutely tuned by his studies to the sounds of the railways, detected an engine approaching. A white cloud of steam appeared in the distance and in a few moments an engine arrived in the engine house. There Babbage went to find Brunel himself covered with smoke and soot. Arriving from Bristol at the farthest point of the line then completed he had missed the last train. But finding an engine with steam up he had driven it the whole way at the then phenomenal speed of fifty miles per hour.

Naturally they considered what would have happened had Babbage's engine not been delayed and the trains met. Brunel declared that he would have put on steam to sweep the opposing engine off the rails. However they concluded that 'If the concussion had occurred, the probability is, that Brunel's engine would have been knocked off the rail by the superior momentum of my train, and that my experimental carriage would have been buried under the iron contained in the waggons behind.' It had been a close shave.[36]

One bitterly cold day Babbage and his team met Dr Lardner carrying out his experiments on the Great Western Railway:

He was drawing a series of trucks with an engine travelling at known velocities. At certain intervals, a truck was detached from his train. The time occupied by this truck before it came to rest was the object to be noted. As Dr Lardner was short of assistants, I and my son offered to get into one of his trucks and note for him the time of coming to rest.

Our truck having been detached, it came to rest, and I had noted the time. After waiting a few minutes, I thought I perceived a slight motion, which continued, though slowly. It then occurred to me that this must arise from the effect of the wind, which

[36] *Passages*, 324–5.

was blowing strongly. On my way to the station, feeling very cold, I had purchased three yards of coarse blue woollen cloth, which I wound round my person. This I now unwound; we held it up as a sail, and gradually acquiring greater velocity, finally reached and sailed across the whole of the Hanwell viaduct at a very fair pace.[37]

On Sunday 3 March 1839 Babbage made his last experimental run. Once again he invited John Herschel to join him when the train stopped at Slough.[38] By that date the fate of the wide gauge had been settled.

The decisive meeting of the proprietors of the Great Western Railway took place at the London Tavern on Wednesday, 9 January 1839.[39] The opponents of the directors had exhibited round the walls the results of whole sets of experiments. Lardner proved with enthusiasm that trains of the wide gauge could not travel at more than forty miles per hour. Babbage then completely demolished Lardner's arguments and wiped the floor with him. He showed without difficulty that Lardner's experiments were meaningless and that his apparatus, consisting of mercury in a tube, could be made to give any result you wished: it should be remembered that while Lardner was dependent on observing the mercury, Babbage had its indications traced out by those remorseless pens. To have been saved by Babbage from the angry crowds at Newcastle was bad enough, but the terrible public humilation at the London Tavern was too much. Lardner never forgave him. Brunel completed the humiliation by destroying Lardner's irritating story about air-resistance.

Babbage had in fact by the time of the meeting conducted sufficient experiments to demonstrate the superiority of the wide gauge, and the evidence was permanently recorded by the untiring pens. The proprietors had never heard anything like it; nor would they again, or their children; or most probably their children's children: it has the ring of the 1940s rather than 1839. The day was won, but it had been a close thing. Northern proprietors holding thousands of proxies had come from Liverpool predisposed to support Stephenson but intending to vote according to the arguments. Hearing Babbage's speech they decided to support Brunel and the wide gauge was given a new lease of life.[40]

[37] Ibid., 325–6.   [38] Babbage/Herschel correspondence, f 293, RS.
[39] *The Times*, Friday 11 Jan. 1839 (2e). Also The *Railway Times*. Lardner's paper on air resistance had been published in the *Monthly Chronicle*.
[40] *Passages*, 321. 'On the discussion at the general meeting at the London Tavern, I made a statement of my own views, which was admitted at the time to have had considerable influence on the decision of the proprietors. Many years after I met a gentleman who told me he and a few other proprietors holding several thousand proxies came up from Liverpool intending to vote according to the weight of the arguments adduced. He informed me that he and his friends decided their votes on hearing my statement. He then added, "But for that speech, the broad gauge would not now exist in England."'

12

○●○

The Analytical Engines: Social Life

In the autumn of 1834 Babbage fell under a potent spell. He saw a vision of a computer, and remained enthralled for the rest of his days. Thus it was Babbage started on the most extraordinary period of rapid invention and discovery even in his long, fertile and creative life.

After eighteen months of stalling and prevarication by his engineer Clement, Babbage at last had the drawings of the first Difference Engine in his hands again. As soon as he had recovered the drawings he began to reconsider the whole plan of the Engine, both in detail and in its general principles. Old ideas which had been germinating began to shoot up under the warmth of new attention. Of these by far the most important was the concept of feeding the results of calculations back into the beginning of subsequent calculations. This concept was to prove capable of unlimited development. Ultimately it has become the stored program of the modern computer, but that was still far away. In its simple form Babbage had had this concept in mind almost from the start of his work on the Difference Engine. Indeed he had referred to it in his 'Letter to Sir Humphry Davy',[1] which had launched the project publicly. Still in a straightforward form the idea had been embodied in the drawings for the Difference Engine itself, and in a rather different form can still be seen as actual cog-wheels in the extant portion of the Engine. At this stage Babbage picturesquely referred to the concept as 'the engine eating its own tail'.

Explaining his ideas to others always helped Babbage to clear his own mind, and explaining the working of the Difference Engine to Herschel, his eldest son, who was living with him at the time, crucially helped Babbage to develop the ideas which had been germinating for more than a decade. He soon embarked on the work which was to lead to the great series of Analytical Engines and to the unique development in the nineteenth century of so many of the basic concepts of the stored-program computer. Common to Analytical Engine and computer are: the separate store for holding numbers and mill for working on them; a distinct control or operations section; punched-card input/output system; and many other features.[2]

Babbage started on his new phase of work from a position of great strength.

[1] *Babbage's Calculating Engines*, 1889.
[2] cf. Anthony Hyman, *Computing, A Dictionary of Terms, Concepts and Ideas*, entries Analytical Engines, Calculating Engines, Arrow, 1976.

He was forty-three years old, financially independent, and one of the leading English mathematicians of his time. He also had behind him the first Difference Engine, which had been a triumphant success in terms of technology. The completed portion worked perfectly, as it still does today. There was no doubt that if the government had chosen to continue financing the project the entire huge engine would have worked equally well.

Babbage's social life had established a routine: he entertained regularly on Saturday evenings during the London season and his company was widely sought in society. Following enforced idleness while he was separated from the drawings of the first Engine his mind was teeming with ideas. Thirty years later, when he wrote his fragmentary autobiography, *Passages from the Life of a Philosopher*, the excitement of those days was still with him:

The circular arrangement of the axes of the Difference Engine round large central wheels led to the most extended prospects. The whole of arithmetic now appeared within the grasp of mechanism. A vague glimpse even of an Analytical Engine at length opened out, and I pursued with enthusiasm the shadowy vision. The drawings and the experiments were of the most costly kind. Draftsmen of the highest order were necessary to economize the labour of my own head; whilst skilled workmen were required to execute the experimental machinery to which I was obliged constantly to have recourse.

In order to carry out my pursuits successfully, I had purchased a house with about a quarter of an acre of ground in a very quiet locality. My coach-house was now converted into a forge and foundry, whilst my stables were transformed into a workshop. I built other extensive workshops myself, and had a fire-proof building for my drawings and draftsmen. Having myself worked with a variety of tools, and having studied the art of constructing each of them, I at length laid it down as a principle—that, except in rare cases, I would never do anything myself if I could afford to hire another person who could do it for me.

The complicated relations which then arose amongst the various parts of the machinery would have baffled the most tenacious memory. I overcame that difficulty by improving and extending a language of signs, the Mechanical Notation, which in 1826 I had explained in a paper printed in the 'Philosophical Transactions of the Royal Society.' By such means I succeeded in mastering trains of investigation so vast in extent that no length of years ever allotted to one individual could otherwise have enabled me to control. By the aid of the Mechanical Notation, the Analytical Engine became a reality: for it became susceptible of demonstration.[3]

At the time and since his death most commentators have assumed that he was trying to build an Analytical Engine and that he failed. In fact, as he was well aware, he did not have the necessary financial resources; but if he was not actually building one it is reasonable to ask what exactly he was doing with such enormous effort and at such high cost. The answer is clear: he was carrying out experiments and preparing designs. This is a common procedure in modern industrial laboratories, and if it was incomprehensible to Babbage's contem-

[3] *Passages*, 112–13.

poraries it should not be difficult to comprehend today. Babbage may have entertained hopes that ultimately he would be able to build an Analytical Engine, but with the bitter experience of the difficulties he had faced, and indeed still faced, in securing government understanding and support for the relatively straightforward first Difference Engine, he had no illusions about the magnitude of the problems which would arise in financing one of his far more ambitious new engines. For the foreseeable future the intrinsic interest of the research would have to be its own justification.[4]

The work proceeded at speed. In less than two years he had sketched out many of the salient features of the modern computer. A crucial step was the adoption of a punched-card system derived from the Jacquard loom. In this type of loom, and in an earlier loom made by another Frenchman, Falcon, the patterns woven were controlled by the patterns of holes in a set of punched cards strung together in the sequence in which they were to be used.[5] In the loom it was necessary to push a set of wooden rods lifting the particular set of threads required to form the pattern designed. Falcon had used separated cards with holes punched in them to select the required sets of rods: a workman pressed the appropriate cards one after another on to the wooden levers. Then one of the greatest French engineers, Vaucanson, had suggested stringing the cards together and using an automatic system for feeding them in sequence.

Many different methods have been used through the ages for storing information. One technique which is quite familiar is the drum with little spikes raised on it, used for example in musical boxes. Vaucanson himself actually made a loom using punched cards moved by a perforated cylinder. He also made many beautiful and ingenious automata. But it was Jacquard, using the punched-card method, who designed the first really successful loom which automatically produced complex patterns and pictures woven in silk.

Babbage's plans separated the store, or storehouse, holding the numbers, from the mill which carried out numerical operations, such as addition, subtraction, multiplication, and division, using numbers brought from the store. The analogy was with the cotton mills: numbers were held in store, like materials in the storehouse, until they were required for processing in the mill or despatch to the customer. Following the introduction of punched cards early in 1836 four functional units familiar in the modern computer could soon be clearly distinguished: input/output system, mill, store, and control.[6]

A great series of engineering drawings, about three hundred in all, among the finest ever made, came from Babbage's drawing office. His principal

---

[4] Babbage to Herschel at the Cape of Good Hope, 18 Sept. 1834. Babbage/Herschel corr. f 287, RS.

[5] An exhibition at the Science Museum, London, shows the working of the Falcon and Jacquard looms.

[6] Anthony Hyman, op. cit.

draftsman was Jarvis, who had earlier worked under Clement on the drawings for the first Difference Engine. An outstanding draftsman with far greater technical knowledge than Clement, Jarvis deserves an honoured place in the history of the computer. In November 1835 Babbage came to a permanent agreement with Jarvis, who had been offered employment abroad with a heavy penalty clause if he should break the contract. The pay was high, as the railway boom had begun and a skilled draftsman could command a high salary. Babbage consulted his mother who gave him encouragement and excellent advice: 'My dear son, you have advanced far in the accomplishment of a great object, which is worthy of your ambition. You are capable of completing it. My advice is— pursue it, even if it should oblige you to live on bread and cheese.'[7] It never came to that, but Babbage paid Jarvis a guinea for an eight-hour day, a very high wage for the time.

Accompanying the drawings are hundreds of large sheets of the corresponding mechanical notations. Together they form a complete description of the Analytical Engines. The detailed study of these Engines is a task of considerable complexity. Each part had to carry out the required logical and arithmetical functions, but it had also to work mechanically. In addition the various parts had to be so grouped that interconnection between them could be made. Even in modern computers, where wires or other conductors can be placed and joined with relative ease, the design of the interconnection system has necessitated a great deal of attention and has often been a limiting factor. In Babbage's mechanical system it was, at the early stages of the Analytical Engines, a problem of the greatest difficulty. A general pattern of development in the work may be observed. As the designs proceeded Babbage was able gradually to simplify each part of the Engines. This in turn permitted him to add more functional units, thus increasing the overall power of the Engines. By repeated simplification and enhancement of computing power the work proceeded, with an interruption between 1848 and 1857, for most of the remainder of his life. Gradually he developed plans for Engines of great logical power and elegant simplicity (although the term 'simple' is used here in a purely relative sense).

The store of an Analytical Engine was formed of a set of axes on each of which was placed a pile of toothed wheels. Each wheel was free to rotate independently of the other wheels on the same axis. In the more straightforward designs each axis stored a number, while each wheel on the axis stored a single digit. The toothed wheels, which were called figure wheels, were marked on their edges with the figures 0 to 9. Only the top wheel on each axis had a special function, indicating whether the number stored on that axis was positive or negative. It was not necessary for Babbage to use the decimal system and he did indeed consider using other number bases, including 2, 3, 4, 5, and 100.

[7] *Passages*, 113–14; see also BL Add. Ms. 37,189, ff 201–3.

Nowadays the use of number bases other than the base 10 of the ordinary decimal system is widely familiar from modern school mathematics. However, if Babbage had adopted a base of 2 (the binary system commonly used in modern computers) the use of storage space would have been inefficient as each figure wheel would only have held a single binary digit. Thus either the numbers which could be held on each axis would have been too small, or the number of discs inconveniently large. On the other hand a number base of 100 made the use of storage space very efficient, but was impractical because arithmetical calculation on the engine would have been too slow.

Babbage used his punched cards for three principal purposes.[8] The numerical value of a constant, such as π, could be introduced into the engine by a *number-card*. A second type of card, called a *variable-card*, defined the axis on which the number was to be placed. The variable-cards could also order the number held on a particular axis in the store to be transferred to the mill when required; or alternatively to transfer a number from the mill to a designated axis. Cards of the third type, called *operation-cards*, were smaller than the other cards. The operation cards controlled the action of the mill. An operation card might order an addition, subtraction, multiplication, or division, or any other operation for which facilities were provided on the Analytical Engine in use. Babbage considered many different arrangements of cards at different times and for varying purposes.

Eighteen thirty-six was Babbage's *annus mirabilis*. It is quite astonishing that within two years, working with Jarvis and two or three other assistants, Babbage should have advanced so far in so short a time. Sequences of punched cards gave instructions to an Engine. Instructions were then decoded by a fixed store, which in turn distributed the detailed instructions appropriate to each particular part of the Engine for the operation being carried out. Both the detailed operations and the order in which the cards themselves were called into action could be modified by the results of calculations previously carried out by the Engine. This is not yet the stored-program computer which plays so large a part in modern life, but rather a versatile and powerful calculator.

However Babbage's ideas did not end at that point. In 1836 he was already thinking of designing an Algebra or Formula Engine. Indeed the famous term Analytical Engine seems to have come into use at this time in contradistinction to the more general Algebra Engine. Whether that is indeed the way the terms were coined, Babbage's plans went far beyond the Analytical Engines which have become widely known. Sometimes his more general plans were thought of as a separate class of Algebra Engines, sometimes as further developments of the Analytical Engines. It is in these more general plans that one can find

---

[8] Babbage planned a card-copying machine, not only for his own use. He was acutely aware of the problems of standardization which would arise when Analytical Engines proliferated.

Babbage's closest approach to the formulation of the general concepts of the modern computer. He even proposed an array processor.

Meanwhile the project for the first Difference Engine remained in suspense. Funded initially in the early 1820s, it now languished, ignored if not forgotten under the indolent ineptitude of Melbourne's government. Babbage often met Melbourne himself in society. We find in Creevey's diary, for instance: 'Tuesday at Charles Fox's, Addison Road—no joke as to distance; 8 shillings coach hire out and back, besides turnpikes! The company—Madagascar [Lady Holland], Allen, Babbage the philosopher, Van de Weyer, Belgian Minister, Hedworth Lambton and wife, . . . and Melbourne.'[9] But even though Babbage might meet government ministers, nothing was done to resolve the problem of the Difference Engine. Handling prime ministers was a far more difficult and delicate problem than dealing with the grant-making bodies of today. On 21 October 1838 we find Babbage writing to Spring Rice, Chancellor of the Exchequer, begging for a decision: 'The question I wish to have settled is whether the government wish me to superintend the completion of the Calculating Engine, which has been suspended during the last five years, according to the original plan and principles, or whether they intend to discontinue it altogether.'[10]

No decision could in fact be made by anyone but the Prime Minister, and Lord Melbourne was not in the least disposed to decide the question. It would have been embarrassing to discontinue the project after £17,000 of public money had been spent, and it would have meant antagonizing Babbage and his friends amongst the Whigs. On the other hand to support the project was far from Melbourne's thoughts. Thus the project remained in suspended animation.

When he had first approached Wellington, Babbage had remarked to the Duke of Somerset that all he sought was understanding. Somerset, who knew the English upper classes, had thought it entirely unrealistic to expect understanding of a scientific problem by any English politician.[11] But in spite of this depressing prognosis Babbage had indeed received both remarkable understanding and steady support from Wellington. Marc Isambard Brunel also looked back on the Duke's government as a benign period from the point of view of the engineers.[12]

Now Babbage met complete incomprehension. The very idea of a calculating engine was, it was felt, more than a country gentleman might reasonably be asked to contemplate. To complicate matters even further Babbage had invented a new type of engine. Did the new supersede the old, making the project on which £17,000 had been spent a write-off? Most people assumed that Babbage was now asking for money for his new project. Further Babbage

[9] H. Maxwell, ed., *The Creevey Papers*, II, 329, John Murray, 1909.   [10] *Passages*, 92.
[11] Duke of Somerset to Babbage, Park Lane, 8 June 1829. BL Add. Ms. 37,183, f 337.
[12] M. I. Brunel to Babbage, 19 June 1834. BL Add. Ms. 37,188, f 392.

was saying that he could make a version of a Difference Engine more economically by starting again than by finishing the original design, even though part of it had been assembled and worked perfectly. Confusion was complete, although Babbage had in fact explained the situation very clearly in successive applications to succeeding governments.

The position was as follows: the first Difference Engine had been technically entirely successful. Furthermore, the project had already repaid the country for the money spent through spin-off in machine-tools and machining techniques developed to make the Engine. Leading engineers testified to this fact.[13] However, after the long time which had passed since the project had first started, new advanced techniques had been developed and the constructional methods so much improved that it might well be cheaper now to start building a Difference Engine from the beginning than to complete the original design. As Babbage pointed out it was not uncommon to find that machinery was obsolete as soon as it was built during that period, so rapidly were manufacturing techniques developing—even without delays such as had attended the construction of his first Engine. A Difference Engine was needed as much as ever to make mathematical tables for navigation and for other purposes. The new engine complemented the old rather than replacing it: the Analytical Engine was far more versatile, but the Difference Engine was more efficient at its defined task—the generation of mathematical tables.

Computer scientists looking back to Babbage's work seek to discover how far he had gone in making the discoveries which at present seem significant in modern computing, but Babbage himself was far more concerned with problems of mechanical detail. The intricacies of design were the bread and butter of his work and they fascinated him, occupying most of his working time. One crucial problem was the carry system. Every schoolchild knows that if you add seven to eight the answer is fifteen. You place a five in the units column and *carry* one to the tens column. Babbage required a special mechanism for carrying the one from the units to the tens, from the tens to the hundreds, and so on as required right up the axes of toothed wheels to the top.[14] Now let us suppose that it takes ten units of time to add the seven to the eight, leaving Babbage's unit wheel indicating 5; then it will take at least one further unit of time to form the carry by adding one to the tens wheel. The difficulty may perhaps be seen if we consider adding one to nine hundred and ninety-nine:

$$+ \ 999$$
$$\underline{1}$$
$$1000$$

First one must be carried to the tens column, and only after that one to the

26   [13] James Nasmyth to Babbage, 22 June 1855. BL Add. Ms. 37,196, f 249.
27   [14] That is to say the top figure wheel; above this was the sign wheel.

hundreds, and only then one to the thousands. Thus the basic addition has taken ten units of time and the carry a further three units of time. As Babbage was planning to use axes with fifty figure wheels to store numbers with fifty digits, it will be seen that the carry would take fifty units of time, or five times as long as the basic addition.

In describing his approach to this crucial problem Babbage gives a picture of how he worked and spent many of his days:

The most important part of the Analytical Engine was undoubtedly the mechanical method of carrying the tens. On this I laboured incessantly, each succeeding improvement advancing me a step or two. The difficulty did not consist so much in the more or less complexity of the contrivance as in the reduction of the *time* required to effect the carriage. Twenty or thirty different plans and modifications had been drawn. At last I came to the conclusion that I had exhausted the principle of successive carriage. I concluded also that nothing but teaching the Engine to foresee and then to act upon that foresight could ever lead me to the object I desired, namely, to make the whole of any unlimited number of carriages in one unit of time. One morning, after I had spent many hours in the drawing office in endeavouring to improve the system of successive carriages, I mentioned these views to my chief assistant, and added that I should retire to my library, and endeavour to work out the new principle. He gently expressed a doubt whether the plan was possible, to which I replied that, not being able to prove its impossibility, I should follow out a slight glimmering of light which I thought I perceived.

After about three hours' examination, I returned to the drawing-office with more definite ideas upon the subject. I had discovered a principle that proved the possibility, and I had contrived mechanism which, I thought, would accomplish my object.

I now commenced the explanation of my views, which I soon found were but little understood by my assistant; nor was this surprising, since in the course of my own attempt at explanation, I found several defects in my plan, and was also led by his questions to perceive others. All these I removed one after another, and ultimately terminated at a late hour my morning's work with the conviction that *anticipating* carriage was not only within my power, but that I had devised one mechanism at least by which it might be accomplished.

Many years later, my assistant, on his return from a long residence abroad called upon me, and we talked over the progress of the Analytical Engine. I referred back to the day on which I had made that most important step, and asked him if he recollected it. His reply was that he perfectly remembered the circumstances; for that on retiring to my library, he seriously thought my intellect was beginning to become deranged. The reader may perhaps be curious to know how I spent the rest of that remarkable day.

After working as I constantly did, for ten or eleven hours a day, I had arrived at this satisfactory conclusion, and was revising the rough sketches of the new contrivance, when my servant entered the drawing-office, and announced that it was seven o'clock— that I dined in Park Lane—and that it was time to dress. I usually arrived at the house of my friend about a quarter of an hour before the appointed time, in order that we might have a short conversation on subjects on which we were both much interested. Having mentioned my recent success, in which my host thoroughly sympathized, I remarked that it had produced an exhilaration which not even his excellent champagne could rival. Having enjoyed the society of Hallam, of Rogers, and of some few others of that delightful circle, I retired, and joined one or two much more extensive reunions. Having thus forgotten science, and enjoyed society for four or five hours, I returned home. About one o'clock I was asleep in my bed, and thus continued for the next five hours.[15]

Although he had established the principle of the anticipatory carry on one heady day its implementation gave Babbage a great deal of work. After designing several devices for implementing the principle, Babbage's next step was to devise a mechanism which would work both for addition and subtraction.[16] But his greatest problem with this carry and borrow system was to make it sufficiently compact. The carry axis was taller than any of the other axes, and as the whole plan required corresponding wheels on the various axes to be at the same level, the height of the carry axis in practice determined the height of the other axes. When it is noted that the axes were at this date about eight to ten feet of precision mechanism the dimensions of the problem can be appreciated. The whole Engine was about fifteen feet tall. Ultimately Babbage was able to simplify the mechanism and compress the design until a carry column for fifty figures was about two feet tall.[17]

It may be remarked that the principle of the anticipatory carry has been used in modern computers and is just one example of Babbage's many detailed inventions which were to be rediscovered in the 1940s. Once the idea had been rediscovered it was child's play implementing it with electronics when compared with the problems which Babbage faced using mechanical techniques. Babbage himself considered using electrical, or rather electromechanical methods. The idea occurred to him after seeing some equipment which Wheatstone had designed for use in his electric telegraph,[18] but Babbage did not pursue it: electrical technology was not sufficiently advanced for Babbage's requirements. He had to be quite certain that no undetected errors could creep into the calculations, and for decades to come it was quite impossible to make electromechanical switching devices sufficiently reliable.

During the first period of work on the Analytical Engines, Babbage's eldest

[15] Passages, 114–16.

[16] Subtraction was actually carried out in the Analytical Engines using an arithmetical dodge.

[17] Using a casting technique. A demonstration column made by Babbage in his last years is usually on view in the Science Museum, London.

[18] It has often been stated that Babbage did not consider using an electromechanical system; and also that if he had used such methods it would have been more practicable for him to construct an Analytical Engine. Even if there were no direct evidence, knowledge of Babbage's work suggests that it would be a far more reasonable assumption that he considered developing an electromechanical system but did not pursue the idea. With only his private resources it would have presented formidable problems at that date. I think those who have written regretting that Babbage did not turn to such methods can have little experience of the problems of reliability in electromechanical digital systems.

son Herschel was living at number one Dorset Street. He was a companion to his father, taking great interest in the Analytical Engines as the plans developed, and they had become very close. Also Herschel was a reminder of the happy days when his mother was alive and the family still together. He seems to have been more similar to Babbage than either of the other two surviving sons, but that may have been largely due to his growing up directly under his father's influence. After leaving Bruce Castle school Herschel went to University College for a period. He then joined the New River Company which supplied old London's water. Herschel would certainly have spent much time in his father's workshops and drawing-office learning both workshop practice and engineering draftsmanship in what was in effect one of the finest schools of its kind in the country. Herschel was also thoroughly familiar with his father's famous mechanical notation.

In 1834 another tragedy occurred in Babbage's life: Georgiana, his only daughter, died. She had been a favourite and he had even been a little afraid of spoiling her. No doubt she reminded him of her mother. Little Georgiana had been living with her grandmother in Babbage's old home in Devonshire Street. The old lady was lonely without Georgiana and after her death Babbage's two younger boys, Dugald and Henry, went to live permanently with their grandmother.[19]

When their mother had died in 1827 Dugald and Henry had stayed with Harriet Isaac, one of their mother's sisters who was married to a country banker with an estate at Boughton near Worcester. The two boys actually lived at the lodge with a Mrs Powell who looked after them and of whom they became very fond. They attended the nearby village school. One day in July 1831 their elder brother Herschel came to fetch them from Worcester, taking them to London. They arrived in the evening, fascinated by the gas lighting in the streets, to be met by their father. After spending the night in Devonshire Street they went as boarders to Bruce Castle school where Herschel was already enrolled. After their sister died the two younger boys left Bruce Castle at the end of the term to attend University College school, then in Gower Street, as day pupils. Life with their grandmother was spartan. Cold water for washing was provided in jugs and in winter it would be frozen over. The Benjamin Babbages were given to early rising and a simple life, and lived to a ripe old age on the strength of it. The two younger boys saw little of their father and went in considerable awe of him.

After University College school they went for a couple of years to University College, promoted to an allowance and frock coats. They also worked with Jarvis and learned the mechanical notation; and they spent a good deal of time

[19] Most of the details of Henry's life are taken from: Major General Henry Prevost Babbage, *Memoirs and Correspondence*, Clowes, London, 1910.

in the workshops doing a variety of jobs. Clearly Babbage was determined that his sons should receive an all-round practical training as well as an academic education: a rare combination in the nineteenth century.

The annual jamborees of the British Association formed part of Babbage's summer holidays. After the first meeting at York the Association went the round of the university cities: Oxford, Cambridge, Edinburgh, and then Dublin. Edinburgh and Dublin were capital cities and Dublin particularly provided an intensive round of entertainments. There were so many visitors that only a fraction of those wishing to attend could be accommodated. Some of the more earnest members thought there was too much frivolity. Babbage thoroughly enjoyed himself.

Although he had several invitations from friends to stay with them Babbage preferred to occupy the rooms which the provost and fellows of Trinity placed at his disposal. After a few days a learned friend advised Babbage that he was giving offence to his hosts by wearing a bright green waistcoat: O'Connell's colours in the heart of the protestant university. Babbage was anxious to remove the source of offence but his friend pronounced every other waistcoat Babbage had with him objectionable on some similar ground. A visit to the tailor yielded only a multicoloured cloth which was unobjectionable on symbolic grounds, but a trifle gay for breakfast. However there seemed to be no choice, so wearing this politically neutral garment Babbage for a time acquired the reputation of a dandy.

Ireland was the source of a story which Babbage was fond of relating. William Edgeworth had an estate at Edgeworth town. Being often approached for alms he decided to discriminate between the lazy and the industrious poor, and also to secure a return for the money he was distributing. The cistern in the upper part of his house was filled by a hand pump. To this pump he added a mechanism to provide a penny after a number of strokes so calculated that continuous work would offer the rate of pay customary in that part of the country. The idle beggars merely passed on. Those who claimed to be seeking only 'a fair day's pay for a fair day's work', that favourite slogan of nineteenth century trade unions, usually went away *cursing* the hardness of their taskmaster. The tale nicely combines several of Babbage's concerns: mechanical ingenuity, economic principle, and the question of the idle and the industrious. He had it from the Marquis of Lansdowne.[20]

The Dorset Street Saturday evening soirées continued as regular events during the season, usually attended by two or three hundred guests: the Fittons, the Lyells, the Somervilles, the Grotes; legal friends, clerical friends, and artistic friends; any visiting practitioner of science with an introduction from one of Babbage's many friends abroad; visitors from Scotland, Ireland, one of the provincial cities or the country. At one of the evenings the young

[20] *Passages*, 438–9.

Cavour first met de Tocqueville[21]. They were lively and intelligent gatherings and Babbage's drawing-room was one of the great meeting places of liberal intellectual Europe. For a decade the Difference Engine in its case of mahogany and glass could be put on display, besides being shown to the many people who came specially to inspect it.

When Babbage was a child his mother once took him to see an exhibition of machinery in Hanover Square. The exhibitor, who called himself Merlin, noting Charles's precocious interest in mechanical detail, took him upstairs to the attic workshop to see some still more wonderful automata. There he saw two silver nudes, about a foot high, the craftsman's masterpieces, still unfinished:

> One of these walked or rather glided along a space of about four feet, when she turned round and went back to her original place. She used an eye-glass occasionally, and bowed frequently, as if recognizing her acquaintances. The motions of her limbs were singularly graceful.
> The other silver figure was an admirable danseuse, with a bird on the fore finger of her right hand, which wagged its tail, flapped its wings and opened its beak. The lady attitudinized in a most fascinating manner. Her eyes were full of imagination, and irresistible.[22]

In 1834 Babbage chanced on the dancing lady at the auction of the contents of Weeke's mechanical exhibition in Cockspur Street where she had lain in an attic, neglected for many years. She was in reasonable condition apart from want of polishing and he secured her for £35.[23] He repaired and cleaned the mechanism himself, and placed her under a glass case on a stand in his dining room. Dressed by some of his female friends she was soon executing her figures at the Dorset Street parties. The dancing lady and the Difference Engine contrast nicely two sides of Babbage's character: the deep earnestness and his sense of fun. However serious Babbage might be, however deeply his interests were involved, he was never a bore.

A story he tells of one of the soirées where both objects were on display shows him in good form:

> A gay but by no means unintelligent crowd surrounded the [dancing lady who was wearing a particularly brilliant dress that evening]. In the adjacent room the Difference Engine stood nearly deserted: two foreigners alone worshipped at that altar. One of them, but just landed from the United States, was engaged in explaining to a learned professor from Holland what he had himself in the morning gathered from its constructor.
> Leaning against the doorway, I was myself contemplating the strongly contrasted scene, pleased that my friends were relaxing from their graver pursuits, and admiring the really graceful movements produced by mechanism; but still more highly gratified at observing the deep and almost painful attention of my Dutch guest, who was

---

[21] Cavour, *Diario*, 23, maggio 1835, 172. [22] *Passages*, 17–18, 365–7, 425–7.
[23] BL Add. Ms. 37,188, f 452. John Richard Lunn to Babbage, 17 July 1834.

questioning his American instructor about the mechanical means I had devised for accomplishing some arithmetical object. The deep thought with which this explanation was attended to, suddenly flashed into intense delight when the simple means of its accomplishment were made apparent.

My acute and valued friend Lord Langdale, who had been observing the varying changes of my own countenance as it glanced from one to the other, now asked me,

'What mischief are you meditating?'

'Look'

said I

'In that room—England. Look again at this—two foreigners.'[24]

In 1836, before the Bristol meeting of the British Association, Babbage stayed with William Henry Fox-Talbot at Lacock Abbey. One of the most important pioneers of photography, Fox-Talbot is one of the many scientific figures who played a part in Babbage's life. Fox-Talbot went to Trinity College, Cambridge, graduating in 1821. In 1822 he was elected a member of the Astronomical Society. He was a mathematician of some ability and originality although this work has been largely overshadowed by his contributions to the development of photography. At that time several people were trying to make permanent records of the action of light. Fox-Talbot developed a workable process as early as 1833 for forming what he called 'photogenic drawings' which were relatively permanent. These early productions could register little range of tone, showing simple pictures such as the outlines of leaves and the pattern of latticing in leaded windows. His later prints, which he called 'calotypes' and 'talbotypes', showed variations of light and shade in considerable detail. Fox-Talbot's process was based on the formation of a negative from which many positives could then be printed. Both Babbage and Sir John Herschel were very interested in all forms of photography. Herschel made the crucial suggestion of using thiosulphate for fixing.[25]

The rival type of photograph was the daguerreotype, invented by Louis-Jacques-Mandé-Daguerre. Made on a metal base, whereas the calotype was formed on paper, the daguerreotype showed far more detail than its rival but made a positive only. Thus it permitted but a single copy of each exposure. Although Daguerre's method was first to be announced, the calotype was immediately championed by Babbage's close friend and ally Sir David Brewster and became the basis for the development of modern photography. Able to make positives in considerable numbers, although the process was tedious, Fox-Talbot established a commercial processing laboratory in Reading. He also made the prints for the world's first book of photographs, 'The Pencil of Nature'. In the book he described both the processes and the way he had come to invent them. As one might guess, early samples of photogenic drawings,

[24] *Passages*, 425–7.

[25] John Herschel made the vital suggestion of using thiosulphate for 'fixing' as a direct result of his early work on the 'sulfosalts'.

16. Charles Babbage, *c.* 1850.

18. Mary Anne Hollier, 1856.

17. Charles Babbage, c. 1860.

19. Henry Prevost Babbage and his family, c. 1870.

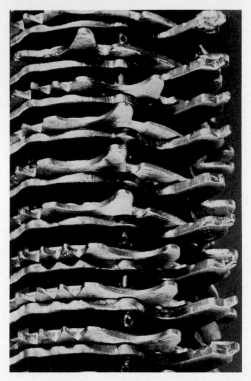

20. Experimental Carry Column for an Analytical Engine, late 1860s. Replica.

21. Mill of minimal Analytical Engine under construction when Babbage died.
In this photograph the racks stick out in positions they could never have reached in use.

calotypes, and daguerreotypes were soon being exhibited at the Dorset Street parties.

One famous visitor at Babbage's was the heiress Angela Burdett-Coutts.[26] They had many friends in common, including Samuel Rogers the poet, Charles Dickens, and the Duke of Wellington. Her father, Sir Francis Burdett was a wealthy man and one of the most colourful figures in radical politics at the beginning of the nineteenth century, sitting as radical MP for Westminster for thirty years. Later in life he became a Tory and was considered a renegade by the radicals, but in his earlier days he had been a popular idol. However Angela's great fortune came from her grandfather on her mother's side, Thomas Coutts the banker. In his old age Coutts made a second marriage to Harriot Mellon, an actress. To everyone's astonishment he left his fortune and control of one of Britain's banks entirely at her disposal. She in turn married the impoverished Duke of St Albans. But when Harriot died there was another surprise: she had left the fortune to shy little Angela Burdett, of whom the worldly old lady was fond and had formed a high opinion. Angela took the name Burdett-Coutts and settled down to the anxious task of handling one of the country's great fortunes. Until he died the Duke of St Albans remained in possession of the Holly Lodge estate in Highgate which had belonged to Harriot; then Angela went to live there, holding her famous garden parties which Babbage often attended.

Deeply religious and believing in the beneficial effects of science, Angela was naturally congenial company for Babbage. She was most appreciative of the Difference Engine and he was glad enough of her moral support in the face of official neglect or hostility. She was a philanthropist on a vast scale and gave some support to science, although Babbage himself would never have dreamed of accepting any financial aid from her: he was far too proud. She took lessons in astronomy from Babbage's old friend Sir James South at his private observatory on Campden Hill and Babbage often escorted her on these visits. During the Crimean War Angela sent Florence Nightingale a drying-chest with capacity for drying a thousand articles of clothing in twenty-five minutes: a practical gesture of which Babbage would have thoroughly approved.

In 1833 a visitor first came to one of Babbage's parties who was to play an important part in his life: Augusta Ada Byron, daughter of the poet.[27] She had encountered Babbage at a party on 5 June of that year, and he had made a deep impression on her. On 21 June she attended one of his soirées with her mother and was profoundly interested in the Difference Engine. Her grasp of its principles was altogether unusual for a young lady of the time. Such was the impression made by Babbage himself that her principal interest in attending the Queen's ball on the same night was the hope of seeing him again. Much

[26] Edna Healey, *Life of Angela Burdett-Coutts*, Sidgwick and Jackson, 1978.
[27] Doris Langley-Moore, *Ada, Countess of Lovelace*, John Murray, 1977.

romance attached to Ada as Byron's daughter. The lines from Childe Harold were very well known:

> Is thy face like thy mother's, my fair child!
> Ada! sole daughter of my house and of my heart?
> When last I saw thy young blue eyes they smiled,
> And then we parted,—not as now we part,
> But with a hope.[28]

The story of the triangle formed by Byron, his wife, Lady Noel Byron, and his half-sister, Augusta, who had been his mistress and after whom Ada was named, has become a legend. When Byron had left for Italy with a parting shot—'When shall we three meet again?'—Lady Byron pursued Augusta with relentless vindictiveness; but until recently Lady Byron's treatment of her own daughter has been less well known. Ada came to play an important part in Babbage's life and work; we owe to her the only clear statement we possess of Babbage's views on the scope of his engines. And the terrible circumstances of her death and her mother's cruelty affected him deeply.

Ada had a disjointed upbringing, without a father, attended by a whole series of governesses and teachers, her domineering, dissembling, hypochondriacal mother detested by the household staff. After she had met Babbage, Ada attended a series of lectures on the Difference Engine by Dionysus Lardner at the Mechanics Institute. With her mother she made a short tour of Midlands industry, probably following Babbage's advice and supplied by him with the necessary introductions. Mary Somerville, that rare figure, a woman with a knowledge of science, became a friend, sometimes taking her to Babbage's parties. In 1835 Ada married Lord King, who had been at Trinity College, Cambridge. No doubt he had heard a good deal there about Byron and Babbage, both rebellious figures in their different ways. King had two estates, the old family seat at Ockham Park in Surrey and a smaller house, Ashley Combe, on the lovely Somerset coast near Porlock. Babbage was to spend much time at both, welcomed equally by husband and wife.

Contemporary diaries give many pictures of Babbage's social life. For example William Charles Macready,[29] one of the most famous actors of the time and a close friend of Dickens, saw a good deal of Babbage. On 25 April 1840, he recorded: 'David Colden came into my room and accompanied me to Babbage's, where I saw Sydney Smith, Professor Wheatstone, the Brockendens, two or three whom I knew but not by name, Harness, Travers, Hawes, Lady Stepney, Dr Arnott, Milman, the Bishop of Norwich (Stanley), who wished to know me.' And again on 9 May: 'Talfourd came into my room ... Dickens came, and we went in his carriage, all three to Babbage's. The room was very

[28] Childe Harold, canto 3, l 1–5.
[29] William Toynbee, *The Diaries of William Charles Macready*, Chapman and Hall, 1912.

much crowded, but I saw a few that I knew there—Wheatstone, Edward Kater, Lady Stepney, and Rogers, who invited Dickens and myself to dine with him on Sunday fortnight. Talfourd left us soon; Dickens set me down.' Macready records meeting Babbage frequently at public dinners and in the houses of the great.

Charles Darwin also gives us a vignette: 'I remember a funny dinner at my brother's, where, amongst a few others were Babbage and Lyell, both of whom liked to talk. Carlyle however silenced everyone by haranguing during the whole dinner on the advantages of silence. After dinner, Babbage, in his grimmest manner, thanked Carlyle for his very interesting lecture on silence.'[30] Babbage occasionally affected a grim aspect. Darwin thought it a pose: underneath was the kindest of hearts.

Babbage's position in society gave him opportunities unique for a man of science. Since his time there have been many bureaucrats, but there has been no British scientist quite in Babbage's position: he was able to approach leading figures both in Britain and abroad on a personal level, which was in no way derived from an official position.

One delightful example of the use of Babbage's influence arose in conection with the old Quaker, John Dalton, pioneer of atomic theory. The two men make a striking contrast: the turbulent, sociable Babbage, and the shy, retiring Dalton. He had been born in the quiet Cumberland village of Eaglesfield, and supported himself from his youth by elementary teaching. Although he had moved to Manchester, and become the idol of the Literary and Philosophical Society, he had continued to teach, and in his old age was still taking private pupils. Dalton had himself observed the contrast between France, where leading scientific figures were officially rewarded and supported, and England, where there was an almost total lack of official support. It was felt to be to the nation's shame that one of the great men of science of his day should have had to eke out a living taking pupils, and Babbage was instrumental in securing a pension for him and making amends.

At the meeting of the British Association for the Advancement of Science in 1832 Dalton was awarded an honorary degree of Doctor of Laws by the University of Oxford. Having no university of its own to award a degree, Manchester decided to have a statue of him carved instead. Two thousand guineas were subscribed and the commission went to Francis Chantrey, perpetrator of so much dull and lifeless iconography in marble. To complete the round of honours it seemed to Babbage that it would be fitting for William IV to say a few words to the venerable philosopher at a levée. Brougham, at the time Lord Chancellor, liked the idea and offered to present Dalton to the King. The problem was what Dalton should wear. Court dress was impossible because, as a Friend, he could not wear a sword. The scarlet gown of a Doctor

[30] Nora Barlow, ed., *The Autobiography of Charles Darwin* (1809–82), 112, Collins, 1908.

of Laws of Oxford also seemed to present difficulties as the colour was forbidden to Quakers. However Dalton was colour-blind, the gown appeared to him the colour of mud, and he made no objection. The velvet cap, as symbol of office to be held in the hand rather than worn, was also acceptable.

One morning when Dalton had breakfasted with Babbage and inspected the Difference Engine, Babbage staged a rehearsal. Dalton played the part of the King while Babbage kissed hands. There was a second rehearsal on the morning of the presentation and the ceremony went forward smoothly. Dalton's scarlet gown attracted some attention and the opinion was that he was some provincial mayor come to be knighted. Never missing a chance, Babbage assured the enquirers that Dalton had no wish to be knighted as his name would be remembered long after the orders of knighthood had been forgotten. Babbage was also glad to take the opportunity of shocking the Bishop of Gloucester by informing him that Dalton was a member of the Society of Friends, rare figures at court.[31]

In February of 1838 Babbage had a long interview with the Duke of Sussex, still President of the Royal Society. Babbage wished to secure the Order of the Bath for John Herschel, recently returned from the Cape of Good Hope. There was a dinner in his honour attended by Melbourne and the young Queen Victoria—altogether a new recognition for science in England. But then Herschel's great catalogue of the southern stars had been made in one of the colonies, and colonial matters were always a popular subject with government. Shortly after the dinner, at a party in Lansdowne House, the Duke of Sussex informed Babbage that during the dinner Melbourne had changed his mind and Herschel was to be offered a baronetcy instead. Herschel declined the offer three times, and then accepted: he would have preferred a non-hereditary honour.[32]

Besides attending the annual meetings of the British Association Babbage visited friends in many parts of the country. He also had many minor scientific interests to divert him. For example on 27 November 1839 George Rennie wrote to Babbage,[33] announcing the invention of a new type of paddle-wheel, which was lighter and cheaper than those in use, entering the water without shock, noise, or vibration, and leaving it without raising a cascade behind. Rennie asked Babbage to be at the West India Docks at 2 o'clock on the following Tuesday, or on some other day at his convenience, to come aboard a small steamer and see the new paddle in action. About that time, or possibly a little later, Hallam invited Babbage to look in the following evening to join M. Guizot and a few friends.[34] Such were the everyday events of his life.

In April 1840 Babbage was urged to stand for election to Parliament in the

[31] *Passages*, Ch. XXIII.
[32] Babbage to the Duke of Sussex, Feb. 1838, BL Add. Ms. 37,182, f 384.
[33] BL Add. Ms. 37,191, f 269.   [34] Ibid., f 303.

Cambridge seat, when an earlier election had been declared void; but Babbage had had enough of elections after the Finsbury campaigns. He never missed an opportunity of visiting an interesting factory, collecting statistics, or inspecting some novel technical process, whether in industry or a laboratory. He took five months holiday to make his extraordinarily modern study of Brunel's wide-gauge railway.[35] But apart from that interlude the Analytical Engines remained throughout this period his primary concern. Moreover he knew that, sooner or later, there would have to be a showdown with the Government over the project for the first Difference Engine and the whole question of support for the calculating engines. He could at that time expect no backing from the Royal Society; the British Association was ineffectual and far from realizing the hopes which he and David Brewster had entertained when it was founded; the more specialized scientific societies carried altogether insufficient weight for his purpose. In all the circumstances, and with his wide Continental network of friends and colleagues, Babbage looked abroad for understanding and support.

In 1839 a meeting of Italian scientists,[36] or as Babbage continued to refer to them, 'philosophers', was arranged at Pisa. In 1828, during his grand tour, Babbage had suggested such a meeting to the Grand Duke of Tuscany, but only now did the Duke feel the time was ripe.[37] Babbage had been in intermittent correspondence with the Duke ever since 1828, sending specimens of British manufactures and receiving on one occasion from the Duke a thermometer from the time of Galileo.[38] However Babbage was not yet ready and declined the invitation. In 1840 a similar meeting was arranged in Turin. By then Babbage did feel ready, and accepted an invitation from Plana to present the Analytical Engine before the assembled philosophers of Italy. As Plana put it: 'Hitherto the *legislative* branch of our analysis has been all powerful—the executive all feeble. . . . Your engine seems to give us the same control over the executive which we have hitherto only possessed over the legislative department.'[39]

In the middle of August 1840 Babbage left England. A few days earlier he had attended the déjeuné at Wimbledon Park given by the Duke of Somerset to Queen Victoria, the first reception, as Babbage was pleased to note, given by a subject for the new Queen. Fox-Talbot sent from Lacock Abbey an album of thirty-five choice calotypes for the Grand Duke of Tuscany and a packet of calotypes for Professor Amici to exhibit at the Turin gathering.[40] Lyell sent Babbage guide books for a trip up the Seine, and he set off.

In Paris Babbage received from Arago and other friends letters of introduction to people in Lyons. In Lyons he purchased from Didier-Petit and

[35] *Passages*, Ch. XXV.
[37] *Passages*, 431.
[39] *Passages*, 129.

[36] BL Add. Ms. 37,191, ff 132, 139.
[38] BL Add. Ms. 37,189, f 401, 37,185, f 363.
[40] BL Add. Ms. 37,191, ff 432, 596.

Co. of Quai de Retz no. 34, Manufacture d'étoffes pour Ameublements et Ornamens d'Église, a silk portrait of Jacquard which had been woven on a Jacquard loom instructed by those punched cards that were also to instruct the Analytical Engines.[41] Babbage already possessed one of the portraits of Jacquard, a gift from a friend, but he wished to obtain a second copy to present to the Duke of Tuscany. The portraits were only made by special favour and Babbage spent many fascinating hours watching it being made.

Turin, then the capital of the kingdom of Sardinia, was a prosperous city of about 165,000 inhabitants, magnificently sited on the rapid Po river, within sight of the Alps. Babbage's arrival was carefully watched by the secret police as he was travelling with Fortunato Prandi, an old revolutionary and friend of Mazzini's in London.[42] Prandi was a good friend of Babbage's and travelled with him at least from Lyons and possibly all the way from London. No doubt Babbage's protection was intended to give Prandi respectability. Certainly the secret police were very dubious both about Prandi and some of his acquaintances in Turin. However the King directed that Prandi was to be left free to accompany Babbage wherever he wished. And when Babbage returned to England Prandi was allowed to stay on and visit his old father. While Babbage was in Turin Prandi was nominally his interpreter, staying in the apartments which Babbage had rented.

Babbage had persuaded his friend Professor MacCullagh of Dublin to abandon a climbing trip in the Tyrol to join him at the Turin meeting. There in Babbage's apartments for several mornings met Plana, Menabrea, Mosotti, MacCullagh, Plantamour, and other mathematicians and engineers of Italy. Babbage had taken with him drawings, models, and sheets of his mechanical notations to help explain the principles and mode of operation of the Analytical Engine. The discussions in Turin were the only public presentation before a group of competent scientists during Babbage's lifetime of those extraordinary forebears of the modern digital computer. It is an eternal disgrace that no comparable opportunity was ever offered to Babbage in his own country.

Babbage remained permanently grateful to his Italian friends, later dedicating his book *Passages from the Life of a Philosopher*

TO VICTOR EMMANUEL II, KING OF ITALY.

Sire,

In dedicating this volume to your Majesty, I am also doing an act of justice to the memory of your illustrious father.

In 1840, the King, Charles Albert, invited the learned of Italy to assemble in his capital. At the request of her most gifted Analyst, I brought with me the drawings and

[41] Ibid., f 446.

[42] Luigi Bulferetti, *Un amico di Charles Babbage*; *Fortunato Prandi*, Instituto Lombardo di Scienze e Lettere, xxx, fasc. 2, 83–166, Milano, 1968.

explanations of the Analytical Engine. These were thoroughly examined and their truth acknowledged by Italy's choicest sons.

To the King, your father, I am indebted for the first public and official acknowledgment of this invention.

I am happy in thus expressing my deep sense of that obligation to his son, the Sovereign of United Italy, the country of Archimedes and of Galileo.

I am, Sire,

With the highest respect,

Your Majesty's faithful Servant,

CHARLES BABBAGE

The problems of understanding the principles of the Analytical Engines were by no means straightforward even for the assembly of formidable scientific talents which gathered in Babbage's apartments in Turin. The difficulty lay not so much in detail but rather in the basic concepts. Those men would certainly have been familiar with the use of punched cards in the Jacquard loom, and it may reasonably be assumed that the models would have been sufficient to explain the mechanical operation in so far as Babbage deemed necessary.[43] Mosotti, for example, admitted the power of mechanism to handle the relations of arithmetic, and even of algebraic relations, but he had great difficulty in comprehending how a machine could handle general conditional operations: that is to say what the machine does if its course of action must be determined by results arising from its own previous calculations. By a series of particular examples Babbage gradually led his audience to understand and accept the general principles of the Engine. In particular he explained how the machine could, as a result of its own calculations, advance or back the operation cards, which controlled the sequence of operations of the Engine, by any required number of steps. This was perhaps the crucial point: only one example of conditional operations within the Engine, it was a big step in the direction of the stored program, so familiar today to the tens of millions of people who use electronic computers.

In explaining the Engines Babbage was forced to put his thoughts into ordinary language; and as discussion proceeded his own ideas crystallized and developed. At first Plana had intended to make notes of the discussions so that he could prepare a description of the principles of the Engines. But Plana was old, his letters of the time are in a shaky hand, and the task fell upon a young mathematician called Menabrea, later to be Prime Minister of the newly united Italy. It is interesting to reflect that no one remotely approaching Menabrea in scientific competence has ever been prime minister of Britain.[44]

[43] These probably included an anticipatory carriage mechanism. See also M. G. Losano, *La Macchina Analitica*, 1973.

[44] I shall not comment on the question of whether a recent appointment invalidates this statement.

During his visit Babbage got on very well with the King,[45] a shy man with whom Babbage was able to communicate, finding a ready listener to thoughts about the electric telegraph and other applications of science. As usual Babbage had brought with him an assembly of instruments, tools, patent locks, and other examples of the arts of manufacture. These were duly exhibited to the young princes, and samples were left as souvenirs. The whole of the time Babbage could spare to stay in Italy was fully occupied, explaining the Analytical Engines, taking advantage of the gathering to discuss his endless other scientific interests, visiting old friends, wining and dining with local notables, and attending court junketings. Thus he was forced to abandon his intention of visiting Florence. However Charles Albert's Queen was the sister of the Grand Duke of Tuscany and Babbage decided to present the silk portrait of Jacquard to the Queen instead of her brother.

During this visit Babbage chanced to mention that he had never seen wine being made. He was soon invited to watch the grapes being pressed at Racogni, a royal estate a dozen miles from Turin. The successive processes were explained and Babbage and his guides enjoyed an excellent dinner which he never forgot: the study of technical processes was by no means always an arduous undertaking. Returning to Turin he noticed that the general with whom they had dined had sent a dragoon as an honorary escort. Though sensible of the honour, the waste of energy offended Babbage's sense of propriety and the dragoon was dismissed from escort duty.

Before Babbage left Turin a courtier remarked on the signal marks of honour Babbage had received: 'The king has done three things for you which are very unusual: he has shaken hands with you; he has asked you to sit down at an audience; he has permitted you to make a present to the Queen. That last is the rarest of all.'[46] On such matters do courts gossip.

After visiting Milan Babbage had a final long audience with the King. Among other subjects they discussed free trade: these were by no means merely social occasions and Babbage never missed a chance of spreading his and Manchester's gospel. Two days later at the beginning of November Babbage left on the mail coach. Before Annecy they crossed a famous suspension bridge, the Pont Charles Albert, named after the King. During his final audience Babbage had mentioned his interest in the bridge and the King had promptly produced a bronze medal struck for its opening. Mentioning the subject to the courier with whom he shared the coupé Babbage found the man already informed of his interest. They descended from the coach a third of a mile from the bridge, instructing the postillions to drive slowly across and wait on the far side. The gorge was deep and the bridge itself partly covered in cloud.

[45] *Passages*, Ch. XXIV.
[46] Ibid., 438–9.

Babbage later recorded the scene in the almost purple prose in which he occasionally indulged:

We were singularly favoured by circumstances. We saw the carriage which had left us apparently crossing the bridge, then penetrating into the clouds, and finally becoming entirely lost to our view. At the same time the dissolving mist in our own immediate neighbourhood began to allow us to perceive the depth of the valley beneath and at last even the little wandering brook, which looked like a thread of silver at the bottom.

The sun now burst out from behind a range of clouds, which had obscured it. Its warm rays speedily dissipated the mist, illuminated the dark gulf at our own side, and discovered to us the mail on terra firma on the opposite side of the chasm waiting to convey us to our destination.[47]

When he returned to England Babbage was delighted to find that his mother at the age of eighty had taken up modelling in clay. The splendid old lady had summoned an eminent teacher of sculpture and taken lessons, presenting Babbage with a portrait head of himself on his return. It was considered not a bad likeness.

Babbage's primary object in attending the Turin meeting had been to secure understanding and recognition for the Analytical Engine. He hoped that Plana would make a brief formal report on the Engine to the Academy of Turin and that Menabrea would soon complete his article. Babbage sent him further explanations to complement the notes he had made during Babbage's exposition and the discussions in Turin. Babbage had certainly little hope of government comprehension or support in England but he was determined not to miss the slightest opportunity of securing recognition for his Engines.

He set down his own thoughts in a letter written at about this time to Angelo Sismoda, whom he had often seen during the Turin meeting:

The discovery of the Analytical Engine is so much in advance of my own country, and I fear even of the age, that it is very important for its success that the fact should not rest upon my unsupported testimony. I therefore selected the meeting at Turin as the time of its publication, partly from the celebrity of its academy and partly from my high estimation of Plana, and I had hoped that a report on the principles on which it is formed would have been already made to the Royal Academy.[48]

But Plana was old and ill: no report was forthcoming.

Meanwhile Babbage continued to work indefatigably at the designs for the Engines. He was seeing more of the Lovelaces—Lord King had been created Earl of Lovelace—and Ada was beginning to show serious interest in Babbage's project. On 12 January 1841 she invited him to stay at Ockham Park. After urging Babbage to bring his skates she continued:

It strikes me that at some future time (it might be even within 3 or 4 years, or it

[47] Ibid., 309.   [48] BL Add. Ms. 37,191, ff 582, 625.

might be *many* years) *my head* may be made by you subservient to *your* purposes ... I wish to speak most seriously to you.

A A LOVELACE

You *must* spend some days with us. Now don't contradict me.[49]

Later, in 1841, Babbage wrote to his old friend Alexander von Humboldt who was going to be in Paris: 'There is no chance of the machine ever being executed during my own life and I am even doubtful how to dispose of the drawings after its termination.'[50] Von Humboldt had invited Babbage to meet him in Paris and Babbage in turn was urging Humboldt to come to London where he could see the drawings and experimental models. One wonders whether Babbage hoped that an Analytical Engine might be made in Prussia. Evidently von Humboldt could not come to London, but Babbage was planning to visit Tuscany for the third gathering of Italy's learned, this time to be held in Florence, and would travel through Paris. He met von Humboldt there.

Before leaving Babbage secured from George Rennie the engineer a list of manufacturers to visit en route: Messrs. Schneider and Co. at Creusot where they were making large marine engines for the French Government; cotton and machinery works at Mulhouse; the cotton and machinery works of Escher Fils at Zurich, and others. Rennie remarked that he had himself been invited to Florence but could not attend.[51] Evidently the Grand Duke's net was cast wide.

Babbage recorded few details of the meeting in Florence, probably because the Analytical Engines did not play much part in the discussions. On 3 October he wrote sadly to Plana: 'I have much regretted that we could not meet at Florence as I had hoped. We should have had more leisure than Turin to discuss the principles of the Analytical Engine. If you had made a report on the subject to the Academy during the last year it might have been of much service to me in the discussion of the question with my own government.'[52]

No doubt Babbage visited his old Bonaparte friends. In particular he would have seen Lucien's eldest son, Charles Lucien,[53] whom he had met in Rome in 1828, now a leader of the radical scientific movement in Italy. In England, where the working class Chartist movement developed, there was little radicalism left among men of science, who were becoming increasingly absorbed into the bureaucracy: certainly there was in the 1840s nothing like the Analytical Society and the 'decline of science' movement. However in Italy at the time of Babbage's visits the whole scientific movement was becoming

---

[49] Ibid., f 543.   [50] Ibid., f 637.

[51] George Rennie to Babbage, 25 Aug. 1841. Whitehall Place to Poste Restante, Geneva. BL Add. Ms. 37,191, f 639.

[52] Babbage to Plana, 3 Oct. 1841; Ibid., f 645.

[53] Charles Lucien Bonaparte made a report on the progress of biology in Europe to the Turin meeting.

increasingly radical and nationalist: it was the period leading up to the upheavals of 1848. There were further scientific congresses in Padua, 1842; Lucca, 1843; Milan, 1844; Naples, 1845; Genoa, 1846; and Venice, 1847. Every subject mentioned seemed to have revolutionary implications: economic discussion implied. a customs league; social discussion implied politics; geography urged a united Italy. The nationalists saw railways not merely as aids to trade and communication but as forces uniting the states of the peninsula: the railways, it was said, would 'stitch the boot'. The Babbages played a part in this development. Herschel with his brother Dugald built railways in the Sardinian kingdom and Charles Babbage for a time was a formidable radical force: Charles Albert's was not the only ear into which he was pouring ideas of free trade and technological development. However the radicalism of Babbage and his friends was fast running out. After 1848 Babbage turned to thundering against socialism. And Menabrea, after having been an early supporter of Garibaldi, turned against him when he later became prime minister of united Italy, and defeated Garibaldi's forces in a decisive battle.

During this visit to Tuscany Babbage made a study of the region of hot springs near Volterra. From notes he made at the time and a letter he wrote in the following year to the Grand Duke we get an interesting picture of his approach to economic development. At the time the hot springs were mainly used as a source of boracic acid. Babbage saw them as a source of power which could serve as the basis for a substantial and widely based industry. The Duke appointed a M. Paliti as his guide. A paraphrase of his proposals shows Babbage approaching a practical economic problem:[54]

Saw a paper mill; efficient and made good quality paper. Saw a leather tannery; could the process be speeded up by using the hot water plentifully available in the locality?
The boracic acid works are very impressive; even so they are still in their infancy.
Heat could be applied to the produce of vineyards to form alcohol which in turn could be the foundation of an extensive chemical industry.
Iron pipes inserted into the ground could yield steam at [according to the ideas of the time considerable] pressure which would be an unrivalled source of power. Steam engines of Savery's type could be used to pump water; this would be the first use of steam and would be very simple to apply.
Perhaps thermo-electricity could also be applied [i.e. on an industrial scale]; this should be investigated.

This is a programme which the modern ecological movement could endorse: an integrated chemical industry based on geothermal power. Babbage then

[54] BL Add. Ms. 37,192, f 104.

went on to make clear that besides being a far sighted and immensely knowledgeable industrial technologist, he was also a liberal political economist:

> It is much to be hoped that the wealth deservedly acquired by two or three foreigners who have brought to light some of the mineral riches of this most remarkable district may stimulate native capitalists [not then a derogatory term] to follow their example and that by establishing a system of working similar to the Cornish mines, in which the labourers themselves participate in the success of the concern, the whole population may be raised in their intellectual character as well as in their bodily comforts.

In conjunction with Babbage's strong Europeanism the programme would be acceptable even now to the British Liberal party. The boracic acid works and associated developments actually became an economic showpiece, particularly proud of its housing and provision for the needs of its workers. How far Babbage's proposals in fact contributed to these developments it is impossible to determine. But it has been noted how very often in Italy at that period some Englishman, quietly working away, played an important part in the country's economic development.

Babbage returned through Milan and Turin, where he doubtless visited Plana, reaching England before the end of November. In Florence the Grand Duke had offered him any product of Tuscany he cared to name and he had requested the 'Grand Cross of your Order of Merit'. He was interested in a decoration at this time for the same reason that he wanted Plana to report to the Turin Academy: a decoration would be a formal endorsement of his work on the calculating engines, useful in dealing with the British government. Babbage was in an awkward position with regard to foreign honours: having turned down the Order of the Guelph when David Brewster and John Herschel had accepted, he could not now accept a comparable honour from the ruler of another country. It was the Grand Cross or nothing, as he had to explain in rather embarrassed letters. He was similarly placed with regard to the offer of an honour from Charles Albert. In the event no Grand Cross was forthcoming from Florence. But from Turin the news was much more embarrassing: Babbage was awarded the Common Cross, which he could not accept. His friend Prandi, writing from Turin, was not surprised. He advised Babbage to accept the Common Cross but not to tell a living soul. Early in 1842 the matter was sorted out. Some courtier, envious of the impression Babbage had made, had told the King that Babbage held a lowly official position in England and that the Common Cross was the appropriate award. Now all was well and Babbage was to be made a 'Commendatore like Plana'. Apart from academic awards it was the only honour Babbage ever accepted.[55]

The North Italian meetings had been a great success and a tonic for Babbage's morale. The contrast with his treatment in England could hardly

[55] BL Add. Mss. 37,191, f 685, 37,192, f 3.

have been more marked. In Turin and Florence he had been received as an honoured adviser by the ruling princes and acclaimed by the philosophers. For the rest of his life Babbage remembered with pleasure the warmth of his reception in Italy.

13

○●○

Analytical Engines and the Circumlocution Office

Babbage returned from the sunny hills and valleys of Tuscany where he had basked in Ducal warmth and the approbation of philosophers to a chilly climate in England. He sent further explanations to Menabrea who in turn entirely rewrote his article. On 27 January 1842 he wrote to Babbage from Turin: 'Je donnerai la dernière main a l'écrit qui vous concerne et j'espère dans quelques jours l'envoyer a Genêve au bureau de la *Bibliothèque Universelle*'.[1] In number 82 of October 1842 the article finally appeared.

Babbage's busy social life continued quite unaffected by his difficulties with the Calculating Engines. On 24 November immediately after returning from Tuscany he dined with Charles Greville, Henry Hart Milman, Sydney Smith, and Macaulay.[2] Although Babbage no doubt had a good deal to say about Italy and the scientific meeting, Greville noted in his diary that the talk had been pretty evenly divided between Sydney Smith and Macaulay. Babbage liked to talk but in those loquacious two he had more than met his match.

Robert Peel having become Prime Minister again, Babbage decided to try and force him into a decision. Late in November 1841 he drew up a statement on the history and present position of his calculating engines, chronicling his relations with successive Governments since his 'Letter to Sir Humphry Davy', published in 1822, and the commencement of the project in 1823. Babbage asked simply that after so long a delay he might at last have a decision:[3] he wanted to know where he stood. Instead he received a stalling letter from the Secretary to the Treasury, Sir George Clark, referring to the great pressure of business prior to the opening of Parliament. When the session closed in August Babbage had still received no reply. He sought help and Sir W. Follett assisted in pressing Peel for a decision.[4]

Actually Peel had not forgotten the Difference Engine, but it presented a problem and he was wondering how best to handle it. £17,000 of Government money had been spent and he considered the project should never have been started in the first place: he had never liked it. However, as was found with the Caledonian Canal and many another project, it was a great deal harder to stop than to start. We find Peel writing to William Buckland: 'what shall we do to

[1] BL Add. Ms. 37,192, f 22.
[2] Lytton Strachey and Roger Fulford (ed.), *The Greville Memoirs*, IV, 428, Macmillan, 1938.
[3] BL Add. Ms. 37,192, ff 13, 19, 29, 147.   [4] Ibid., f 179.

get rid of Babbage's calculating machine ... worthless to science in my view. If it would now calculate the amount and the quantum of benefit to be derived to science it would render the only service I ever expect to derive from it.'[5]

Peel had his set of scientific advisers, of course: one was the Astronomer Royal, George Biddell Airy, so deferential to his superiors, prototype of the scientific bureaucrat. Airy was a competent calculator, fond of arithmetic. In the course of his astronomical work he carried out a large amount of computation. A pedantic unimaginative man he simply could not see the point of calculating machinery and thought the Engine useless in principle. But the choice of advisers reflected Peel's own opinions. Respectful academics and the retiring Faraday were acceptable, but Babbage's industrial interests and the possibilities of commercial application of his calculating engines were quite unattractive to Robert Peel.

At the beginning of November 1842 Babbage finally received a letter from Goulburn, now Chancellor of the Exchequer once again dealing with the Calculating Engine. In it he stated that both Sir Robert Peel and himself regretted the need for abandoning the completion of a machine on which so much scientific labour had been bestowed, but the expense needed to render it satisfactory to Mr Babbage and generally useful appeared to be such that they felt they had no alternative. They offered to place the Difference Engine at Babbage's disposal: an offer which he promptly declined.[6]

Babbage had no intention of being fobbed off with a letter from a subordinate. On 6 November he demanded an interview with Peel and on 11 November was accorded one. The interview was entirely unsatisfactory. The project had been abandoned by the Government without consulting professional engineers about the actual cost of completing it and in spite of the fact that the finished portion worked perfectly satisfactorily. Moreover Peel would not admit that Babbage had any claim on the Government whatsoever for his years of labour. In the House of Commons when Peel announced the decision, Benjamin Hawes, I. K. Brunel's brother-in-law, alone spoke up for the Difference Engine.[7]

Undoubtedly Peel's character and personal interests played an important part in the unhappy end of the project for the first Difference Engine, but to assign this as the basic reason is to trivialize the question. In the 1840s there was very little backing for the Engines which in the 1820s had aroused obvious enthusiasm and support. The underlying cause for this change was the developing Victorian academicism. A wide gap was opening between the superior pure science of the universities and inferior applied science, the bread and butter of the engineers. Babbage's calculating engines fell through this gap into a century's oblivion.

[5] BL Add. Ms. 40,514, f 223.
[6] BL Add. Ms. 37,192, ff 138, 146, 172, 174, 176, 178, 189; *Passages*, 92–5.
[7] *The Times*, 28 March 4b, 1843; *Phil. Mag.* 23, 235.

After his interview with Peel the Somersets pressed Babbage to stay with them at Stover,[8] near Newton Abbot and his native Totnes. He went thankfully to their sympathy and warm hospitality, which played such a large part in his life. The Duke of Somerset was by then married to his second wife, Margaret, daughter of Sir Michael Shaw-Stuart of Blackhall, Renfrewshire. She became a very close friend of Babbage, on one occasion even writing to him as 'Dear Charles',[9] an intimacy otherwise claimed in the whole of the voluminous correspondence only through childhood friendship. While at Stover Babbage received a letter from William Falk, son of the former tenant of Babbage's farm at Dainton and now a cider manufacturer. Interestingly it refers to the copy of *On the Economy of Manufactures* which Babbage had kindly presented to Falk's late father. One would like to know more about Babbage's tenants[10] but scarcely any documentation appears to remain: he would not have presented a copy of his book to a fool. Babbage never lost his warm feelings for Totnes and the surrounding countryside.

Two or three years later Somerset and Lovelace discussed the possibility of securing some honour for Babbage as compensation for all his work, but Babbage was not interested.[11] If taking some honour could have helped the Calculating Engines no doubt he would have been glad to accept: otherwise he was little interested. He turned down a knighthood for himself when John Herschel and David Brewster accepted; later he was to decline to become a member of the Privy Council unless the gesture was clearly made as a recognition of the importance of his work on the Calculating Engines.

Babbage had chosen to leave the Difference Engine as Government property in order to force them to make proper provision for housing and displaying it. He gave careful instructions for its preservation and display:

It should be kept in a warm well-ventilated room. It stands on a stout mahogany table, height from the floor 30 inches, length of table 36 inches, breadth of table 27 inches, height of machine above table 29 inches. The machine is covered on all sides and on top by thick plates of glass. It should be placed where the public can see it for example the British Museum.[12]

The Engine was soon placed on view in the museum at King's College in the

---

[8] BL Add. Ms. 37,192, f 197.   [9] BL Add. Ms. 37,194, f 505.

[10] I am indebted to Mr Fisher Maddicott and Mrs Tozer (née Maddicott) for information about their grandfather, Robert Maddicott (1827–1905), who bought the farm from Babbage's sons after his death. Robert Maddicott was a 6′ 4″ giant of a man. He came from Wolborough Barton near Newton Abbott and his grave can be seen in the Ippelpen churchyard. The property included rights on 'Babbage Common' and Storeycombe Quarry. Babbage is remembered as a good landlord who left his property in excellent condition. The gateposts which particularly struck the young Fisher Maddicott were erected by Babbage. Made of granite with the iron hangings fixed into holes in the granite with molten lead poured in, they were virtually indestructible.

[11] Lovelace to Duke of Somerset, 1845. DCRO.   [12] BL Add. Ms. 37,192, f 224.

Strand. Later, with all its drawings and mechanical notations it was transferred to the Science Museum in South Kensington. There it can be seen by the enquiring visitor, still in perfect working order, without doubt the finest achievement of the precision engineering of its time.

The remaining practical details were cleared up with the Board of Woods and Forests. Babbage purchased from the Government the uncompleted parts of the Difference Engine at the cost of the metal by weight. The small house in East Street, and the sites of the specially constructed buildings which had been planned and built to house the Engine and associated tools and equipment, had been leased by the Government until the termination of Babbage's lease on the rest of the property, that is to say on his house and garden in Dorset Street. Thus the Government had no option but to continue the lease. However Babbage in turn leased back the fireproof building to use as his own workshop and drawing office. The Government in due course rented out the house in East Street, doing, as Henry Babbage noted many years later, very well on the deal.

In 1843 Babbage published a 'Statement of the Circumstances respecting Mr Babbage's Calculating Engines' in the *Philosophical Magazine*. Although published anonymously it was well known that it came from him.[13] After telling the sorry story of succeeding prime ministers after the reform government, Babbage turned to the current state of the Analytical Engine. Upwards of one hundred drawings had been made and the mechanical notations covered four to five hundred large folio sheets of paper. 'No part of the construction of the Analytical Engine has yet been commenced. A long series of experiments have, however, been made upon the art of shaping metals, and the tools to be employed for that purpose have been discussed, and many drawings of these prepared. The great object of these enquiries and experiments is, on the one hand, by simplifying as much as possible the construction, and on the other, by contriving new and cheaper means of execution, at length to reduce the expense within those limits which a private individual may command.'

In public Babbage kept up a bold front but his private letters tell a different story. He had little hope of an Analytical Engine being built in his lifetime. The government of the Duke of Wellington and the reform government had come and gone, and with them all hope of securing the necessary support for such a project in nineteenth century Britain. Babbage had no illusions whatsoever about the position of the engines. With no expectation of public support or backing, the courage and determination with which he pursued the work in the face of enormous difficulties is very impressive.

It has hardly been noticed that Babbage's treatment by successive administrations inspired some years later the creation of the most famous

[13] *Phil. Mag. Sept.* 1843, 235.

government department in the history of literature, the HOW NOT TO DO IT OFFICE in *Little Dorrit*.[14] It would be unreasonable to suggest that Babbage's experience was the sole source of inspiration for that venerable institution: other examples of its activities occur daily, hourly, continually, and on an enormous scale, but certainly the treatment of Babbage for daring to develop a calculating engine was a primary inspiration. It is not only that Dickens knew Babbage very well; the people twisted up by the OFFICE give the clue: 'Mechanicians, natural philosophers, soldiers, sailors, petitioners ...'. The first two categories can hardly be accidental, even if the mechanician is Daniel Doyce himself.

In the world of fiction one should not look for an exact reflection of the real world, but the picture Dickens draws is recognizably that of Babbage who had been seeking an answer from the Government since 1834:

A dozen years ago, he perfects an invention ... of great importance to his country and his fellow creatures. I won't say how much money it cost him, or how many years of his life he had been about it, but he brought it to perfection a dozen years ago ...

He addresses himself to the Government. The moment he addresses himself to the Government he becomes a public offender! Sir, he ceases to be an innocent citizen, and becomes a culprit. He is treated, from that instant, as a man who has done some infernal action. He is a man to be shirked, browbeaten, sneered at ... he is a man with no rights in his own time or his own property, a mere outlaw, whom it is justifiable to get rid of anyhow, a man to be worn out by all possible means.

And what action of Doyce's raised such horror in the young Barnacle? Why, Babbage's own: he wanted to know.

Daniel Doyce is certainly not a direct portrait of Babbage: Doyce is a character with a life of his own in the novel. He has 'a certain free use of his thumb never seen but in a hand accustomed to the use of tools.' Babbage was certainly accustomed to the use of tools, though not perhaps possessed of lifelong familiarity with them likely to give the suppleness of thumb which Dickens has noticed. Although Doyce is a working man his character has other traits in common with Babbage's. As Mr Meagles puts it of Doyce, 'He is the most exasperating fellow in the world, he never complains!' Babbage's patience and resilience were remarkable. He never bored his companions with his problems. 'You see, my experience of these things does not begin with myself. It has been in my way to know a little about them from time to time. Mine is not a particular case. I am not worse used than a hundred others, who have put themselves in the same position—than all the others, I was going to say.' This is Doyce speaking. In another passage we have the following exchange:

'Disappointed?' he went on, as he walked between them under the trees, 'yes. No doubt I am disappointed. Hurt? Yes. No doubt I am hurt. That's only natural. But what

[14] It was actually noted by Dickens's biographer, John Forster, *The Life of Charles Dickens*, 542, London, 1928.

I mean, when I say that people who put themselves in the same position are mostly used in the same way—'

'In England,' said Mr Meagles.

'Oh! of course I mean in England. When they take their inventions into foreign countries that's quite different. And that's the reason why so many go there.'

Mr. Meagles very hot indeed again.

'What I mean is, that however this comes to be the regular way of our government, it is its regular way. Have you ever heard of any projector or inventor who failed to find it all but inaccessible, and whom it did not discourage and ill-treat?'

'I cannot say that I ever have.'

'Have you ever known it to be beforehand in the adoption of any useful thing? Ever known it to set an example of any useful kind?'

'I am a good deal older than my friend here,' said Mr. Meagles, 'and I'll answer that. Never.'

'But we all three have known, I expect,' said the inventor, 'a pretty many cases of its fixed determination to be miles upon miles, and years upon years, behind the rest of us, and of its being found out persisting in the use of things long superseded, even after the better things were well known and generally taken up?'

They all agreed upon that.[15]

After describing Doyce's experience and portraying the Barnacles, Dickens generalizes, spelling out the consequences of the OFFICE for the country's future. Although Babbage was articulate enough most of the engineers who thought like him were not. Here Dickens seems to be making himself their spokesman, as he did for so many oppressed or neglected groups in society: 'As they went along, certainly one of the party, and probably more than one, thought that Bleeding Heart Yard was no inappropriate destination for a man who had been in official correspondence with my lords and the Barnacles—and perhaps had a misgiving also that Britannia herself might come to look for lodgings in Bleeding Heart Yard, some ugly day or other, if she over-did the Circumlocution Office'.

During the period of his visits to Northern Italy and the final cancellation of the Government project to build the Difference Engine Babbage was seeing more of the Lovelaces, and Ada suggested that she might be able to help him in his work. She had the mathematical knowledge to understand some of his plans for making use of the Engines, sufficient leisure to devote time to the project, and she was longing to find some useful outlet for her talents. When so few cared about his engines, and fewer still understood them in the slightest degree, the combination of intelligent interest and charm was seductive. Cemented by her interest Ada and Babbage established a close friendship lasting until her death.

He was a frequent visitor to Ockham Park, to East Horsley where the Lovelaces moved after leasing Ockham, and more especially to Ashley Combe,

[15] Dickens, *Little Dorrit*, 1, ch. X.

their smaller house near Porlock. Babbage in turn executed many small commissions for her in London and discussed with her husband taxation, social reform, and plans for improvement of his estates. At a meeting of the Institute for Civil Engineers in 1849 Lovelace described a remarkable design for the roof supports of the new hall he built at East Horsley. They consisted of a set of 24 foot span laminated wooden arches. The younger Brunel considered the method superior to his own at the Temple Meads station in Bristol, in which the hammer-beam effect is merely decorative.[16] One suspects the Babbage touch: if not, Lovelace deserves more credit than he has been given for engineering skill.

Following a suggestion by Wheatstone, Ada Lovelace translated Menabrea's article on the Analytical Engine. Babbage suggested to Ada that she should add some explanatory notes. She liked the idea and her notes, which are much longer than the original article, give a unique insight not only into some simple mathematical calculations possible with the Engine and how to go about programming it, but also into Babbage's general views on its powers.[17] The translation and notes were due to be published in *Taylor's Scientific Memoirs* and by the end of July they were in the hands of the printer. At this point Babbage had second thoughts. The notes were so good that it seemed to him that it would be better if Ada were to redraft them, no doubt with detailed guidance by Babbage, and publish them as a separate article or booklet. He wrote to her and suggested the idea. Ada was furious: she had no intention whatsoever of having her first publication interrupted at this late stage; moreover she was afraid of endless delay. Material had been going backwards and forwards between her and the printer and she was busy with the final details of proofreading and correction. Worse, she found that Babbage had been altering her sentences. Byron's daughter was not going to tolerate that. She wrote him a sixteen page letter in protest. Her letter is a delightful mixture of firmness, cajolery, and good sense. Ada knew her Babbage, and knew quite well that he adored her. She made it perfectly clear that she had no intention whatsoever of permitting publication to be delayed. She said very firmly that she had Babbage's interests at heart; and he knew it too.[18]

She then went on to make two suggestions. Firstly, that she would place her pen at his disposal for some years, but he must undertake that there would be no repetition of the incident in which he had asked her to break ('be released from' was his phrase) her agreement with the publisher. Secondly, she said, she and Lord Lovelace had some scheme, unknown to Babbage, for financing the

[16] Stephen Turner, William Earl of Lovelace, *Surrey Archaeol. Col. LXX*, 108; *Proc. Inst. Civ. Eng. 8*, 282 (1849).

[17] Reprinted in *Babbage's Calculating Engines*.

[18] Ada Lovelace to Babbage, Ockham Park, Monday, 14 Aug. 1843. BL Add. Ms. 37,192, f 422.

construction of an Analytical Engine. Possibly she had in mind the fortune she would inherit when her mother died. As her mother never stopped complaining of ill health the idea is not implausible. But if the scheme were to go forward Babbage must undertake to place practical decisions in the hands of an agreed set of his friends. Who she had in mind is not clear: possibly the Duke of Somerset, who was also a friend of her husband, and an engineer such as Whitworth. Babbage would become effectively technical adviser. Again there was good sense in this: Babbage himself was to make similar proposals some years later for his plan for a second Difference Engine, although the proposal then was for the Government to finance construction.

She also had something to say about their different motives:

My own uncompromising principle is to endeavour to love *truth* and *God before fame* and *glory* on *each just appreciation*, and to believe generously and unswervingly in the good of human nature, (however dormant and latent it may often seem).

Yours is to love truth and God (yes, deeply and constantly); but to love *fame, glory, honours, yet more*. You will deny this; but in all your intercourse with *every* human being (as far as I know and see of it) it is a practically *paramount* sentiment. Mind I am not *blaming* it. I simply state my belief in the *fact*. The fact may be a *noble* and *beautiful fact*. *That* is another question.

The two last sentences seem to be a reference to Babbage's justification of the quest for lasting fame in his *Ninth Bridgewater Treatise*.

She went on to soften the blow:

Far be it from *me* to disclaim the influence of *ambition* and *fame*. No living soul ever was more imbued with it than myself.... And my own view of duty is, that it behoves me to place this *great* and *useful* quality in its proper *relation* and *subordination*; but I certainly would not deceive myself or others by pretending that it is other than a very important motive and ingredient in my character and nature.

Her slightly exalted tone is attributable to the effect of opium or alcohol or both. Babbage would have known how to discount such influences on his beloved interpretress. But on the following day he was apparently in Porlock and declined both offers. He knew the problems far too well to involve his friends in financing a calculating engine. It may also be that her work on the notes had been a strain for Ada. Babbage had had to carry out a major part of the calculations to help her. None the less it is a thousand pities that she did not write a book on the Analytical Engines. She had barely completed the notes when she herself realized that she could do far better starting again from the beginning: a common enough discovery. What a difference a coherent book on the engines published at that time would have made, especially one written under Babbage's guidance. None has ever been written and his work has remained greatly misunderstood to the present day.

The article and notes together give an outline of the concept of a straightforward Analytical Engine, showing in some detail how to proceed in constructing programs to work out a few simple mathematical examples selected by Ada from a number suggested by Babbage.[19] Babbage and his assistants had earlier carried out such calculations themselves. She repeatedly emphasized the great generality of the Analytical Engine, but the full significance of this generality has not always been understood. Its scope covered the whole of the science of operations, in which she included any process, excepting formal logic, altering the relation between two or more well-defined abstract entities. A happy analogy she made has become more widely known: 'We may say most aptly that the Analytical Engine *weaves algebraic patterns* just as the Jacquard-loom weaves flowers and leaves.' One example she gives of the Engine's powers is musical composition: 'Supposing, for instance, that the fundamental relations of pitched sounds in the science of harmony and musical composition were susceptible of such expression and adaptions, the engine might compose elaborate and scientific pieces of music of any degree of complexity and extent.'

Popular writers on science of the time were given to wild flights of fancy and most readers of Ada's notes would doubtless have considered such statements entirely fanciful. They were, however, strictly true. To an uncomprehending scientific world, which included the budding scientific bureaucrats of the time, the machines were toys or humbug. To Babbage and a handful of friends the scope of the computer, so evident in the modern world, was gradually becoming clear. Ada's notes give us some idea how far Babbage's mind was reaching even at that relatively early stage in the development of the Analytical Engines.

On 27 March 1844 Menabrea wrote to say that he had met Babbage's son, presumably Herschel, and had been astonished to find that the excellent notes added to the translation of his paper on the Analytical Engine were by Ada Lovelace.[20] She had signed them A.A.L. and he had been unable to think of an English mathematician whose name fitted. Beautiful, charming, temperamental, an aristocratic hostess, with the romance attached to her as Byron's daughter, mathematicians all over Europe thought her a splendid addition to their number.

The first phase of work lasted for about thirteen years, from 1834 until the end of 1847 or the beginning of 1848. Throughout the period Babbage kept up

[19] Babbage, probably some of his assistants, and possibly Herschel Babbage, had carried out similar exercises about five years before, and Babbage carried out the calculations for the Bernoulli numbers to save Ada the work. Ada was probably the third or fourth person in the world to write simple programs, if that is the right word for them. She worked under Babbage's careful guidance and they are student exercises rather than original work. Indeed there is not a scrap of evidence that Ada ever attempted original mathematical work; if she did it would probably have been in her mathematical scrap-book, long since lost. Ada's importance was as Babbage's interpretress. As such her achievement was remarkable.

[20] Henry Prevost Babbage, *Memoirs and Correspondence*, 9, London, 1910.

a wide range of scientific interests although he had little time to pursue them in detail. He continued to see a wide circle of friends but what remained of his family was breaking up. After the experiments on the G.W.R. had been completed Herschel Babbage became engaged to a girl, Laura Jones, of whom his father disapproved. Herschel and his younger brothers had met her with her family at Clifton in 1836 when they were there for the meeting of the British Association. Apparently she was from a lower middle-class family with no fortune of her own, or at any rate a very small one. Probably Charles found her dull. But he was reacting much as his own father had reacted to Charles's marriage to Georgiana a quarter of a century before.

Herschel persuaded James Maclaren, an old friend of Charles and Georgiana to mediate. Maclaren was a wealthy man, principal proprietor of the Carron iron works, much favoured by Nelson and Wellington as a supplier of artillery and famous for the carronade. Letters show Maclaren intervening tactfully, and they give us some idea of what was happening:[21]

Hursley August 11, 1839.

I am very glad you have agreed to be one of Herschel's trustees; it would have been too serious a charge for me without some member of his family. What do you mean to do about going to his wedding or asking them to your house? I told him I would stand his friend with you in the matter but I can only be so because I wish him to be happy. I cannot find any arguments in this case. You have undoubted cause to be angry with him. I am sorry when I look back a year or two and remember how very much you were attached to each other & now you are like to separate I fear for long. Of course we must receive them here for they have done nothing to offend us and we will your family. I suppose I believe M.J. [probably Laura's father] wished it from the first. Pray do what you can to keep up affection between H & yourself & believe me

Yours ever
J MACLAREN.

Evidently Charles declined to take the initiative in writing to Herschel for on 27 August Maclaren wrote again:[22]

I think you are justified in acting as you have said you will until the marriage, but you surely do not mean to cut H off for ever thereafter. A young man who makes a foolish match has no right to expect pecuniary assistance from his father, but he should not be treated as if he has made a disgraceful one nor his wife as if she were an improper character. There is not such a disproportion in rank or circumstance as would warrant such a course in the eyes of the world or any individual in it. Besides I do not think there is any very great chance of their becoming actually a burden on you . . . for his own sake and for yours pause, before you lose him, or at least his affections, for they cannot be recalled after a long estrangement. You cannot believe how painful and vexing this whole affair is to me.

Yours truly
JAS MACLAREN.

[21] BL Add. Ms. 37,191, f 214.   [22] Ibid., f 241.

Herschel Babbage and Laura Jones were married in Bristol on 10 September 1839. Relations were maintained but Babbage never liked his daughter in law. In 1840 Brunel took Herschel to make a survey for a railway in the Sardinian Kingdom. In the spring of 1842, when work on a railway from Genoa to Turin was planned, Herschel returned for three months to make a more detailed survey. When it was decided definitely that he should stay and work on the railway he was followed by his wife, sister, and a maid, and also by his second brother Dugald Babbage who went as Herschel's pupil. The party departed from Dorset Street to travel in a light carriage better suited to a trip in the park. Babbage helped stow packages in the carriage, regarding the unsuitability of the conveyance and general muddle with evident contempt. However they arrived safely beyond the Alps.

While working on the North Italian railways Herschel met some of his father's friends in Turin. Such introductions were valuable and it would appear that Herschel was a competent engineer. He became Brunel's representative and resident engineer for several Italian projects. He asked Babbage's old friend Sir John Herschel to be godfather to one of his children.[23] In 1844 Dickens met the Herschel Babbages in Italy.[24] Returning to England Herschel became an engineering inspector for the board of health. In 1851, on the recommendation of Sir Henry de la Beche, founder of the British Geological survey and an old acquaintance of Babbage, Herschel was appointed to conduct a geological survey for the colony of South Australia. After visiting engineering works in England to secure the latest information, Herschel and his family embarked on the *Hydaspes*, arriving in South Australia on 27 November 1851.[25]

Henry, the youngest son, would willingly have gone with Herschel to Italy in Dugald's place if the opportunity had been offered. Left behind in England while the other two brothers were abroad, he decided to enter the Indian service and start on a military career. Early in 1843 he left for India. His father bade him farewell in the library and did not see him even as far as the cab. No sentimentality! Henry remembered it all his life. His grandmother was heartbroken to see the last and youngest of the children going abroad. Well in her eighties she could never expect to see them again. Her daughter, Mary Anne Hollier, visited her frequently in Devonshire Street and was in the house when she died on 5 December 1844.

Babbage was continuing his round of scientific and social engagements. On 8 June 1842 Charles Wheatstone wrote to him about the atmospheric railway which Brunel was to attempt to use on the G.W.R. extension past Exeter:[26] 'I think you have not yet seen the Atmospheric Railway. There is to be another, and the last trial at Wormwood Scrubbs on Friday next at one o'clock. Professor

---

[23] Babbage/Herschel correspondence, f 308; Florence, 21 May 1844, RS.
[24] M. House and G. Storey *The Letters of Charles Dickens*, 4, 231, Clarendon.
[25] See *Dictionary of Australian Biography*.   [26] BL Add. Ms. 37,192, ff 98, 122.

Daniell and myself will be there. As Brunel has almost decided on introducing this novel means of locomotion into the Sardinian dominion your son will probably have something to do with it'.

The idea of using air pressure directly to drive locomotives had been patented by George Medhurst in 1810, before the steam locomotive had established itself. Medhurst's idea was to make the entire train a large circular piston thirty feet in diameter and drive it through a huge tube by compressed air. The effect on the passengers, if it had worked at all, would have been traumatic. Other inventors considered the more practicable plan of attaching a cylinder to the bottom of a railway carriage and placing the cylinder in a tube under the rails. The great problem with this plan was to keep the slot in the top of the tube air-tight while allowing the arm joining engine to cylinder to pass freely. In 1840 an experimental railway was laid out in which the seal was a continuous hinged flap of leather backed by plates of iron. Under the track, and sealed from the atmosphere by the leather flap, was a nine inch diameter tube which could be evacuated by a small steam-engine ahead of the train being driven.

There were obvious difficulties with the whole system. For example it was not easy to see how lines could cross each other, although Brunel claimed to have solved that problem with a simple mechanical contrivance, and shunting would have been difficult. At that time, however, the small trucks in use were simply manhandled on sidings and turntables so that shunting may not have appeared to the engineers to present any particular problem. Another difficulty was that there was no way for the driver to control a train in motion, because the power was in the hands of the pumping station ahead and the inefficient brakes then in use would have been hopelessly inadequate. However, the great power of the atmospheric propulsion system meant that trains could travel up far steeper gradients than with steam-engines, thus saving enormous sums when constructing railways through hilly country like South Devon. Brunel was impressed and decided to go ahead.

A line of atmospheric railway was also laid between Kingston and Dalkey near Dublin and it worked quite successfully, if occasionally dramatically, for a time. Brunel decided to use the system for the extension of the G.W.R. from Exeter to Plymouth. The line of the railway passed Babbage's quarry at Dainton and the South Devon Railway actually purchased a small piece of his land. Brunel had to reassure Babbage that it was so designed that he could remove rock if he wished to work the quarry: apparently he was not doing so at the time. There was actually a pumping station at Dainton, conveniently placed between Newton Abbot and Totnes, and sidings there were conveniently placed for the lime quarry if Babbage should at any time have wished to work it. For nearly a year atmospherically powered trains ran successfully between Exeter and Teignmouth. Unfortunately the leather rotted, and the rats nibbled at it until the entire seal needed renewing. Brunel faced the reality without evasion and returned to steam.

On 20 March 1844 Babbage was informed that he had been elected a corresponding member of the section for Political Economy and Statistics of the Academy of Moral and Political Sciences of the French Academy. Foreign academic honours poured in upon Babbage until they became almost a nuisance. Augustus de Morgan the mathematician wrote plaintively: 'Do set conf[ounded] long tail of yours to rights. I never saw such a conger of curt[esy] tit[les]. It is for the Astronomical Soc[iety] Mem[oirs].'[27]

The same year the Duke of Wellington asked Babbage to show the Difference Engine to the Prince of Prussia.[28] After the Prince had seen the Engine Babbage sent some further information, writing sadly: 'The Analytical Engine is too much in advance of my countrymen, scarcely one of whom comprehends its results or sympathizes with the author. The solitary exception—the Duke of Wellington—may however console me for their neglect.' Even so the extent to which the Analytical Engine entered the intellectual consciousness of the time can be seen in a letter dated 17 February from Elizabeth Barrett to Robert Browning:[29] 'Do you know Tennyson? that is with a face to face knowledge?... That such a poet sh^d submit blindly to the suggestions of his critics is much as if Babbage were to take my opinion & undo his calculating machine by it.'

One day Babbage received two letters from an agitated Ada Lovelace.[30] They concerned an unnamed Italian Countess. There has been some suggestion that she was Teresa Guiccioli,[31] Byron's last mistress, whom Babbage had met earlier. It would have been daring for Ada to seek her acquaintance, but not unlike her.

12½ o'clock
I am at 52 George St. & shall be with *you* in a very short time. Lovelace wants *you* to take me to call on Countss Italia-Italia; so I hope you know her address.

Then later:

3 o'clock
My dear Babbage,
I am in despair about not finding you this morning. I don't know what to do about Countess Italia ...

ever yours
A.L.

[27] 7 Jan. 1845, 7, Camden St. BL Add. Ms. 37,193, f 166.
[28] 1 Sept. 1844; Ibid., f 110. The date when the Engine was moved to King's is uncertain.
[29] E. Kintner ed., *The Letters of Robert Browning and Elizabeth Barrett Browning*, Harvard, 1968.
[30] 18 June 1864, Add. Ms. 37,193, ff 286, 287.'
[31] Doris Langley Moore, *Ada, Countess of Lovelace*, John Murray, 1977.

Certainly Countess Italia-Italia was a woman of the world. The Marquis de Boissy trailed after Countess Guiccioli, as he said, like a poodle. When she eventually married him he continued to boast of her liaison with Byron, even introducing her as 'Madame la Marquise de Boissy, autrefois la Maitresse de Milord Byron.'

We catch occasional glimpses of Babbage's social life in the reminiscences or diaries of his contemporaries. He met at Macready's on different occasions both Rachel and Jenny Lind. For 4 June 1846 Macready records:[32]

Went to Hampstead: on my way called on Stanfield who was just setting out for Cockerell's with May Stanfield. [Charles Robert Cockerell was a well known architect.] Arrived at North End, Cockerell's place—a very pretty villa, very charmingly situated to command a view of Hendon, Mill Hill etc. There was an elegant and beautiful assemblage on the lawn. Music and dancing—talking and promenading. I saw Cockerell and his wife who were glad to see me. Eastlake, Bezzi whom I was delighted to see, Milman and Mrs. Milman, Babbage, Chalon, Martin, Edwin Landseer, Knight, Lady Essex and Miss Johnstone, Maclise, Wyon, Rogers, Charles Landseer, Hart, Mulready, Campbell the sculptor, Thornburn, Blake etc. The scene was very brilliant.

Babbage's great friend MacCullagh, one of the few people who may really have understood his more general concepts for Analytical Engines, wrote approvingly from Trinity College, Dublin:[33] 'I suppose you are going on pretty much as usual—devoting your mornings to hard work [it should be remembered that mornings at that time went on until dinner] ... and your evenings to Society. It is a good mixture I am sure and the one gives relish to the other.' Babbage's son Henry noted in February 1842 that his father had at least thirteen invitations to dinners and parties for every day of that month, including Sundays.

In 1843 a short-lived society of authors was formed: there were several such attempts in the nineteenth century. Among its early members were Babbage's friends Hallam and Rogers, but they soon dropped out. It was probably at this time that Babbage, who was much addicted to founding societies, drew up his 'Outline of a plan for an Authors' Publishing Society':[34]

The principle of the Authors' Society is entirely commercial. Its object is to convey the manufactured article to the consumer as cheaply as possible. It will in no case enter into any speculations for the purchase of copyright nor will it advance money on credit to enable its members to print their works.

[32] William Toynbee *The Diaries of William Charles Macready*, 340, 4 June 1846, Chapman and Hall, 1912.

[33] Trin[ity College] Dublin. 25 Feb. 1842. BL Add. Ms. 37,192, f 53.

[34] BL Add. Ms. 37,195, f 54. Included among the Mss. for 1852. The earlier date is more probable and the Ms. was probably shifted on some occasion when it was being considered in conjunction with the later discussion of retail price maintenance. See John Clapham to Babbage, 30 April 1852, 142, Strand, London; Add. Ms. 37,195, f 52. For a discussion of this subject see *Authors by Profession* Vol. 1 by Victor Bonham Carter.

It will give every assistance to its members in the way of information and advice relative to the Ms, the printing, the paper and illustrations etc. It will be directed by a Committee who will employ a manager and establishment for the warehousing and sale of works of its members.

At that time it was common for an author to deal directly with the printer.

Next Babbage produced a more comprehensive draft:

Plan for a Society of Authors united for the general *protection* of *literary property*.

The want of union amongst authors as well as their want of capital has hitherto prevented them from deriving that full benefit from their works which under better management they might obtain.

The real expenses and risk of their agents in distributing intellectual property are small whilst the charges for agency are unusually heavy as compared with those for other manufactures.

The remedy rests with authors themselves. By union—by confining themselves entirely to the judicious management of their own property, to the maintenance of their admitted rights and to the steady but temperate application to the legislature of their own and other countries for such additional protection as is at once consistent with justice and sound policy, they may enrich themselves by a larger sale and supply the public at a greatly reduced price.

Increased wealth will have the same effect upon this as on all other classes of Society in England. It will give it greater respectability and a much more powerful influence. The stimulus thus supplied to the best intellects of the age will become productive of a higher standard of literary excellence and will make it the *interest* of authors to bestow upon their works that careful and laborious revision which can alone render them objects of enduring interest.[35]

However nothing came of Babbage's plan. Authors are far harder to organize than astronomers, statisticians, or civil engineers. Dickens himself had chaired a meeting of the abortive authors' society on 8 April, but he had castigated the whole plan, distrusting its founders, and considering authors unorganizable. It was possibly Babbage's plan that provoked Dickens to reply to him on 27 April:[36] 'Authors and Publishers must unite, as the wealth ... and business habits of the latter class are of great importance to such an end.' The Society of Authors which flourishes today was not founded until 1884.

Babbage occasionally attended concerts and the theatre, although he was never greatly devoted to either. In Berlin in 1828 he had felt daring in attending a concert of Mendelssohn's[37] and in May 1844 we find Mendelssohn himself invited to one of the Dorset Street parties.[38] On that occasion Mendelssohn had a headache after a six hour rehearsal and did not attend. After his experiences as parliamentary candidate for Finsbury Babbage had written a playlet, typical of

[35] BL Add. Ms. 37,195, f 56.
[36] Walter Dexter (ed.), *The Letters of Charles Dickens*, I, 516, Nonesuch, 1938.
[37] *Passages*, 202. Babbage attended the concert with Alexender von Humboldt.
[38] BL Add. Ms. 37,193, f 76.

the election squibs of the time: 'Politics and Poetry or the Decline of Science'.[39] It gives a picture of what Babbage imagined life would have been like if he had been elected. Turnstile is obviously Babbage, thinking of how he would have idled his time away in political life. The tory Lord and Lady Flumm tempt him from the House of Commons on the occasion of crucial divisions with offers of the pleasures of society, while his supporters at Finsbury demand government contracts in repayment for their support. Government appointments are given out by jobbing. Finally he loses his seat after wasting two years away from science. In the theatre his rather puritanical upbringing made it difficult for Babbage to enjoy scenes of strong emotion. He disliked tragedy and could not stand the reconciliation scenes in comedy, but he liked light entertainment and enjoyed the spectacle of the dance. Even so he was more interested in the mechanism of the theatre than in the plays themselves.

One evening he joined a party of friends in their box at the opera for a performance of Mozart's Don Giovanni. Bored with the performance he wandered behind the scenes where one of the stage hands, discovering his considerable technical knowledge, offered to show him round the theatre. They examined the ventilation system, the huge water tanks near the roof for use in case of fire, wandered through a maze of passages and steps, coming at length to a large dark area lit by dim rushlights and topped by a flat wooden ceiling. A bell rang summoning Babbage's guide to his post. The man gave directions, indicated a door, and disappeared into the gloom.

Calmly Babbage mounted an oblong platform six feet above the floor to survey his surroundings. There was a flash of lightning and he looked up to see a brightly lit opening in the ceiling the size of the platform on which he stood. Suddenly two devils with long forked tails jumped out, one at each end of the platform, and the platform began to rise: 'What do you do here?' asked one devil, a tricky question to answer on the spur of the moment. 'You must not come with us', said the second devil; to which Babbage's reply was 'Heaven forbid!' He saw a beam a few feet higher and at a moderate distance. As they came level it was still too far away. Waiting till they were above the beam he bent low to avoid the approaching ceiling and jumped, landing safely. He often wondered what would have been the correct thing to do if he had appeared on stage next to the Don.[40]

In 1845 Babbage had a stall at the German Opera House. At that time most London theatres were illuminated by gas, which had been used for stage lighting in the Lyceum as early as 1817. Gas lighting was amenable to effective and sensitive control by the prompter, and it can give quite subtle effects, but the light is subdued. One evening Babbage noticed that his companion's bonnet was tinted pink by the stage lights, as also were the programmes. The effect gave Babbage the idea that spectacular stage lighting effects could be achieved

[39] *Passages*, ch. XXII.   [40] Ibid., Ch. XX.

by using the brightest sources of light then available and modifying them to produce alterable colours. The two intense light sources available were limelight and a primitive form of electric arc-light. Limelight was invented in 1816 by Thomas Drummond but it had not come into regular use. An oxyhydrogen flame was directed onto a cylindrical block of lime which was slowly rotated to provide the flame with a fresh surface. Humphry Davy had shown an experimental carbon arc in 1808 but it too had failed to come into use, primarily because stable sources of electricity were not easily available. It seemed to Babbage that if either of these lights were focussed through coloured liquids in glass cells bright coloured lighting could easily be produced and used to form most varied effects. He conducted experiments and soon found a method that worked.[41]

A friend of Babbage approached Bernard Lumley, the lessee of the Italian Opera House (Her Majesty's Theatre), and the leading impressario of the time. He was interested and provided facilities for Babbage to make a series of experiments in the main theatre. Michael Faraday offered his counsel and assistance. It must have been quite the most scientifically high-powered stage lighting team in the history of the theatre. Not content with merely solving the technical problems, Babbage consulted with the *chef-de-ballet* and sketched out a vehicle to demonstrate his system.[42] This was his ballet:

ALETHES AND IRIS

Scene 1

The Temple of the Sun. Alethes, a Priest of the Sun determines to search for the book of Fate. An illuminated Missal descends on a golden cloud. Alethes learns that, with constancy and perseverance through danger, he may find in a distant sphere a spirit worthy of his love and that he may give to her renewed life.

Scene 2

Alethes descends into a glacier to search amongst the frozen waters for the object of his admiration. The dance of the frozen by the gnomes of the ice-berg in which Alethes the Priest of the Sun joins. NB This dance must be slow and gliding. After this Iris appears in a vision.

Scene 3

The ice-gnomes having disappeared and the vision of the Nymph having occurred in the last scene, Scene 3 opens with the conversion of the ice into fiery rock, the interior of the crater of a volcano. Varied lights will accomplish this. Gnomes of Fire appear and the Fire Dance on red hot lava is joined by Alethes. The gnomes disappear—the vision of his love reappears and Alethes gradually falls asleep whilst the fires cool amidst slow music.

41 Ibid., Ch. XX.; BL Add. Ms. 37,193, ff 204, 237, 249, 250, 254.
42 *Passages*, Ch. XX.; BL Add. Ms. 37,193, f 196.

Scene 4

Alethes appears asleep in a white marble palace—In the background is a tomb on which are black velvet cushions and a female figure in pure white is reclining. Alethes awakes and by magic powers revives the lady—Iris appears dead, but the blue gradually being turned off and pink light as gradually turned on she becomes warmed with life— Alethes then respectfully raises her; she descends from the marble and they dance, the marble palace having become also coloured.

Scene 5

Alethes and Iris enter accompanied by a train of nymphs in pure white. They form a group and dance round in the rainbow—Alethes and Iris in the centre—suddenly they form four groups each of a different colour—with each a central dancer.
These groups break up and again rejoin. This is the rainbow dance.

Scene 6

Alethes and Iris and their Nymphs dance—The latter retire. A small circular rainbow appears in the horizon. It opens and enlarges always touching the ground—The central space then opens disclosing the Nymphs arranged in order. Alethes and Iris join them and the rainbow again becomes a small circle.

Iris opens the book of fate and disappears. Alethes wakes from the spell and reads in the book of fate appropriate bad verse.

It would appear that a set of limelights were to be used as footlights. For one dance four lights, blue, yellow, red and purple were placed behind four large urns of flowers. Four groups each of fifteen danseuses in white costumes performed, one group to each colour. Occasionally a dancer would spring from one light source to another, resembling, as Babbage put it, a shooting star. To end the dance the sixty dancers formed a large ellipse, changing colours as they passed around. In the rainbow dance Alethes sees a light projected onto a back screen which appears to be approaching him. Starting as a small blue centre surrounded by yellow which was in turn enclosed by a red ring, the whole grew larger until it appeared as a huge rainbow.

This is all commonplace nowadays, but it must have been most impressive at the time and capable of much development by an imaginative impressario. After the rainbow dance there was to be an epilogue in the shape of a large diorama showing Alethes' travels:

1. A representation of all the inhabitants of the ocean, comprising big fishes, lobsters, and various crustacea, molusca, coralines, & c.
2. A view of the antarctic regions,—a continent of ice with an active volcano and a river of boiling water, supplied by geysers cutting their way through cliffs of blue ice.
3. A diorama representing the animals whose various remains are contained in each successive layer of the earth's crust. In the lower portions symptoms of increasing heat show themselves until the centre is reached, which contains a liquid transparent

sea, consisting of some fluid at a white heat, which, however, is filled up with little infinitesimal eels, all of one sort, wriggling eternally.

The wriggling eels were merely added to make fun of divines who preached the eternity of bodily torment in hell. Babbage was embarrassed because an ancestor of his, Dr Burthogge, a friend of John Locke, had written a book to prove the continuation of physical torment to eternity.[43] Babbage himself had advanced a much more subtle form of torment in the *Ninth Bridgewater Treatise*, a theory nicely based on Newton's physics. Some parts of the story can be related without difficulty to Babbage's own life: Iris to the lost Georgiana, and the volcano to his exploration of the crater of Vesuvius. The scientifically educational diorama at the end is interesting.

At the great experiment two fire-engines stood by while the danseuses cavorted in front of the bright and novel lights. Lumley was impressed but felt that he could not run the risk of fire: even if the theatre could be insured, the audience was irreplaceable.[44]

The date of Babbage's experiments, the beginning of 1846, is interesting. That year the Paris Opéra used a carbon arc to form a beam of sunlight. The arc was in the auditorium and would have needed a mirror or lens to focus it. However the limelight, which Babbage probably used, though he may have experimented with both systems, needs a mirror in any case and Babbage certainly had a focussing system. It was also about that time that limelight began to be used to produce effects like sunlight, moonlight, fire, water, and cloud. These ideas would derive directly from Babbage's work. Lumley went often to Paris and Babbage's experiments would certainly have been known in the London and Paris theatres at the time. A diorama using limelight for projection had been used by Macready to show views of the Continent for his pantomime of 1837–8, but he had abandoned it because the inventor wanted 30 shillings a night, which Macready considered extortionate. It seems probable that Babbage's foray into stage lighting helped to bring both carbon-arc and limelight into more common use in the theatre.

Towards the end of 1847 Babbage felt that his work on the Analytical Engines was drawing to a close. The main principles had been established as early as 1836 and were worked out in detail by the time of the Menabrea paper. Since that time Babbage had greatly simplified the whole system until he felt he could go no further. Using the novel techniques he had developed for the Analytical Engine he prepared a new design and set of drawings, together with the associated mechanical notations, for a second Difference Engine, lighter, simpler to make, and more rapid in action than the first. Jarvis was well able to work out the details of the notations and of the design, freeing Babbage from the work. At the end of 1848 he even started on what was to be a third

[43] *Passages*, 5.   [44] Ibid., 258.

Difference Engine, using some interesting methods of sequence control. These ideas were incorporated in a modified version of the Second Engine.[45] Babbage felt he had completed all that he had set out to do: indeed he had done far more. It has often been said that he failed to construct an Analytical Engine, but that is not correct. He was not attempting to do so: merely to carry out experiments, prepare plans and notations, and establish manufacturing techniques sufficient to satisfy himself that an Analytical Engine could be constructed if anyone chose to do so.

It would not be appropriate to discuss in this book the technical details of the Analytical Engine, but we may glance very briefly at the course which developments had followed. The main components of the Engines were described by Babbage in a paper written in 1837–8, a year and a half after he had introduced a punched-card input system (see Appendix B). The control system was most ingenious, using several barrels, or fixed stores, in conjunction. To alter the arithmetical operations performed by an Engine it was necessary to do little more than use different barrels, or what was in most cases the same thing, replace the rows of studs on each barrel. In modern terminology each row of studs was a word. Addition and subtraction took a few seconds, multiplication and division, some minutes. The other two operations principally considered were the extraction of square roots, and the method of finite differences on which the Difference Engines were based.

Babbage went to enormous lengths to make calculation as rapid as possible with his mechanical system, even developing a complex form of arithmetic with a carry at 5 as well as 0. He called this 'half-zero' carriage. He developed plans for a double engine, that could carry out operations simultaneously on two numbers each with half the number of digits which could be handled by the whole engine. Stepping numbers up or down the columns was carried out by pinions, but in 1843 he thought of the more elegant method of stepping by raising an entire column of wheels. The latter technique was later used in the 'minimal engine' on which he was working when he died. He planned a whole range of peripherals, including card reader and punch, line printer, and curve plotter. These could work locally or remotely (in the next room). He could effect several operations simultaneously, including transfer of numbers between store and mill while operations were being performed in the mill, thus taking one or two steps towards the modern technique of pipelining. At several stages he developed a 'great engine'—in modern terms a number-cruncher—and a simple engine. The achievement was matchless: the detailed drawings and notations reveal incomparable logical power combined with a mechanical ingenuity worthy of the great Da Vinci.

Inevitably the question is asked: if someone had chosen to construct an Analytical Engine at that time, could it have worked? The first Difference

[45] Ibid., Ch. XII.

Engine had worked perfectly when assembled and engineering techniques had improved greatly in the intervening years, so that what was novel in 1828 was not uncommon by 1848. Moreover Babbage was immensely experienced by then. So also was his head draftsman, Jarvis. Certainly Babbage, a good judge, was in no doubt that with the great strides that had taken place in engineering techniques construction of an Analytical Engine was entirely practicable. Careful consideration suggests that Babbage was right, that it could have been built if the necessary funds had been forthcoming.[46] The problem was financial. It may well be that such a project could only have been financed in the middle of the nineteenth century if Babbage had had at his disposal a great personal fortune, like that of Cavendish in the eighteenth century.

Be that as it may, the first phase of work on the Analytical Engines[47] had ended. This time, he told an American visitor, if the Government wanted his designs, by God they should pay for them.[48] Babbage was free to turn his attention to other interests.

[46] The question as to whether a working Analytical Engine could have been built is very difficult indeed to answer. Perhaps a cumulative tolerancing study of the manufacture of an Engine following one of the sets of plans would go some way towards answering it. However not only would such a study be difficult but it would not be easy to decide the values of the tolerances appropriate. We have little detailed knowledge of the studies which Babbage carried out in the period 1834–48, or of the models which he made; and in any case he would certainly have modified the engineering techniques as manufacture proceeded. Perhaps the best evidence that manufacture of an Analytical Engine was practicable is Whitworth's offer of liberal help in constructing an Analytical Engine as a contribution to pure science. Whitworth knew Babbage and his work very well since the time when the young Whitworth had worked on the first Difference Engine. It is highly probable that they had discussed plans for one or more of the Analytical Engines in some detail. Thus Whitworth's offer is strong *prima facie* evidence that manufacture of an Engine was technically feasible. See Babbage to Whitworth (draft) 9 July 1855, Add. Ms. 37,196, f 366.

[47] The generality of the powers of the Analytical Engine which Babbage already envisaged during this period may be illustrated from Ada Lovelace's notes to the Menabrea paper: 'Many persons who are not conversant with mathematical studies, imagine that because the business of the engine is to give its results in *numerical notation*, the *nature of its processes* must consequently be *arithmetical* and *numerical*, rather than *algebraical* and *analytical*. This is an error . . .. The engine might develop three sets of results . . . *symbolic* results . . .; *numerical* results . . .; and *algebraical* results in *literal notation*.' (Note E). The latter merely indicates Babbage's initial approach to non-numerical powers for his Engines. It is the 'symbolic' results that are important in connection with the question of the relation between the Analytical Engines and the stored-program computer. Ada had already observed (Note B): '. . . each circle at the top [of a column] is intended to contain the algebraical sign + or −, . . . In a similar manner any other purely *symbolical* results of algebraic processes might be made to appear in these circles.' See also discussion on pp. 242–5.

[48] P.C. 574, Francis L. Hawks papers, North Carolina State Archives.

14

○●○

The Great Exhibition

The highlights of the forthcoming years were the great exhibitions: London in 1851; then New York, Paris, and London again. They focussed the attention of the world on the products of science and the rapidly developing commerce of the industrializing nations. The exhibition of 1851 was a landmark in Britain. After a long period of economic difficulty; after the revolutions of 1848; after the agitation surrounding the poor law, the factory acts, and chartism; the great exhibition opened the era of mid-Victorian prosperity with a general celebration of the arts of peace.

The discovery of a new planet a few years earlier in 1846 had aroused much public interest in science. Planets were part of people's lives. Astrology was even more popular than it is today. With poor street lighting, and in the absence of radio and television, far more people knew the night sky. The position and apparent size of the planet later called Neptune had been predicted by John Adams in Cambridge and Urbain Jean Joseph le Verrier in France. Adams had tackled what was at the time a well-known problem; the anomalies in the orbit of Uranus, the remotest of the then known planets. The orbit had certain irregularities, or perturbations, which Adams was able to show could not have been due to the gravitational attraction of the known planets nearest to Uranus: that is to say the irregularities could not have been caused by Jupiter and Saturn. He proved, after the most extended and laborious calculations, that the irregularities were caused by a hitherto unknown planet; and he concluded with a prediction of the location and apparent diameter of the new planet. After twice failing to see the Astronomer Royal, Airy, Adams left his paper for Airy to read on 21 October 1845. Airy criticized the paper and delayed instituting a search for the new planet.

Meanwhile quite independently in France Le Verrier had carried out his own set of calculations and in his third memoir, which appeared on 31 August 1846, also predicted the position and size of the new planet. He sent the results to several astronomers who had powerful telescopes at their disposal. J. G. Galle in Berlin received Le Verrier's letter and two days later wrote to say that he had identified the new planet. This dramatic achievement of scientific theory and mathematical analysis, clear in its consequence to the widest range of people, did a great deal to make both government and the public conscious of the possibilities of scientific research in general. Le Verrier's discovery was acclaimed while Adams' earlier work remained unknown. It was first revealed publicly by Sir John Herschel in a letter to the *Athenaeum* at the beginning of October 1846, when a controversy about the priority of discovery broke out. Young and cautious, Adams had hesitated to publish the results of his calculations until his prediction had been confirmed.

Babbage saw immediately that the calculations would have been straightforward had an Analytical Engine existed, and moreover that automatic computation would leave little room for careless error. He wrote to Adams drawing attention to the idea: 'I ... cannot ... help half suspecting that if you had felt that confidence in your *arithmetical* results which the fact has proved they deserved even the most discouraging circumstances would not have prevented you from publishing the results of a theory of which you entertained no doubt.' Babbage also sent Adams a copy of the Menabrea paper with Ada's notes.[1] Adams replied: 'It would be difficult to overestimate the value of such a machine, even if it were confined to the particular case you mention, the elimination between any number of linear equations, as nearly all the great problems of astronomy ultimately reduce themselves to that form.'[2] Babbage also wrote to Le Verrier in a similar vein. However Le Verrier did not like the idea of automatic computation, too proud perhaps of his own laborious work. It is pleasant to be able to record that on 10 July 1847 Adams and Le Verrier met on the lawn of Sir John Herschel's house at Collingwood and became good friends.

In 1848 Babbage published a pamphlet on the principles of taxation. A small but characteristic work, it shows some of his opinions with particular clarity. A second edition of the pamphlet followed in 1851 and a third in 1852. All three editions are prefaced by a quotation, which Lovelace had apparently pointed out to Babbage, from Thiers' *History of the Consulate and the Empire*[3]—a tribute to Napoleon's financial wisdom. No sooner had Napoleon seized power than he proposed to the *Conseil d'État* the reimposition of the unpopular property tax; and he made sure that the tax should fall at the same time on all property and all industry to give the tax as wide a base as possible. It was the latter point that Babbage was concerned with: that there should be as few exceptions to taxation as possible. Once again one notes the influence of the Bonapartes.

Babbage was primarily concerned with Income Tax, however, a question much discussed in Britain at the time, and he was worried that too many exceptions were being allowed. His basic argument was simple. If there is an electoral system of one-man-one-vote, with the poor in a great majority, the

---

[1] 12 Feb. 1847; BL Add. Ms. 37,193, f 410.

[2] St John's College, 13 March 1847, Ibid., f 437.

[3] In reprints of the pamphlet on taxation (Kelley, 1971) the quotation from Thiers has been omitted.

electorate will vote for low taxes on themselves and high taxes on the rich, thus destroying private enterprise. Later, in 1865, he was to spell out this point in detail in his pamphlet, *Thoughts upon an Extension of the Franchise*. Babbage was by no means the only person in the nineteenth century to hold that democracy and capitalism were mutually opposed, but he put his ideas with unusual logic and clarity.

The utilitarian background of Babbage's ideas is as obvious as the ideas themselves now seem quaint. He attempted to establish a moral basis for taxation, concluding: 'We have now arrived at the principle, that each person ought to be taxed *annually* for the protection of his personal liberty and property during that year.'[4] So remote is this from modern concepts of social engineering, let alone from the problems of a Chancellor of the Exchequer desperate both to raise sufficient in taxes and to keep the economy afloat, that one marvels at the spacious and luxurious way in which the matter was discussed at the time. However Babbage was also very much a realist and he soon turned to practical considerations which sound modern enough:

Another evil resulting from a high rate of tax upon income is perhaps of more dangerous consequence [he had previously discussed tax evasion], from its being less open to observation. Its effect will be, the *transfer of capital to other countries*. No laws however stringent can prevent this consequence [Babbage knew his capitalists well], nor follow the transported capital to its adopted home, and there tax its annual produce. The injustice of a government taxing capital which it does not protect, would remove from the minds of its possessors the impediment of moral wrong; and the sagacity of commercial enterprise would soon place that capital far beyond the grasp of the most rapacious chancellor of the exchequer, even with all the aids which legal ingenuity could devise.

The evil effects of such an abstraction of capital might at first be almost imperceptible; they would be slow, but certain and cumulative. The impact would fall more heavily than before on the capital remaining at home, crippling the manufacturing enterprise of the country ... [5]

In the 1870s the export of capital became a dominating factor in the British economy, and its effect on British manufacturing is only too well known.[6] After 1848, the year of revolutions, Babbage, like his friend Menabrea in Italy and many others, abandoned any remnants of political radicalism: in the preface to the third edition of the pamphlet Babbage was calling for steps to 'check the advance of our career in the direction of Socialism'.[7] His stern demand that

---

[4] *Thoughts on the Principles of Taxation*, 10.   [5] Ibid., 16.

[6] Such, at least, has been the conventional wisdom according to J. A. Hobson and Lenin. Recently it has been questioned whether the net export of capital from Britain after 1870 was indeed significant. Even for before 1870 British investment overseas has been exaggerated (cf. D. C. M. Platt, *Econ. Hist. Rev.* 2nd. ser. **XXXIII**, 1). However, the flight of capital caused by punitive levels of taxation has certainly been a problem for recent chancellors of the exchequer.

[7] *Principles of Taxation*, Preface to the third edition, 15.

economic laws should be allowed freely to run their course was accompanied by an uncompromising fight against injustice. In the pamphlet he instances some horrifying examples of the miscarriage of justice against men unable to afford legal advice or even to bring witnesses before the courts, commenting: 'but who ever heard of such calamities happening to a rich man?'[8] A few months before the pamphlet was written a convict who had suffered transportation to Australia and served part of his sentence was found to be innocent. On his return to England the Government had presented him with ten pounds in compensation. About the same time some grossly overpaid offices connected with the Chancery Court were abolished: the weak government, anxious to conciliate powerful interests, had awarded the former holders of the offices pensions of more that £6,000 a year. Babbage repeatedly drew attention to such injustices.

Although he had much sympathy for individuals who fell victim to economic laws, his attitude to beggars was almost Brechtian. But what strikes one most is the thoroughness with which he investigated a whole series of cases.[9] When he had first come with Georgiana to live in London, he had inevitably encountered many street beggars with tales of distress. One bitter winter he began to make enquiries, soon forming two conclusions: firstly that wherever Babbage happened to be when accosted, the beggar always lived in a district of London remote from that place; secondly that while the beggars professed to want work, not charity, they always belonged to trades in which employment would have necessitated entrusting valuable property to them. Babbage decided to investigate. One man accosting him in Belgrave Square claimed to be a watchmaker with a starving family, unable to find work. Babbage obtained his address, which was in Clerkenwell, and the next day visited the house, finding no trace of the man. Three months later Babbage was accosted by the same beggar, this time on the Portland Road. The man had forgotten Babbage and told him the same story, giving the same address. Confronted with Babbage's evidence the man concocted another story, and a few days later Babbage followed that one up as well. It was of course no more true than the first.

He investigated many cases with extraordinary perseverance without finding a single example to disprove his conclusions. Professional begging was endemic in nineteenth century London and not something the innocent well-meaning middle-class citizen could cope with. His advice to people with a small amount of charity to bestow was to place it in the hands of a sensible and kind-hearted magistrate. Babbage was a friend of Dickens and of many reformers and could always be relied upon to support charity for the deserving poor. However it is likely that he believed in original sin, that he would have seen justice in the pronouncement that the sins of the fathers shall be visited on the children, even unto the third and fourth generation: people had free will and it was always

[8] Ibid., 6.   [9] *Passages*, Ch. XIV.

open to any man to improve his lot by perseverance and hard work. One must not forget that Babbage was also a friend of Sir John Guest and many a tough Midlands industrialist, and in the class battles that raged through the factories of Britain's growing industry Babbage's sympathies were engaged on the side of the industrialists. To the workmen he recommended savings banks, life insurance, placing the labour of the several workers of one family in different industries to mitigate the impact of fluctuations in trade, and similar palliatives. Improvement in the lot of the working man, he considered, was dependent on the play of the inexorable laws of the market, and in the long run on the march of science.

Babbage's household in Dorset Street included as a rule two servants, a married couple. There may also have been daily help. Inevitably in running a household with little time to attend to the details he had servant problems. On one occasion his sister wrote from Lynmouth, where she had moved after her husband's retirement: 'I am most truly sorry to hear of your losses and the shameful conduct of your servants, and I sincerely trust your present ones will prove more faithful; indeed in *former* days I found so *many* unpleasant circumstances in regard to servants, that I have long since ceased to regret that I have only one to trouble me ... '[10]

Babbage remained on the warmest terms with his sister and her family. Letters only begin to appear in the collection after his mother's death. She was no doubt the centre of family communication before that time, but many family letters have disappeared. He also remained on good terms with Georgiana's large family. For some years, from about 1834 until 1843, his brother-in-law, Wolryche Whitmore, made number 1 Dorset Street his London home, paying Babbage rent. In January 1844 Wolryche Whitmore wrote from his country house to end the arrangement, as he had few friends left in London. A story has been handed down in the Whitmore family that a slight coolness developed between the two men as a result of sharing a residence. Possibly so; certainly they were both growing older and travelling less. But when Henry Babbage returned from India on furlough and stayed with his father, he and his family went happily to stay with Uncle Wolryche over Christmas.

In the spring of 1849, as work on the second Difference Engine drew towards an end, Babbage sought to find other work for Jarvis, his senior assistant and draftsman. Babbage was in a position to approach virtually all the senior engineers for assistance: even so it was not a simple problem. He wrote to, amongst others, Robert Stephenson, who replied:[11] 'I shall bear Mr Jarvis in mind, but I dare say you are fully aware of the state of our profession just

[10] Lynmouth nr. Minehead, 2 July (probably 1853), BL Add. Ms. 37,195, f 319.
[11] 14 April 1849, 24 Gt. George St., BL Add. Ms. 37,194, f 256.

now—many leaving it, and many half starving in it.' So many romantic stories have been written about Victorian engineering that such statements serve as a valuable corrective. Some years earlier, when Babbage had enquired of I. K. Brunel about the prospects for a young engineer, Brunel had replied:

I would counsel you to dissuade (as I always do ...) your friend from making an Engineer of his son—starvation is quite as plentiful as with any other profession.... If he will be an engineer I believe he must go through *some* engineer's office— ... I do what I can to deter people and amongst other methods I adopt the plan (to keep them out of my office) of making them pay a large premium—£800—half down—and as this still leaves more applicants than I want I think next year of doubling it. There are fools enough still to pay.[12]

In August 1849 Babbage visited Paris,[13] staying at the Hotel Castiglione, to see the French industrial exhibition. While he was there he visited old friends and dined with De Tocqueville, who had an English wife. Babbage had known him many years earlier in England and they had many mutual acquaintances. One feels that the two men would have got on well together. They were both liberals with an aristocratic turn of mind, strongly anti-socialist after 1848, and deeply sceptical about the effects of mass democracy. Returning from Paris after a month, he went to stay at Ashley Combe. He continued to see the Lovelaces both in the country and in London. Both he and they were frequent guests at Dickens's table.

The great exhibition of 1851, was held in the specially constructed crystal palace. England invited the civilized world to meet in London and display in friendly rivalry its raw materials, the products of industry, and some of its arts. Lyon Playfair, one of the better early scientific bureaucrats, who played a leading role in organizing the exhibition, suggested that Babbage should be in charge of the industrial commission which was to be responsible for an important part of it. He would have been the ideal person for the task, but the old scientific radical was not wanted: he was alienated from much of the growing scientific bureaucracy, and the Government, still at odds with him over the calculating engines, would not have him on the commission. A mediocrity was appointed instead and, as Playfair later wrote, 'The Industrial Commission necessarily sank in importance and use'.[14]

In arranging the exhibition, Playfair divided manufactures into twenty-nine different classes. His plan was submitted to leading manufacturers and modified to meet their criticisms. Playfair wrote in his memoirs: 'This classification, the first attempted of industrial work, met with great success, and had the good fortune to be highly commended by Whewell and Babbage, both masters in classification ... France alone made some objections, as the French Commission

[12] BL Add. Ms. 37,192 f 471.   [13] BL 'Add. Ms. 37,194 ff 298, 301, 304, & 314.
[14] BL Add. Ms. 37,199, ff 162, 173, & 217.

had drawn out a logical and philosophical classification for itself.... I suggested that we should fix upon any common object, and see who could most quickly find it in an appropriate division. My French colleague had a handsome walking stick in his hand, and proposed that this should be the test.' Playfair found it under 'Objects for personal use'; the French commissioner under a subsection 'Machines for the propagation of direct motion'. It was agreed to work under the English classification.[15]

Playfair tells a characteristic story of Babbage during this period:

Another philosopher whom I frequently visited was Babbage, ... Babbage was full of information which he gave in an attractive way. I once went to breakfast with him at nine o'clock. He explained to me the working of his calculating machine, and afterwards his method of signalling by coloured lights.[16] As I was engaged to lunch at one o'clock, I looked at my watch, which indicated the hour of four. This appeared obviously impossible so I went into the hall to look for the correct time, and to my astonishment that also gave the hour as four. The philosopher had in fact been so fascinating in his descriptions and conversation that neither he nor I had noticed the lapse of time.[17]

Playfair records that he suggested to the Prince Consort that scientists should be made Privy Councillors, proposing Babbage and Faraday as the first scientific members.[18] However Babbage was as usual only prepared to accept if the award were directly related to his work on the calculating engines, and the idea lapsed. At the time he was in fact unaware that his name had been proposed and rejected for the leadership of the Industrial Commission: he merely felt ignored. Bitterly resentful of the slight, he decided to publish a book putting before the general public the ideas which he would have put before the official body. His work, *The Exposition of 1851*, was intended to appear before the exhibition opened; it actually appeared shortly afterwards.

In it Babbage gives an interesting history of the origin of the exhibition.[19] In France national industrial exhibitions had been held about every five years since 1798, and it was probably the French example which gave Babbage the idea of the industrial exhibitions held at the British Association meetings in 1838 and 1839. During the preparations for the Paris exhibition in 1849 the Minister of Commerce had actually sent out circulars to several chambers of commerce throughout France in order to test opinion about the feasibility of making the exhibition international. It had been felt that the time was not ripe. In England even a national exhibition was a novelty, and the proposal for a

[15] Wemyss Reed, *Memoirs and Correspondence of Lyon Playfair*, 116, Cassel, 1889.
[16] This is a mistake—not the only one in Playfair's memoirs: the whole point of Babbage's signalling system was the use of occulting or flashing lights with an easily memorized code.
[17] Wemyss Reed, op. cit., 155.
[18] Ibid. and BL Add. Ms. 37,194 ff 601, 617, & 622.
[19] *Exposition of 1851*, 23–5.

great international exhibition raised strenuous opposition. Babbage's description of the protests is still entertaining:

> The ministers could not of course commit themselves by publicly avowing their disapprobation of an undertaking commenced under such high auspices [i.e. under Prince Albert's patronage]. It might readily have been forseen that they would be averse to such a scheme, because whilst it was sure to give them a good deal of trouble; it would afford them no compensation in the shape of patronage.

> Those, however, who usually reflect and retail the opinion of the Government, were by no means silent; at first it was said to be Utopian, then ridiculous, then in the slang of official life it was '*pooh-poohed*'; at a later period, when great public meetings had been held and when public dinners began to give it an English character, the best speech which has yet been made on the subject, containing the far-sighted views of a statesman, was ridiculed as full of *German* notions, by coxcombs whose intellect was as defective as their foresight, and whose selfishness was more remarkable than either.

> Another class of persons, the Belgravians, though actuated by the same motives, were induced to join the outcry for other reasons. As soon as it became known that the locality of the building would be the southern side of Hyde Park, they represented that the park would be destroyed, and become utterly useless. As if a building covering twenty acres out of above three hundred and twenty, could prevent the people from enjoying air and exercise on the remaining three hundred . . .

> The Belgravians found other causes of complaint. They could not tolerate the mass of plebeians of all nations who would traverse their sacred square, and they threatened to destroy the London season by going out of town. When it was suggested to them, that in these days of agricultural distress, if they left town they might console themselves by letting their houses at a high price, they refused to be consoled.

> The Belgravians next consulted their '*medicine men*' who, seeing they wanted to be frightened, suggested to them that *some* foreigners were dirty,—that dirt in *some* cases causes disease. The Belgravian mind immediately made the inference that the foreigners would bring with them the plague; then they dwelt on sanitary measures, and on the danger to the public until they themselves became nearly insane.

> It was then suggested that the foreigners might become assassins by night,—or take military possession of London by day.[20]

There were also objections on the part of a protectionist lobby, afraid of the effect on British commerce, although in retrospect these objections were difficult to understand. But whatever the objections, once a large number of people in distant parts of the world had begun to participate it was difficult to draw back. Babbage makes the interesting observation that in England the working classes had been in favour of the exhibition from the start, and that an understanding of its advantages advanced slowly in society from below upwards.

Babbage had held a high opinion of Prince Albert ever since the time, a decade earlier, when he had inspected the Difference Engine in the fire-proof room in Dorset Street.[21] In 1842 Count Mensdorf, the Queen's uncle, who was visiting London had told the Duke of Wellington that he would like to see the

[20] Ibid., 27–9.   [21] *Passages*, Ch. XI.

Calculating Engine. Prince Albert had asked to join the party. Babbage had been none too pleased, expecting just another stupid aristocrat. Much to his surprise he found the Prince a receptive and intelligent listener knowledgeable enough to have some appreciation of what he was doing. Faced with the almost unanimous opposition to his work on the part of the British Government, support from anyone holding a high position came to Babbage as greatly needed encouragement. The fact that the Great Exhibition had as patron Prince Albert, the only man then in high position able and willing to back applied science, made Babbage doubly annoyed at being excluded.

A major feature of the plan of the exhibition as it was first announced was the promise of large prizes. One was to have been for £5,000, and the total amount £20,000. This munificent offer had raised high hopes, and many a working man made heavy sacrifices to complete a model of some instrument or scheme close to his heart. Unfortunately the offer of prizes was withdrawn after it had become widely known. From the inventions described in his book it would appear that Babbage had himself planned to enter two models in the competition:[22] one of his system of occulting or flashing lights for lighthouses; the other of a versatile machine tool which he designed as equipment for building the Analytical Engine. Such a multipurpose machine tool would have been of great value. In large workshops it was usually practicable and economic to have a specialized machine for each of the principal processes: drilling, routine lathe work, planing and others. Specialized machines are generally cheaper than more versatile machines and require less time to set up for a particular job. On the other hand small workshops might not be able to afford or accommodate more than one or two machine tools, and the advantage of being able to complete all, or nearly all, the operations required within a single workshop would justify the expense of a versatile machine tool. Babbage later planned to have two machines, a high-precision machine sufficiently accurate to meet the most exacting requirements, and a larger machine for heavier work. So crucial is machine tool design to engineering that the spin-off from this development would by itself have amply repaid the cost of developing an Analytical Engine. Had prizes been offered at the Great Exhibition Babbage might well have had such a machine constructed at his own expense. As it was, Babbage's plans for versatile machine tools went unrecognized for more than a century, like the many uncomprehended plans for Analytical Engines.

The greatest exhibit of all was the building which housed the exhibition. Designed by Joseph Paxton like an enormous greenhouse, light and airy, it was made by mass-production techniques and assembled rapidly on site. Paxton had been experimenting with conservatories and forcing houses, and a particularly fine example, the Victoria Regia house, had been completed at Chatsworth in 1849. Thus he came to design the Crystal Palace with an already

[22] *Exposition of 1851*, Ch. IX.

proven constructional system. The building was four times the length of Saint Paul's Cathedral, housing undisturbed some large elm trees. It covered eighteen acres of ground and was supported by 3,300 cast iron columns. There were 2,224 cast and wrought iron girders, and 1,200,000 square feet of glass were used. Perhaps most remarkable of all was the speed of construction: the tender was accepted on 26 July 1850 and the contract signed on 31 October; the first column was fixed on 26 September and the building finished, painted, and ready for the official opening on 1 May 1851. Babbage had one cavil: the building was erected half a mile from Hyde Park Corner. Had it been erected on a site between Hyde Park Corner and what is now Marble Arch a great deal of time and energy would have been saved, and he went on to estimate what the saving might have been.[23]

During construction there were terrible prophecies of disaster. Rumours circulated suggesting that condensed vapour, well-known in greenhouses, would damage the precious contents; that hailstones, as every farmer knew, would smash the glass; that such a building could never stand against a gale: it was at once sieve, lightning-conductor, and house of cards. Amongst the Jeremiahs, as one might have expected, was the Astronomer Royal, but all were proved wrong. Brunel and Stephenson both vouched for the technical soundness of the building, even though it had competed with Brunel's own design; and it was a triumphant success. System built, with its elegant simplicity and clean lines, the Crystal Palace, more than any of the exhibits it contained, was symbolic of the machine age.

Babbage made suggestions for facilities required in such an exhibition: a place to deposit umbrellas, sticks, coats, and bags; the provision of soap, water and clean towels. What is surprising, looking back, is that such facilities were not automatically considered. One of his proposals was more unusual: he suggested the installation of a light railway,[24] quietened by rubber insets in the wheels, both to give visitors a bird's eye view of the whole exhibition and to transport visitors from one part of the huge building to another. The concept has now been adopted for some really large exhibitions and would no doubt have been useful in the crowded Crystal Palace.

The question of attaching price-tags to objects displayed was a subject of much controversy. The idea had been mooted and then abandoned in the face of hostility on the part of the manufacturers. Some foreign countries published such information in their catalogues in any case. Babbage felt that much of the value of the exhibition was lost without the display of prices: certainly he could not obtain the statistical information so dear to his heart. Playfair claimed that hostility from the manufacturers was so strong that showing prices was quite impossible, but Babbage dismissed this plea as mere pusillanimity on the part of the commissioners.

[23] Ibid., Ch. XII.   [24] Ibid., 39–40.

Although the exhibition was intended to show the finest products of industry, Babbage's Difference Engine, arguably the finest product of precision mechanical engineering to date, although made twenty years earlier, was not on display.[25] That fact alone shows the extent of the prejudice which Babbage had to face. He contented himself with reprinting in his book both chapter XI from Weld's *History of the Royal Society*, which gave the story of the project, and also Augustus De Morgan's review of Weld's book in the *Athenaeum*. Babbage also devoted a few pages to outlining some of the mechanisms used in the Analytical Engines, a topic which had not been mentioned in the Menabrea paper or Ada Lovelace's notes.

In his book Babbage dealt with the long-standing quarrel between Sir James South and the Revd. Richard Sheepshanks, a man of remarkably persevering vindictiveness, which had started over the mounting for South's telescope (see p.146). Such was Sheepshanks's tenacity that almost everyone in the scientific community was involved to some extent. Although he had been warned that the repercussions might have an indirect effect on the first Difference Engine, Babbage had remained aloof, despising Sheepshanks and disdaining to notice his antics. However, Sheepshanks's friend the Astronomer Royal was brought into the controversy and Babbage eventually came to suspect that there might have been some connection between this extraneous affair and the government attitude towards the calculating engines. It does not seem probable. Lord Melbourne was not likely to have taken much notice of Sheepshanks or Airy, but neither was he inclined to support Babbage, any more than was Peel.[26] For South the consequences of the affair were unpleasant, alienating him from most of his scientific friends. For Babbage, Sheepshanks's effusions were sound and fury signifying nothing.

London took the fairy-tale palace to its heart.[27] When the exhibition finally opened its impact was enormous. Its sheer size, and the huge variety of produce

---

[25] Babbage also noted: 'In 1847, Mr Dangan nobly undertook at a vast expense to make an Exhibition in Dublin to aid in the relief of his starving countrymen. It was thought that the exhibition of the Difference Engine would be a great attraction. I was informed at the time that an application was made to the Government for its loan, and that it was also unsuccessful.' *Passages*, 149.

[26] Lord John Russell indeed consulted Babbage's friend Beaufort at the Admiralty, but it had little effect.

[27] *Exposition of 1851*, 62–3. Babbage wrote: 'the present beautiful building arose as if by magic. Among all the curious and singular products which the taste, the skill, the industry of the world, have confided to the judgment of England, there will be found within that crystal envelope, few whose manufacture can claim a higher share of our admiration than that palace itself, which shelters these splendid results of advanced civilization. The building itself was regularly manufactured. Simple in its construction, and requiring the multiplied repetition of few parts, its fabrication was contrived with consummate skill. The internal economy with which its parts were made and put together on the spot was itself a most instructive study.'

coming from all parts of the globe, were overwhelming and tended to still criticism. The exhibition was evidently the main subject of conversation, even though it plays quite a small part in the correspondence and memoirs of the time. Dickens noted that whenever he was asked if he had seen some exhibit he always said he had, to avoid having to listen to people's boring explanations. Many people had season tickets and visited the exhibition repeatedly. The Duke of Wellington went daily and Babbage, much in demand as expert guide, often accompanied him.[28]

The exhibits from abroad were excessively ornamental, and although raw materials and goods made by recently developed manufacturing methods were to be found it would have been difficult to see the Exhibition as a whole as a celebration of the machine age. The emphasis was on craftsmanship and decoration. Among the exhibits from many countries were: from Greece a bas-relief from the Parthenon, a carved cross in the old Byzantine style, and embroidered goods; from Russia, gilt-bronze candelabras, malachite orna-ments, a case of jewels; from North Germany musical instruments, embroidered portraits of Queen Victoria and the Prince of Wales, a plaster relief illustrative of northern mythology; from Holland, a diamond corsage, a silver ornamental tea-table, a case of military ornaments; from Belgium, statues and wood carving; from Austria, an elaborately carved bedstead, an eau-de-Cologne fountain, marble carvings, Radetzky in bronze, more marble and glass; from the German states of Zollverein, yet more ornaments and statuary. The large French section had a few industrial products, including a case of goods electroplated by Elkington's process, but the main effect also in the French section is of elaborate ornament: Göbelin tapestries, Sèvres porcelain. Portugal sent a tabernacle for a cathedral, baby linen made for the prince of Austria, barrels of snuff; Switzerland, a model of Strasburg cathedral, ornamental clocks and musical boxes. Italy sent some classical antiquities, a figure of Bacchus reclining, and—from the country of Leonardo!— a landscape in needlework. From Tunisia came tents, carpets, blankets, dresses and cushions; from Turkey, carpets, bottles of scent, dried flowers from Mount Hebron, embroidered saddles and trappings; from China, screens and ornaments. The United States sent a Sioux saddle and hunting-belt, and a model of the Niagara Falls, but also some reminders of the industrial age: a case of cheap American newspapers, a vulcanized rubber trophy, a model of the 'Rider' Iron Truss bridge; and the Colt revolver.

Outstanding in the colonial contribution was the huge Indian exhibition, which included a stuffed elephant with trappings, hunting saddles, arms, an ivory throne, a crown, jewels, the Durri-a-nor diamond, a silver bedstead, groups of illustrative Indian figures, pottery, rude tools of natives, and a statue of the Queen by a native artist. The other colonies showed ornaments in shell-

[28] *Passages*, 173.

work, palm leaves, a collection of minerals from South Australia—loaned, one imagines, by De la Beche. Canada sent a birch-bark canoe, sledges and carriages, linen and woollen goods, and chairs presented by the Ladies of Montreal to Her Majesty the Queen.

Dickens declared he had been twice and was completely 'used up' by the fusion of so many sights into one. He was not sure he had seen anything but the large fountain and an Amazon; but he managed to have the best exhibition story. A party of school-children had been taken to the exhibition:

> The school was composed of a hundred 'infants', who got among the horses' legs in crossing to the main entrance from Kensington Gate, and came reeling out from between the wheels of coaches undisturbed in mind. They were clinging to horses, I am told, all over the park.
> When they were collected and added up by the frantic monitors, they were all right. They were then regaled with cake etc., and went tottering and staring all over the place; the greater part wetting their forefinger and drawing a wavy pattern on every accessible object. One infant strayed. He was not missed. Ninety and nine were taken home, supposed to be the whole collection, but this particular infant went to Hammersmith. He was found by the Police at night, going round and round the turnpike, which he still supposed to be a part of the Exhibition. He had the same opinion of the police, also of the Hammersmith workhouse, where he passed the night. When his mother came for him in the morning, he asked when it would be over? It was a great Exhibition, he said, but he thought it long.[29]

Even in the British section there was far too much emphasis on luxury items. In the industrial section, playing a smaller part in the Exhibition than is often supposed, there was a remarkable lack of system amongst the mass of disparate items which had been brought together, from railway engines to Babbage's friend Snow Harris's lightning-conductor for ships, from Clayton's brick, tile and pipe-making machine to Jones's twin semi-cottage pianoforte. Whitworth's machine-tools stand out as purposeful, logical, and of decisive importance.

Babbage wrote contemptuously to Alexander Dallas Bache, superintendent of the United States Coast Survey, that the machinery exhibited was driven 'at railroad speed. Through this mistake it was not easy even for professional mechanicians to understand the structure of new machines:—the dust raised by their rapid motion rendered it difficult to *see* their structure and the noise made it almost impossible for the few who understood them to convey information to the many who were so anxious to be informed.'[30] In the hands of mediocrity the Industrial Commission had failed to give the clear scientific direction which industry badly needed.

During the next two decades of the mid-Victorian prosperity Britain could look back to the Great Exhibition with a feeling of satisfaction, confident in her

[29] Charles Dickens, *Letters*, 2, 325–6, Nonesuch, 1938.
[30] Babbage to Alexander Dallas Bache, 8 Aug 1854: collection of A. W. van Sinderen.

commercial superiority, the strength of her industries, and in their ability to take on the world. Only a few, like Babbage and Playfair, could see that all was not well and that fundamental weaknesses were developing. The slow development of labour-saving machinery for many industries, when compared with the United States, for example;[31] the lack of systematic application of science to technology; and also the glaring weaknesses in the provisions for scientific and technical education;[32] these were the signs which should have caused the greatest concern. But in the sunshine of those years it seemed wholly unreasonable to worry about a few little clouds. Over the drawing-rooms of the middle classes, unchallenged, and not unduly concerned about the ocean of misery upon which his affluence floated, Podsnap reigned.

[31] H. J. Habakkuk, *American and British Technology in the Nineteenth Century*, C.U.P., 1962.

[32] Professor Gowing has told the extraordinary story of Britain's failure to provide adequate technical education during the nineteenth century in the 1976 Wilkins Lecture: *Notes and Records*, 32, no. 1, July 1977. Her lecture might well have been subtitled 'The Road to Bleeding Heart Yard'.

15

○ ● ○

The Death of Ada : A Family Reunion

During the Great Exhibition Babbage placed in a top-storey window of his house a working model of a new signalling system which he had devised for lighthouses, flashing coded messages. Passing friends would write down the numbers they had seen being flashed and drop cards through the letter-box so that Babbage could see if they had read the messages correctly. The system of occulting or flashing lights, now used by lighthouses all over the world, was of all his inventions the most rapidly adopted: not in Britain but first in the United States.[1] The origin of Babbage's idea is a nice example of his systematic approach to discovery. He was continually formulating general principles, and then, stepping back, he sought to use them as working tools of thought. The principle which led to his signalling system was this:[2]

Whenever we meet with any defect in the means we are contriving for the accomplishing a given object, that defect should be noted and reserved for future consideration, and inquiry should be made—

Whether that which is a defect as regards the object in view may not become a source of advantage in some totally different subject.

Babbage described the actual development of the signalling system as follows:

I had for a long series of years been watching the progress of the electric, magnetic, and other lights of that order, with the view of using them for domestic purposes; but their want of uniformity seemed to render them hopeless for that object. Returning from a brilliant exhibition of voltaic light, I thought of applying the above rule. The accidental interruptions might, by breaking the circuit, be made to recur at any required intervals. This remark suggested their adaption to a system of signals. But it was immediately followed by another, namely: that the interruptions were equally applicable to all lights, and might be effected by simple mechanism.

Within weeks of the first night when the little lighthouse was flashing its messages from Babbage's window he received an enquiry asking whether the mechanism was secret. Quite the contrary, he replied, he was solely interested in having his idea investigated and applied to the common good. He prepared a short paper on the system and had twenty copies printed. A dozen were sent to the appropriate authorities in the leading maritime countries.

One copy of the pamphlet was sent to Louis Napoleon, the elected President of France. Babbage may well have met him in Rome during the winter of 1827-

[1] *Passages*, 452-6.   [2] Ibid., 452.

8, and they had certainly met in London during Louis Napoleon's visits. The letter accompanying the pamphlet was dated 30 November 1852. Three days later the coup took place and Louis Napoleon seized permanent power. Babbage felt that his timing had been unlucky as the President would have more pressing matters on his mind. However on 5 December he was delighted when a note arrived from an *aide-de-camp* saying that the President had lost none of his former interest in applied science. It was part of his imperial heritage. But whereas his uncle had successfully applied science to military ends, Louis Napoleon talked, and achieved remarkably little.

In September 1853 Babbage spent several weeks in Brussels at an international congress of naval officers held to discuss maritime signals and codes of practice. One evening the whole party visited Babbage's apartments to witness and examine his signalling system. The portable flashing light was exhibited from the first storey verandah while the party trooped along the boulevard to observe the signals from several distances. Later a paper which Babbage read on the statistics of lighthouses was published in the report of the Brussels congress.

Writing at a time of jealous military secrecy it is strange to observe how easily ideas of military importance passed from country to country in the nineteenth century. Although Babbage had sent a copy of his paper on occulting lights to Trinity House and exhibited his system in Dorset Street, and although it must have been known to virtually every scientist in the country, it was not taken up in Britain. The Russian representative at the Brussels congress, however, was interested and with Babbage's permission copied the whole of his pamphlet for transmission to St Petersburgh. Later, during the siege of Sebastapol, the Russians used Babbage's ideas to good effect, sending messages by flashing reflected beams of sunlight. The British forces, in contrast, suffered from the lack of an effective signalling system.

After the Brussels congress Babbage went to Paris for two months. During his stay in Paris he explained his system to the Ministry of Marine, the Ministry of Roads and Bridges, and to many of his friends. A year or two later Prince Albert heard of Babbage's system and asked to see the model lighthouse. However by that time it had been taken to pieces and the mechanism applied to other experiments.

The copy sent to the United States was considered by the North Eastern Lighthouse Board and Babbage's plan approved in principle. Congress allocated $5,000 and experiments were made to investigate the system in practical use. Henry Reed, Professor of English Literature in Philadelphia, conveyed a pressing invitation to Babbage to return with him to the United States and assist in the practical establishment of an occulting or flashing lights system. Babbage had long wanted to visit the United States and this seemed a good occasion. He investigated the possibility of giving a series of public lectures to publicise the Calculating Engines and defray the cost of the trip. They planned to travel in October when the weather was likely to be good, on the finest steamship in the United States merchant fleet, the *Arctic*. A month before he was due to leave Babbage changed his mind. The trip did not seem worth while for less than a year and he decided that he could not spare the time as his son Henry was returning from India on furlough. On 20 September 1854 Professor Reed and his sister embarked. On 27 September the *Arctic* collided with another steamer and sank. Professor Reed and his sister were both drowned. Babbage mourned their loss and congratulated himself on a lucky escape.

In the 1840s and 1850s the electric telegraph was being developed very rapidly. I. K. Brunel commissioned the first commercial installation along the line of the Great Western Railway between the terminus at Paddington and West Drayton. Thus the first telegraph was being made and installed while Babbage was conducting his railway experiments on the same line. This telegraph, which used a set of five needles to indicate the letters, was opened in July 1839. Seeing some of Wheatstone's apparatus gave Babbage the idea that he might use electromechanical switching instead of mechanical techniques for the Calculating Engines.[3]

Businessmen and the general public feared the lack of security of the telegraph. They soon realized that their fears were exaggerated, but governments required codes for telegraphy and they also needed methods for cracking other governments' codes. For one thing telegraphs could be tapped. Writing in code must be almost as old as writing itself: no sooner is there a method of recording or transmitting information which can easily be read than there is a need to conceal the information from prying eyes. A small coded cuneiform tablet, for example, exists dating from about 1500 B.C. which gives a formula for making glazed pottery: the scribe was at once recording and guarding his secret. The Arabs began cryptology proper and since then, although codes have often been associated with magic, many great names have been associated with cyphers and code-breaking: Roger Bacon, Chaucer, Leon Battista Alberti, and Francois Viète, pioneer of algebra.

Babbage was the outstanding cryptologist of his age, wholly without rival: he developed and used mathematical techniques far ahead of those in use at the time.[4] His creative approach to the mathematics of ciphers is reminiscent of his early work on the theory of functions, carried out when he was in his twenties. He obviously found breaking ciphers great fun. In its elementary form Babbage practiced deciphering at school, particularly enjoying reading the ciphers of

[3] Notebook, 18 May 1839. 'Yesterday saw Wheatstone's model for telegraph & his drawings for Multi[plication] Engine ... It would therefore require seventeen times as many units of electrical time as it does of engine time—But the unit of Elec[trical] must really be limited by the unit of escapement time.'

[4] No adequate technical discussion of Babbage's work on cryptology has so far been given. See however David Kahn, *The Codebreakers*, New York, 1967.

older boys, to their annoyance. He thought that any cipher could be broken given sufficient time and patience. In theory this is true, but the time needed can be extended to such a degree that even with modern computers ciphers can be devised which are in practice unbreakable. Babbage and his friends used to vie with each other in cracking codes which they had devised. He became an accepted authority, and so many people sent coded messages, challenging him to read them, that he was forced in self-defence to decline further challenges.[5]

Deeming analysis of the language of great assistance in breaking codes, Babbage had constructed a set of code-breaking dictionaries. One is again struck by the energy and thoroughness with which Babbage pursued his every undertaking. He had a good English dictionary remade as a set of twenty-six dictionaries. The first comprised all the one letter words; the second the two letter words, and so on to the few twenty-six letter words. Plurals of nouns, comparatives, and superlatives of adjectives, forms of verbs and so forth, were then added and a new edition of the twenty-six dictionaries was prepared. These dictionaries were arranged, as is customary, in alphabetical order as determined by the first letter of each word. He then had further editions made, arranged according to the alphabetical order of the *second letter* in each word. Further dictionaries were prepared according to other classifications. The work continued for many years and was never finished: indeed such a task is virtually unending. Any member of the family was liable to be roped in to help in deciphering. Babbage passed on his interest in ciphers to his youngest son, Henry, and a niece of Babbage's was still addicted when she was grown up.[6]

All forms of disguise fascinated Babbage. He was intrigued to meet the French detective, Vidocq, who had trained himself to walk with his height reduced by an inch and a half, a great help to disguise. Babbage even wrote a paper on picking locks. In the paper he also proposed a lock which he thought could not be picked. During the Great Exhibition he noticed a particularly splendid lock, exhibited with its mechanism displayed by the American lock-maker Hollis. The next time Babbage accompanied the Duke of Wellington to the Exhibition they discussed the mechanism with Hollis. He too thought he had invented a lock which could not be picked, and Babbage was pleased to find that they had both had the same idea. The paper on picking locks was never published. Possibly he feared that it would have become required reading for every literate thief and its techniques standard practice in the criminal world.

From time to time Babbage was consulted by members of the government and senior civil servants on a wide range of technical problems: although he held no official post and his views were considered by many to be heterodox and disturbing, he remained one of the greatest technical authorities in the country. On 16 January 1854 we find Sir Charles Edward Trevelyan, Assistant Secretary

5 *Passages*, Ch. XVIII.
6 Louisa to Babbage, 29 Portland Sq. 1 Nov. 1854, BL Add. Ms. 37,196 f 5.

to the Treasury, writing on behalf of the Lords Commissioners of Her Majesty's Treasury to ask Babbage's views on the selection of the scale for the detailed ordnance survey maps. Three proposals were being considered which may be paraphrased:

1st 24 inches to the mile for rural, and 10 feet to the mile for urban districts.

2nd 3 chains to the inch or 26⅔ inches to the mile for rural, and 10 feet to the mile for urban districts.

3rd Scales having a definite numerical proportion to the linear measurement of the ground to be mapped i.e. the scale of 1/2500 for rural districts and 1/500 for towns, as recommended by the Statistical Society of London, and the Statistical Conference at Brussels—these being the fractions then in use on the continent of Europe: that is to say, approximately 2′ ⅓″ for rural, 10′ 6⅔″ for urban districts.

Babbage recommended the third proposal. His reply has its usual modern ring: 'Doubtless each country would prefer a scale capable of being expressed in round numbers in its own smaller measures', but the use of a standard system of measures which could be common to all countries in his opinion outweighed the admitted inconvenience. 'There is one point I wish *strongly* to press upon their attention', he wrote. 'Whichever scale they may decide upon adopting, *every* part or single portion of each map ought to have engraved upon it a scale explaining English feet and also French metres.'[7] The advantage which would have been gained if Britain had adopted European standards and the metric system in the middle of the nineteenth century would have been very great indeed.

At this time the question of free trade in books, which had been such a controversial topic in Babbage's *The Economy of Machinery and Manufactures*, again became a centre of interest: in modern terms it was the question of retail price maintenance.[8] Babbage's friend Grote was appointed to a commission of arbitrators headed by Lord Campbell, and the Booksellers' Association agreed to abide by the commission's decision. A meeting of authors, politicians, and the more independent booksellers was called to prepare the case against price maintenance. Amongst the sponsors of the meeting were Charles Dilke, Babbage, Herbert Spencer, Richard Cobden M.P., and Charles Knight, the bookseller who had published *The Economy of Manufactures* more than twenty years before. Lord Campbell handed down a decision in favour of free trade: the Booksellers' Association collapsed and was not revived until near the end of the century.[9]

[7] BL Add. Ms. 37,195, f 432.

[8] Printed letter over the name of John Chapman, 142, Strand, London, 30 April 1852. Amongst the early signatories supporting a meeting to protest against the system of maintained profits were: Charles Dickens, Charles Knight, Babbage, Herbert Spencer, Richard Cobden M.P. and W. J. Fox M.P.

[9] V. Bonham Carter, *Authors by Profession*, Society of Authors, 1979.

In 1852 there was a last attempt to persuade a British Government to construct one of Babbage's Calculating Engines.[10] The initiative was taken by the Earl of Rosse, President of the Royal Society since 1848. An astronomer of some note, discoverer of spiral nebulae, he had been interested in Babbage's Engines from the start and was well acquainted with the recently made drawings and notations for the second Difference Engine. It was lighter in weight and faster in operation than the first, and, as a result of advances in mechanical engineering which had taken place, there would have been no great difficulty in finding an engineering firm to construct it.

Well aware of the importance of having a working Engine to make accurate tables, the Earl of Rosse asked Babbage if he would make a gift of the drawings and notations of the second Engine to the Government on condition that they would have the Engine built. Unwilling to spend more of his own time on the practical problems of implementing the project, Babbage munificently agreed. Lord Rosse advanced a sensible plan: he suggested that the Government should approach the President of the Institute of Civil Engineers to determine firstly whether it was possible to estimate the cost of constructing the Engine; and whether a mechanical engineer could be found willing to undertake the construction.

Babbage wrote a long letter to Lord Derby, prime minister at the time, outlining the history of the Engines. Lord Rosse forwarded Babbage's letter accompanied by one of his own, and enclosing letters of support from Sir John Herschel, then Master of the Mint, Adams, co-discoverer of Neptune, and James Nasmyth, one of the leading engineering manufacturers of the time. Nasmyth's point was that manufacturers had benefited through the advance in machining equipment and techniques resulting from the first Difference Engine by many times the cost of the project.[11] Lord Derby referred the question to his Chancellor of the Exchequer, Benjamin Disraeli. The author of *Sybil* was not likely to be sympathetic to a proposal to apply mechanism to activities hitherto reserved to the mind. Besides, Babbage was the founder of the London Statistical Society and one of Disraeli's often quoted phrases is: 'There are lies, damn lies, and statistics.'

Disraeli replied that the project was indefinitely expensive, the expenditure impossible to calculate, and its ultimate success problematical. He was ignoring the advice of the highest technical authorities and the reply was nonsense, merely indicating his dislike of the whole idea. Before Lord Rosse could take further steps Lord Derby's short administration was at an end. In his autobiography Babbage remarked in passing that the Analytical Engine could calculate the millions the ex-Chancellor had squandered. Babbage's final

[10] *Passages*, Ch. VII.
[11] James Nasmyth to Babbage, 22 June 1855, Bridgewater Foundry, Patricroft, Nr. Manchester, BL Add. Ms. 37,196 ff 249 & 251.

comment on Disraeli was: 'The Herostratus of Science, if he escape oblivion, will be linked with the destroyer of the Ephesian Temple.'[12]

In June or July of 1855 Babbage received an offer from Joseph Whitworth, who had worked on the first Difference Engine in Clement's workshops, to help in constructing an Analytical Engine. By then Whitworth was a wealthy and immensely successful engineering manufacturer, Britain's and the world's leading maker of machine tools. His firm could have made an Analytical Engine and Whitworth also offered liberal terms amounting to generous financial assistance as a contribution to pure science. However it would still have required government support and Babbage declined to approach the Government, writing to Whitworth: 'I have long given up hope of seeing an Analytical Engine constructed.'[13] Although certainly generous, Whitworth's offer may not have been entirely altruistic. When working on the Difference Engine in 1831-3 he had seen the effect of that project on the development of tools and he may well have anticipated a spin-off from an Analytical Engine. To see the full significance of this one must look at the dominating position which Whitworth had achieved in British industry.

Joseph Whitworth was born in Stockport, Cheshire, son of a schoolmaster. Thus Whitworth, unlike most of the working engineers, came from an educated background. He learned his trade in the leading workshops of the day, with Maudslay, Holtzapffel, and Clement. In 1833, when the project on Babbage's Difference Engine was stopped, Whitworth went to Manchester and opened his own workshops. A writer in the *Manchester City News* of 1865 recalled:

When Mr Whitworth started business in Manchester, in 1833, the best mechanics were accustomed, in common parlance to speak of a 'full eighth' of an inch, or a 'bare sixteenth,' the latter term expressing to their minds ... something like perfection in mechanical finish. Mr Whitworth's foot rule, on which he had the thirty-second parts of an inch marked, was regarded as a curiosity, and many did not hesitate to affirm that to work to such a standard was unnecessary refinement. How very different the case is now, may be understood ... when we state, that at the works of Messrs. Whitworth and Co. in Chorlton Street, the self-acting machines are made, adjusted, and fitted to the ten thousandth of an inch.[14]

During the Crimean war the use of interchangeable parts enabled Whitworth to complete the engines for ninety gunboats in ninety days because the individual parts could be made in different factories. This was something he had learned from the Difference Engine. With his lathes, planing machines and other equipment, many of which were shown at the Great Exhibition of 1851, he achieved complete domination of the machine tool industry. However, after the early 1850s Whitworth became very conventional in design and exerted a conservative and even stifling effect on industry. This is where the development

---

[12] *Passages*, 111.   [13] Babbage to Whitworth (draft), BL Add. Ms. 37,196, f 366.
[14] Quoted in Joseph Whitworth exhibition catalogue, Inst. Mech. Eng., 1966.

of the Analytical Engine could have had a decisive effect: Babbage was the only man in the country to whom Whitworth would listen on machine tool design. Thus making the special tools which Babbage had designed in the late 1830s and 1840s could have changed the central section of the British machine tool industry, the heart of engineering manufacture. What the Difference Engine had done for the industry in the later 1820s and early 1830s, the Analytical Engine could have done in the late 1850s and early 1860s.

Like Babbage, Whitworth had a healthy appreciation of the significance of developments across the Atlantic, and he made a remakable study of the rapidly developing industry of the country. His general conclusion, published in 1854 was:

The vast resources of the United States are now being developed with a success that promises results whose importance it is impossible to estimate.

This development, instead of being, as in former cases, gradual and protracted through ages, is by the universal application of machinery effected with a rapidity which is altogether unprecedented.

Upwards of thirty establishments visited in different parts of the States, and employing in the aggregate from 6,000 to 7,000 men, afforded direct evidence that the greatest energy and attention are brought to bear upon the manufacture of machinery.[15]

Like Babbage and I. K. Brunel before him, Whitworth despised officialdom and had his rows with the Government. In 1854 the Board of Ordnance requested Whitworth to design and construct machinery for manufacturing the Enfield rifle, then recently adopted by the army. He insisted on first conducting a thorough series of experiments in rifle design and ballistics. It is interesting to note that Babbage was thinking along the same lines and also urged a scientific approach to guns and gunnery on Benjamin Hawes, former M.P., now permanent civil servant at the War Office.[16] One sees again and again how small was the group leading the campaign for a scientific approach to technology and industry: the Brunels, Babbage, to some extent Playfair, Whitworth, to a lesser extent Fairbairn and Nasmyth, and Benjamin Hawes, who was married to I. K. Brunel's sister.

Whitworth designed a rifle with 0·45″ calibre and a curious hexagonal bore to give more effective rifling surfaces. It was found greatly superior to the Enfield rifle in accuracy, range, and penetrating power during tests in 1857. The Board of Ordnance rejected the 0·45″ calibre as too small for military use. In 1869 a new committee adopted the 0·45″ calibre as standard, but the hexagonal bore did not marry with breech loading which was also adopted. However the fame of Whitworth's rifle had spread and he received large orders from the French government. Whitworth also designed cannon with an hexagonal bore. In the late 1850s he made a gun which fired a shell capable of

---

[15] Joseph Whitworth, *New York Industrial Exhibition*, Ch. 1, London, 1854.
[16] Hawes to Babbage, War Office, June 1860, BL Add. Ms. 37,198 f 74.

piercing $4\frac{1}{2}''$ ships' armour, before deemed invulnerable. In the official trials on the Southport sands—the controversy was famous at the time as 'the battle of guns'—Whitworth's cannon was found in every way superior to conventional guns. However the Ordnance Board rejected them also as too novel.[17]

Whitworth contributed large sums of money to technical education and he was prominent on educational committees. In 1868-9 he founded the Whitworth Scholarships, open to young engineers for their further education, to help in 'bringing science and industry into closer relationship with each other than at present obtains in this country.'[18] This was all very much in the Babbage tradition.

In 1854 an article was published describing the recently invented opthalmoscope, the instrument through which an optician looks at the back of the eye.[19] The author mentioned that Babbage had shown him a working opthalmoscope in 1846 or 1847, which would make Babbage among the first to devise such an instrument. He probably made it at about the same time as his experiments on stage lighting. In Babbage's device a bit of plain mirror was fixed within a tube at such an angle that light falling on it through an opening in the side of the tube was reflected into the eye to which one end of the tube was directed. The observer looked down the tube and through some clear spots where the silvering was scraped off the mirror. It was simple but effective, just another of the brilliant ideas which Babbage scattered around in such profusion.

During the period between 1848 and 1856, although he was not working on it, Babbage kept the Analytical Engine very much in mind. He told friends he was seeking some way of earning sufficient money to finance construction of an Engine, and indeed considered several schemes. After taking expert advice he invested £10,000 in United States and Virginia stock, and lost the better part of it.[20] He considered writing a couple of novels, only to be advised that he would probably end in losing money in the venture. Instead he wrote some children's stories to amuse friends. He thought of building an automaton to play tic-tac-toe, and then exhibiting it; but he never made such a machine, although he did investigate strategies for winning games, including nine-men's-morris. Confronted with these desultory efforts it is difficult to take seriously the idea that he was trying to finance construction of an Analytical Engine. Rather he was seeking to keep up his courage, when he was well aware that in the absence of government finance construction of the engine was quite impracticable; though, of course, extra finance would always be welcome to fund his own research.

At the end of November 1852 Ada Lovelace died. Always attracted by a combination of beauty, intelligence, and *savoir faire*, Babbage had many women

---

[17] Joseph Whitworth exhib. cat.   [18] Ibid.
[19] H. G. Lyons, 'Charles Babbage and the Opthalmoscope', *Notes and Records*, 3, 1941.
[20] Babbage to Alexander Dallas Bache, 8 Aug. 1854, coll. A. W. Van Sinderen.

friends after his wife's death, but Ada, so devoted to the Analytical Engines, had played a quite special part in his life. Ada's last years had been a story of mounting tragedy, and her death chamber was a scene of horror. Over it presided the vulturine figure of her mother, Lady Noel Byron. As in the Byron saga itself, Lady Byron systematically destroyed all the evidence she could and the events have only recently and very partially been disentangled.[21]

Fairly soon after she had finished the translation and notes for Menabrea's paper on the Analytical Engine Ada's life began to change. What would have happened if Babbage had accepted the offer of her pen is necessarily a matter of speculation, but it is clear that she was beginning to feel that her life was dull and meaningless. After living a secluded life for some years she began to seek excitement in dubious directions. She was becoming bored with her husband, a kind and honourable man, but hardly romantic; and disastrously he was allowing himself to be dominated by her mother.

Babbage had introduced Ada into a wide circle of scientists and he had stimulated her interest in many branches of science besides the calculating engines. One of the men she met was Andrew Crosse, an inventor much interested in the practical applications of electricity and magnetism, such as the development of loudspeakers and telegraphy. He lived in an ill-kept house not very far from Ashley Combe: Fyne Court, Broomfield, also in Somerset. In October 1843, when Crosse stayed with his eldest son John at Ashley Combe for a few days, Ada was away comforting her mother in one of her imaginary illnesses. However Ada soon established a relationship with John Crosse.

The second complication in Ada's life was betting on horses. She had been fond of riding since childhood and was quite fearless on horseback. She adored going to the races and the first steps in betting are innocent enough and easily taken. Probably the habit grew slowly at first but she became in time a compulsive gambler. John Crosse was a gambling man and may have encouraged her. There is a story that Babbage was involved with her in developing an infallible gambling system based on mathematical analysis but there is not a scrap of evidence and it is entirely out of character. Even the rumour of Ada following such a system is based on unsupported testimony written down long afterwards.[22] Possibly she had a betting system like many others, but it was incidental: she was gripped by gambling fever. The 'book' which has been cited was probably Ada's mathematical scrapbook.[23] By 1848, whether through gambling, family expenses or both—she had a tiny allowance for a woman in her position—Ada was seriously in debt.

The company Ada was keeping at the races was thoroughly sordid: although

---

[21] Doris Langley Moore, *Ada, Countess of Lovelace*, John Murray, 1977.

[22] Mary, Countess of Lovelace, epilogue to *Life of Lady Byron*, by E. C. Mayne, Constable, 1929.

[23] Doris Langley Moore, op. cit.

her contact with low betting men could only be furtive and indirect, even that was dangerous for a woman in her position. How exactly she came into contact with these people is unclear, but gradually she became entangled with a low group who were placing bets for her: no lady could place bets herself at the time. The central figure was probably John Crosse. He lived a disreputable life, possessing a family whose existence he kept secret. Ada had become his mistress. She paid for some of the furniture for his family house and Crosse had compromising letters from her. At the time Lovelace knew nothing of all this. He had himself been gambling a little, but that was of no consequence for a country gentleman.

Babbage had recommended Mary Wilson, Ada's private maid, to her service. Naturally Mary Wilson would have been involved in most of Ada's private affairs, including the gambling: a trusted maid has to do as she is told and keep her mistress's secrets. It is possible that Babbage put Ada in touch with a man called Padwick who took some of her bets but behaved honourably, and he did all he could to control her gambling. It is also possible that as her difficulties mounted she turned in desperation to the capable and experienced Babbage: for there were few others to whom she could turn. In particular she dared not let her husband get wind of her dealings with John Crosse: her *affaire* and her gambling were inextricably mixed. By 21 May, Derby day in 1851, she was £3,200 in debt.

At the same time Ada was mortally ill with cancer of the womb. The malignancy was felt in 1850. Her health recovered sufficiently for her to enjoy a few weeks riding and climbing that year but there was soon a relapse. By the spring of 1852 she was in intense pain. At the same time her gambling debts were pressing. Lady Byron paid some of the debts and seized the opportunity thus provided. She turned bitterly on Lovelace for encouraging Ada's gambling. At the same time, almost unbelievably, she tried to stop Ada from taking opium as a pain killer. Lady Byron was prepared deliberately to countenance her daughter's pain, taking pleasure in increasing her remorse and repentance.

Desperate for money to pay her gambling debts Ada pawned the family diamonds. Her mother found out and redeemed them for £900. The gambling men must have been pressing hard: during a visit from John Crosse Ada managed to pass him the diamonds which he pawned a second time; and her mother again redeemed them. All this was unknown to Lovelace. The two men Ada wanted to see were Crosse and her old companion Babbage. On 12 August Ada wrote him a letter which she placed in his hands:

Dear Babbage,
    In the event of my decease before the completion of a will I write you this letter to entreat that you will *as my Executor* attend to the following directions;
1stly you will apply to my mother for the sum of £600; to be employed by you as I have elsewhere privately directed you. [Probably as bequests to her servants.]

2ndly you will go to my Bankers, Messrs Drummond's and obtain from them my account & balance (if any) and also all my OLD DRAFTS.

3rdly you will dispose of all papers & property deposited by me with you, as you may think proper *after full examination*.

Any *balance* in money at my bankers you will add to the £600 above named to be similarly employed.

In the fullest reliance on yr faithful performance of the above, I am

Most sincerely and affectly yours

Augusta Ada Lovelace.[24]

This document had no legal significance and Babbage begged her to make a legal will.

Thereafter Lady Byron took complete control, and dismissed Ada's servants including the faithful Mary Wilson. She forbade Babbage to enter the house, although Ada begged to be allowed to see him and the doctor also urged that Babbage should see her. Babbage told society what Lady Byron was doing. She even forced Ada to change her bequests. She had left Babbage twelve multi-volume works of his choice as a memento; drugged, dying and helpless she was made instead to leave bequests to her mother's friends, and to Babbage a single book. On Lovelace, distraught with Ada's suffering and deeply upset by his discovery of her *affaire* with Crosse, Lady Byron turned with remorseless bitterness. There remained many weeks of agony for Ada while her mother cruelly rejoiced in the time which Ada was gaining to repent of her 'errors'. One evening while she could still walk a little Ada made Lovelace promise that she would be buried at Newstead Abbey next to her father. In the end she repudiated her mother. It was fitting. Ada had the Byron temperament, and Lovelace, summoning some courage at last, saw that her remains went to Newstead where they were placed in the family vault.

After Ada's death there was a sorry mess of affairs. Compromising letters in Crosse's hands had to be purchased by her mother's lawyer Woronzo Grieg, Mary Somerville's son by her first marriage. Lady Byron ignored Ada's requests to Babbage and gave the servants tiny sums after making them sign humiliating statements. Mary Wilson alone refused to sign, no doubt on Babbage's advice, and received nothing. He must have given her a pension of £100 a year, as he later left as much for her in his will. Lady Byron, against the advice of her lawyers, asked Babbage to return Ada's letters. She received a cold statement in reply. Babbage declared that Lady Lovelace had been unjustifiably kept in her own house under surveillance and restraint, and that after the sixteenth of August she had been deprived of control even of her own servants; that in the circumstances no word or writing of hers was in any way binding on her friends; that she had foreseen such an eventuality and taken some precautions

[24] Lovelace Papers 138, ff 3 & 4, Bodleian Library.

against it—that is to say by writing the letter to Babbage on 12 August; that Lady Noel Byron's conduct had released him from any feelings of delicacy towards her; and that he intended to keep the letters.[25]

Such a row was bound to leave its marks. One of Lady Byron's coterie was Augustus de Morgan's wife, and de Morgan felt that he could no longer see Babbage. Lovelace· too, after all that Babbage had seen and knew, could no longer feel at ease with him and the two men ceased to see each other. Lady Byron humiliated Lovelace himself in every way she could. She refused to see him and turned the two younger children against him. Thus when his older son died little of his family remained to poor Lovelace. The more one learns of Lady Byron's behaviour the more sympathy one feels for Byron in that earlier and much more famous quarrel. For Babbage it was a sad end to his 'fairy lady', his beloved interpretress. Ada's death left a big gap in his life.

In 1854, after ten years in India, Babbage's youngest son, Henry Prevost Babbage, returned to England on a three year furlough. He had married in India and had one child. Babbage had made preparations to accommodate them, establishing a comfortable nursery with many conveniences on an upper floor, and they made Number 1 Dorset Street their home while they were in England. When Henry was a child living with his grandmother in a separate house, Babbage had been a rather remote and awe inspiring father. Henry had got to know him better during the month before leaving for India: returned from ten years military service, he could now meet his father on more equal terms. Babbage took to the whole young family. Slyly he placed a mirror on the sideboard so that he could watch Henry's wife Min without seeming to stare.

The years of the furlough provided a happy time of renewed family life in Dorset Street. Henry was promptly swept into his father's way of life, alternating between scientific activity and a busy social round. After dinner on the first day Babbage took Henry to an artists' soirée at University College. A few days later they dined with Sir Edward Ryan at his house, Garden Lodge in Addison Road. Captain Locke, a retired sea captain of the East India Company and an old friend of Babbage's, paid a visit to Dorset Street. Sir David Brewster was also there, and a few days later, after dining at Dorset Street, he went with Babbage to a Queen's ball. Henry accompanied his father to a soirée at Lord Rosse's; also on an admiralty barge for a visitation to the Royal Observatory at Greenwich. Then the Babbages returned by rail to another soirée at Lord Rosse's.

Henry does not appear to have been a particularly devoted scholar in his youth but he had only been back in Dorset Street for a short time when he started studying algebra, taking a book of problems from his father's library. A Mr Thwaites submitted a cipher to the Society of Arts in the hope of obtaining their medal and the Society asked Babbage for his opinion. To his great

[25] Doris Langley Moore, op. cit., Ch. 11.

pleasure Henry succeeded in breaking it. Henry also tried electroplating aluminium. The range of activities in the Babbage household was remarkable.

Babbage and Henry went on 13 October to see Brunel's *Great Eastern* being built: this leviathan was one of the sights of London. When Brunel had completed the *Great Western*, which established the Atlantic steamship crossing, he went on to design and build two more steam vessels, the *Great Britain*, and the *Great Eastern*, originally planned for the Australia run.[26] The *Great Britain* was an iron hulled, propellor driven vessel and its design established the main principles of ship construction for a century. But the *Great Eastern* was in a class of its own. The displacement of the *Great Western* was 2,300 tons and of the *Great Britain* 3,700 tons; the *Great Eastern* displaced no less than 27,000 tons. The *Great Western* was driven by paddles and sail, the *Great Britain* by screw and sail; the *Great Eastern* had all three. The entire country followed the progress of this monster ship. Unfortunately construction was placed in the hands of a plausible rogue called John Scott Russell,[27] who did much damage, wrecked Brunel's hopes, and in the end helped to kill him. The sideways launch in the constricted water of the Thames had always presented a difficulty, but it was held up because the directors would not allow Brunel to obtain the special hydraulic equipment he needed for the unprecedented task of launching a vessel of that size. When the great ship set out on its maiden voyage to New York in 1860 Isambard Kingdom Brunel was dead.

Over Christmas 1854 Henry took his family to stay with his uncle Wolryche Whitmore at Dudmaston, one of the few warm country houses of the time thanks to the central heating system, copied from Babbage's house in Dorset Street, which warmed the hall and passages. Early in January the Henry Babbages went on to Boughton where Henry had lived during the first years after his mother's death. Cousin John Isaac was now living there with his mother, Henry's aunt Harriet. Henry also saw again the old woman who had looked after him at the lodge a quarter of a century before. John Isaac was married to a sister of Edward Holland, one time member of parliament for Evesham, who lived at Dumbleton. Edward Holland in turn had married John Isaac's sister Sophy. After Boughton the Henry Babbages went on to Dumbleton to stay for a few days, doing the rounds of an extensive but closely knit family. Sophy herself had died and her eldest daughter presided over the household. Edward Holland was a farmer on a large scale and one of the main promoters of the Cirencester Agricultural College. Back again in London there was the usual round of soirées, parties and visits, but Henry was also working hard at trigonometry and geometry. Under his father's eye he seems to have felt a little guilty if he stopped working for a single day.

[26] J. B. Caldwell, 'The Three Great Ships', in *The Works of I. K. Brunel*, ed. Sir Alfred Pugsley, Inst. Civil Eng., 1976.
[27] L. T. C. Rolt, *Isambard Kingdom Brunel*, Longmans, 1967.

In the spring of 1855 an opportunity arose for Henry to play a substantial part in the story of the calculating engines. On 4 April he went to Somerset House to see a Difference Engine which had been made by a Swedish printer, George Scheutz, and his son Edward.[28] The Scheutzes had visited Babbage at Dorset Street in the previous year and seen Babbage's first Difference Engine at King's College in the Strand. George Scheutz's interest in calculating engines had begun when he had read Dionysus Lardner's article on Babbage's engine twenty years earlier in the *Edinburgh Review*. Fascinated he had begun to make a small difference engine himself, and nearly twenty years later he and his son had completed one that worked. Compared with Babbage's compact and elegant engineering, the Scheutz engine is like a piece of Meccano. Even so it was a fine achievement which had cost Scheutz heavy sacrifices and many years of unremitting toil. Now the engine was completed: it both calculated tables according to the method of finite differences and directly produced a mould in which type could be cast to print the tables. Although their machine worked on quite different mechanical principles, the Scheutzes were afraid that Babbage would be jealous, resenting their use of his idea. Instead they were warmly welcomed to Dorset Street. The greeting he gave to the Scheutzes' achievement and the enthusiasm with which he worked to ensure that their work received proper support was quite splendid.

Using his father's system Henry made some large mechanical notations of the Swedish engine. Drawn on Babbage's dining-room table they were intended as illustrations for public lectures. One drawing was $7\frac{1}{2}$ feet by 3 feet 4 inches; the other $13\frac{1}{2}$ feet by 3 feet 4 inches. They are now in the library of the Science Museum at South Kensington and would be creditable work for a professional draftsman. Henry attended the meeting of the British Association at Glasgow in September 1855. He was put on the committee of the mechanical section as a compliment and met Playfair, Tyndall, and others. In a lecture to the Association he explained both the mechanical notation and the operation of the Scheutz engine.

Meanwhile Babbage was visiting Paris for the Great Exposition of 1855 where the Swedish Engine was on display. The commission for the Crystal Palace exhibition was responsible for choosing the official British representatives, and Playfair had written to Babbage inviting him to be a member of the British delegation. After being overlooked during the exhibition of 1851 Babbage had no intention of being patronized by the commission now and he declined the invitation. He asked Fairbairn to make quite clear in Paris that the action was not in any way motivated by lack of respect for the French government. As a member of the Institute of France he could make his weight felt, and his intervention was instrumental in securing a gold medal for the Scheutz difference engine. The contrast between the warm reception of the

[28] The Scheutz Engine is in the Dudley Observatory, Albany, N.Y., U.S.A.

Scheutz engine in Paris in 1855 and the neglect of Babbage's engine in 1851 was striking. Babbage returned to Dorset Street on 15 November. The British Commissioners appointed to report on the Paris Exhibition made no mention of the Scheutz engine, nor indeed of any of the arithmetical machines which were impressively displayed.[29]

The Scheutzes were deeply grateful for Babbage's support. They wrote:

Inventors are so seldom found that acknowledge the efforts of others of identical aims, that your liberality in this respect has, as we hear, made éclat in the French scientific world. Respecting me and son we would not have been so much surprised, having had occasion before, during our stay in London, to learn at your house the true character of an English gentleman; although our admiration of it can only be surpassed by our deep sense of gratitude. We came as strangers; but you did not receive us as such: conforming to reality you received us but as champions for a grand scientific idea. This rare disinterestedness offers so exhilarating an oasis in the deserts of humanity that we wished the whole world should know of it.[30]

In May of 1856 Henry's large notations were exhibited at the Institute of Civil Engineers and he explained them to a meeting. The Swedish Engine was later purchased by an American and went to the United States. Before that a copy with minor improvements was made by the engineer Bryan Donkin which was used by the Government to make tables for civil purposes for some years. This copy is now on view in the Science Museum.

Henry brought news of the relatives whom Babbage, now they were growing old, saw less and less often. But the long furlough was drawing to a close. On 19 November they had a sad dinner at Dorset Street. Henry had had the chance to get to know his father well, and throughout his life the need to earn paternal approval was a powerful influence on him. On 20 November Babbage took Henry and Min to Waterloo Station, and stood on the platform watching the train until it was out of sight. For Babbage it had provided a happy family interlude in what was becoming, for all his social activities, a lonely life.

[29] *Passages*, 150.
[30] George Scheutz to Babbage, Stockholm, 22 Feb. 1856, BL Add. Ms. 37,196, f 418.

16

○ ● ○

Final Passages in the Life of a Philosopher

Ada Lovelace was dead, Henry and his wife were sailing back to India, the other two sons were in Australia, and life in number one Dorset Street offered a sad prospect. Babbage was nearly sixty-five and could have little hope of seeing any of his sons again. Undismayed he responded with characteristic vigour and launched a new project: within weeks of Henry's departure Babbage began the second phase of work on the Analytical Engines.[1] But it was quite different in character from the earlier work on his Engines. From 1823 to 1833 he had worked systematically on the Government-sponsored project to construct a Difference Engine—and come within an ace of complete success. From 1834 to 1848 he had worked on plans and feasibility studies for the new versatile Engines, and achieved everything that he had set out to do and more. In 1848 and 1849 he had completed plans for a simplified Difference Engine which could certainly have been constructed at the time if finance had been forthcoming. But after 1856 there was no such clear objective: Babbage simply could not stop working, and the Engines became a hobby for his old age.

Whilst flirting with the idea of constructing an engine, he was well aware of the difficulties that faced him: after his experience with the first Difference Engine and knowing the problems which the Scheutzes had faced, no illusion remained. A few years later he noted: 'The *constructor* of the *navy* might as well be required to *pay* for the building of a new ship he has devised as the inventor of the Anal[ytical] Engine to manufacture it.'[2]

The work follows his usual procedures: simplify and generalize. After a decade when the Engines lay fallow the simplification was particularly elegant and the generalization powerful. During this time Babbage produced the plans which come closest to the modern computer, embodying a range of extraordinarily advanced concepts. After the first eighteen months the work becomes much less systematic and unfortunately the drawings are not accompanied by sheets of mechanical notation to make interpretation unambiguous. It is not clear who was Babbage's draftsman: probably not Jarvis, as the style differs from the earlier drawings. The loss of Jarvis must have been a severe handicap.

The very first sketch shows an interesting new layout. The highways

[1] Detailed work, as shown by the notebooks, did not begin until well into 1857.
[2] Notebook, 6 April 1868.

connecting the columns of discs in the store are designed on an X–Y grid, a plan used in printed circuit boards of modern computers since the famous IBM system 360.[3] Babbage's new design was based on a main trunk highway which could be connected to any of the branch highways meeting the main highway at right angles. The layout is reminiscent of the plan of the passages in the great exhibition of 1851 and it is tempting to think of the idea dawning during one of Babbage's many visits to the Crystal Palace. The new layout led Babbage to make one of his most remarkable advances in computing techniques. He realized that each row of discs comprising the store could function as a separate engine while remaining under the same central control. On 6 April 1859 he observed:[4] 'thus it was seen that 18 separate Difference Engines each having ten differences and two adding wheels and two carriages could be worked at the same time ... If the op^{ns} [i.e. the small operations in the main mill] were required then 14 such Diff[erence] Machines might work in conjunction with them ... Next day I observed that instead of making differences these 14 racks and wheels might be used for any combination of the small operations.'

Although great caution is necessary in relating Babbage's Engines to modern computers this is clearly what is now called array processing, a standard technique for tackling problems, such as weather forecasting, which require enormous computing power. Those few lines are all we have on the subject and yet it is rare to have so precise a statement of the circumstances of an important technical development. In the general plan of 1858 from which this idea derived two 'great barrels' working together each control a set of 'small barrels', one set controlling the specialized axes of the mill, the other controlling the rows of store axes. No doubt in an array processor this arrangement would have been further developed: Babbage's engines make subtle use of distributed forms of microprogramming.

During this period of work Babbage was juggling with logical structures and word-lengths with much of the freedom of a modern computer designer, but assessment of Babbage's more general concepts for calculating engines is not straightforward. The sketchbooks, from which, in conjunction with the detailed plans and mechanical notations, we draw most of our information on the subject, were almost exclusively concerned with his detailed practical work on the Analytical Engines, containing merely occasional asides on more general themes. Indeed, but for the chance of Ada's notes we should have very little idea of his views on the general powers of Calculating Engines, and Ada's notes were written at a comparatively early stage in the development of the Analytical Engines.

[3] Anthony Hyman, *Computing, A Dictionary of Terms, Concepts and Ideas.* Arrow 1976.
[4] Notebook.

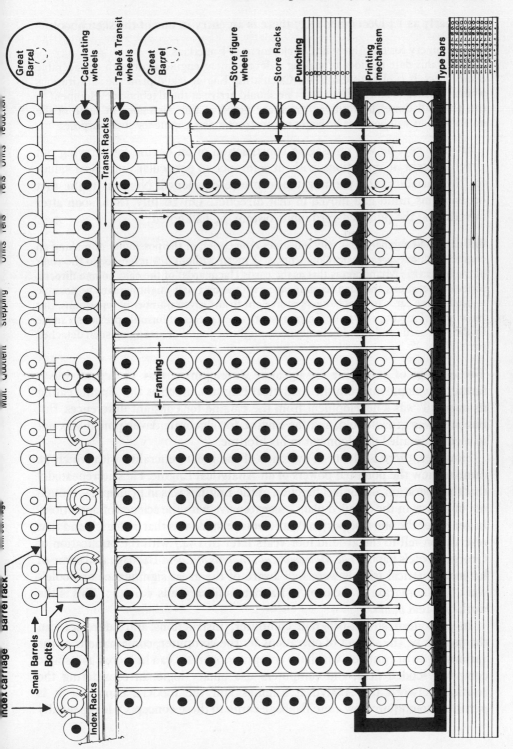

Fig. 2. Plan of Analytical Engine with grid layout, 1858. Redrawn.

As early as 13 December 1837 there is an entry in one of the sketchbooks:

On machinery for the Algebraic Development of Functions

About this date the idea of making a development engine arose with considerable distinctness. It is obvious that if the Calculating Engine could print the substitutions which it makes in an Algebraic form we should arrive at the algebraic development 'it can print all those substitutions which are noted in the *composition* in some of the notations'—it will however be better to construct a new engine for such purposes.

That is to say, he was envisaging construction of an engine for carrying out general algebraic operations, without reference to particular numerical values for the variables, and printing the formulae developed. Nor was it the first occasion his mind had moved in that direction. On 10 July 1836, soon after introducing the punched-card input system, he had noted:

this day I had a general but only indistinct conception of the possibility of making an engine work out *algebraic* developments . . . I mean without *any* reference to the *value* of the letters . . . My notion is that as the cards (Jacquards) of the calc. engine direct a series of operations and then recommence with the first so it might perhaps be possible to cause some cards to punch others equivalent to any given number of repetitions. But these hole[s] might perhaps be [i.e. represent] small pieces of formula previously made by the first cards and possibly some mode might be found of arranging such detached parts.

The concept of an engine for algebraic development may be regarded as a further development from the Analytical Engines, much as the 'Engine eating its own tail' was a development from the Engine for constant differences. But Babbage was growing old, and the Engine for algebraic development was not to receive detailed development in his lifetime.

However, there is plenty of evidence of the great generality with which he came to view the potential powers of an Analytical Engine. He said repeatedly that the operation, or control function of the Engines was in principle—though not of course in the detailed plans—coextensive with the science of operations. In her notes the one reservation Ada Lovelace made was that of 'logic'; and that reservation fell to the ground four years later in 1847 with the publication of George Boole's paper, *The Mathematical Analysis of Logic*. Babbage and some of his mathematical friends were well aware of the great significance of Boole's work. In the margin of the copy in Babbage's library was written: 'This is the work of a real *thinker*.'

In a modern stored-program computer, instructions and addresses can themselves be the subject of operations by the computer, combinations of the various types of entity on which the computer operates can be stored as a single compactly-coded field, and computers use binary logic. In considering the relation between the Analytical Engines and the modern computer it is pertinent to enquire how close Babbage came to these concepts. He repeatedly proposed using the Engines to make calculations on store addresses, sometimes with, sometimes without the use of variable cards, but one plan dating from about 1860 is particularly interesting. It shows the top ten wheels in each column used for purposes other than holding the main number. The top wheel was as usual to be a sign wheel, but the next six wheels held a store address, and the last three wheels of the ten were left undefined. Three ten-tooth wheels had the capacity to hold instructions, or to implement a floating-point representation, or indeed, with a restricted instruction set, to do both. After the introduction of the punched-card input-system in 1836 Babbage had developed methods by which computation could influence the program, both by advancing and backing the string of cards, and hence the sequence of instructions to the Engine, and also by controlling the number of times a given operation was repeated. If it were further admitted that Babbage thought of holding instructions in store, and then operating on them, we should have the basic principles of the modern computer, except for the use of binary logic. However, Babbage also stressed the fundamental importance of what he called 'the principle of the chain'. This principle is defined only by examples, of which the chain in his anticipatory carry system is the most straightforward, but it is clear what he was getting at: even if the principle of the chain is not the general binary operation realized in mechanical form, it certainly points in that direction. That Babbage's thought went much further in these matters than we have detailed proof of there can be little doubt, but he never spelt out his ideas on the subject; and in consequence his friend Wilmot Buxton, who later wrote on the Engines, felt debarred from attempting to do so. Thus the question of just how close Babbage came to the modern concept of the stored-program computer remains, tantalizingly, a matter for speculation.

Richard Wright was Babbage's head workman and there were several other men in the workshops. In the autumn of 1859 Wright fell ill, seriously interrupting the work:[5] both master and man were growing old. Sometimes Babbage himself put the work aside. Even so, besides conceptual advances of great interest, the practical achievements were substantial, particularly in the field of pressure die-casting.

Babbage decided to try and make a small working Analytical Engine to show at the Exhibition of 1862. Knowing he could not afford standard machining techniques to build the Engine he investigated new methods including stamping and moulding. But he was now in a new realm of difficulty: developing such technologies is an altogether more intractable problem than using machining methods. Babbage did indeed make some remarkable advances in pressure casting techniques but the problems of developing them sufficiently to make an engine were not entirely solved. Whereas it can be argued that Whitworth

---

[5] R. Wright to Babbage, 25 Sept. 1859, BL Add. Ms. 37,197, f 440.

and Babbage could have made a working Analytical Engine by machining methods in the 1850s had funds been available, it may be more reasonable to claim that Babbage's methods of the 1860s were technologically premature. Even so he made a successful carry-column with die-cast wheels in his last years which may be seen in the Science Museum. This was the method he was using to build a minimal Engine when he died. But first he investigated stamping and there was no Analytical Engine ready for the Exhibition of 1862. The Difference Engine assembled in 1833, although it had been ignored in 1851, was however exhibited. Babbage was furious to find the Engine placed in a corner where it could only be seen by two or three people at a time. It was demonstrated by the leading engineer Gravatt, and when he fell ill by Babbage's friend, Harry Wilmot Buxton, a Chancery barrister who had written a book on marine law but had known some mathematics in his youth. Babbage remarked: 'I remained incognito behind the Engine whilst he expounded it. But after two or three trials I came to the conclusion that he could explain it in a much more popular manner than I could so I gave him my key and had another cut for myself.'[6]

Later Babbage entrusted to Buxton the task of writing his biography, which was to include both the story of the Engines and some technical detail. An early draft of the biography survives, and together with papers Babbage lent him whilst writing the book form the Buxton collection of the Museum of the History of Science in the old Ashmolean building in Oxford. After the exhibition King's College refused to receive the Engine back. Thus it was transferred to the Science Museum where it remains on display together with the uncompleted Mill of the Analytical Engine on which Babbage was working when he died.[7]

During this period Babbage was conducting his celebrated battle with street musicians. Barrel-organs and other entertainments had come to be a public nuisance. They formed a particular problem for Babbage as the quiet neighbourhood into which he had moved in 1829 had degenerated and a nearby public house became a centre of trouble. He could not move because of the extensive workshops on which he depended, and besides it would have been out of character for Babbage to allow himself to be driven from his home. A

[6] *Passages*, 166.

[7] The Mill was partly constructed under Henry Babbage's direction after his father's death. This fragment, designed and built when Babbage was nearly eighty, is a tribute to Babbage's extraordinary continuing energy and intellectual vitality. However, its construction is irrelevant to the question of whether Whitworth and Babbage, with the formidable array of engineering techniques available, could have successfully constructed an Analytical Engine ten or fifteen years earlier had the necessary funds been available. See however, H. K. Barton, 'Charles Babbage and the beginning of die casting', *Machinery and production engineering*, 27 October 1971.

letter drafted by Babbage to the authorities after he had had a public row with a mob gives some picture of the problem:

The nuisance from Organs and Brass Bands is of much more recent origin and has continued to increase without intermission. About two years ago a cab-stand was placed at the Eastern end of Dorset St against the strong remonstrance of myself and all the householders. That stand is opposite a public house before which mountebanks by day and harp and fiddles in the evening go on sometimes until after eleven at night. Since the change several most respectable tradesmen long resident like myself in the district have been obliged to leave it and many houses ... have been converted into shops and low lodging houses. Some of my friends have even found the passage through it so disagreeable that they have ordered their coachmen when conveying them to my house not to pass along Dorset St.

If Babbage sought a constable to order an organ grinder to move on, it took half an hour to find one and the organ grinder was in any case soon replaced by another source of noise:

Amongst some thousand nuisances comprising: Organs; Brass-bands; Fiddlers; Harps; Punch [and Judy shows]; Pantomime, Monkeys, Military; Dancing and Musical; Athletes; Ladies and Gentlemen walking on stilts and looking inquiringly in at the Drawing Room windows; Hindu or Mohammedan impostors beating monotonous drums, or showing insanity; troups of Scotch imposters dancing with bag-pipes, even more inharmonious than the genuine instrument; it is obviously impossible for the householder to enjoy any quiet.[8]

The authorities offered to station a constable outside Babbage's door,[9] but there was little the law could do to help and Babbage's work was continually interrupted.

On one occasion an unpleasant neighbour paid a musician to perform outside Babbage's house. Babbage had some troublemakers jailed and received in consequence a letter threatening his property and his life. A recognizable figure, some of his antagonists even troubled him in distant parts of London. But Babbage was a formidable opponent even in old age. He fought back by every means at his disposal: the long chapter on Street Nuisances in *Passages from the Life of a Philosopher* was merely one stroke in his campaign. On 25 July 1864 'Babbage's Bill' became law: 'An Act for the better regulation of Street Music within the Metropolitan Police District'. Clause one stated: 'Street musicians may be required to depart from neighbourhood of house in case of illness or interruption of occupation of inmates, etc.' Offenders were liable to a forty shillings fine or three days imprisonment. The Bill was a real help to harassed residents. Soon we find Augustus de Morgan writing to John Herschel:

---

[8] Draft of letter to Sir Richard Mayne, 12 July 1860. BL Add. Ms. 37,198, f 90.
[9] BL Add. Ms. 37,198 f 95.

'Babbage's Act has passed, and he *is* a public benefactor. A grinder went away from before my house at the first word.'[10]

Feeling the end approaching Babbage began settling accounts with the world. He drafted his delightful reminiscences, *Passages from the Life of a Philosopher*. Too fragmentary properly to be called an autobiography, it is difficult to think of memoirs of another man of science which make such entertaining reading. Babbage gave a rather cool justification for writing the book: 'I have no desire to write my own biography, as long as I have strength and means to do better work ... The remarkable circumstances attending those Calculating Machines, on which I have spent so large a portion of my life, make me wish to place on record some account of their past history. As, however, such a work would be utterly uninteresting to the greater part of my countrymen, I thought it might be rendered less unpalatable by relating some of my experience amongst various classes of society, widely differing from each other, in which I have occasionally mixed.'[11] Actually he greatly enjoyed writing *Passages*, adding story after story until friends feared it would never be finished;[12] and Babbage's own enjoyment comes through clearly to the reader.

The work caused a couple of minor rows. John Murray was going to publish the book but on 21 June 1864 he wrote to Babbage:

My Dear Sir,
   When you did me the honour first to consult me about your book, you recited to me a certain epigram & asked me what I thought of it.
   I told you that I thought it grossly offensive ... I had hoped to have heard or seen nothing more of it but ... I regret to find this offensive epigram in print at page 370.
   I do not expect and I will not condescend to ask an author to alter any passage of his book to please me. I have therefore no alternative [but] to decline publication of the book, which I am glad has not proceeded so far as to prevent you making other arrangements.[13]

The offending section appears in the chapter on 'Wit':

The clever and eccentric member for East Surrey, the late Henry Drummond, who founded a professorship of Political Economy at Oxford, made in the House of Commons a most amusing, though rather strong speech against the modern miracles of the Roman Catholic Church, in which he spoke of 'their bleeding pictures, their winking statues, and the Virgin's milk.' On this some profane wag wrote the following couplet:

'Sagacious Drummond, explain, with your divinity:
Why reject the milk, yet swallow the virginity?'

[10] S. E. de Morgan, *Memoir of Augustus de Morgan*, 323–4, London, 1882.
[11] *Passages*, Preface.
[12] Countess Teleki to Babbage. BL Add. Ms. 37,199 ff 38, 66 & 263.
[13] Ibid., f 79. Babbage next offered *Passages* to Blackwood, who turned it down.

Probably some clever fellow of that faith was at the bottom of this mischief; for I have observed that the cleverest fellows seem to think that the merit of adhering to a cause entitles them to the right of quizzing it.

I was particularly struck with this idea when I saw, for the first time, at Cologne, the celebrated picture of St Ursula and her eleven thousand virgins. The artist has quietly made every one of them more or less matronly.

Fairly harmless stuff, one might have thought. However *Passages* appeared under the imprint of Longman, Green. The other minor snag occurred when Sir James South raised a fuss over some of Babbage's comments about Humphry Davy.[14] De Morgan thought South was piqued because he was not mentioned in the book.[15]

It must have been about this time that Babbage asked Wilmot Buxton to write the biography and the story of the Calculating Engines. Babbage was himself preparing a 'History of the Analytical Engines.' In *Passages* it is referred to as 'in the press'. The 'History of the Analytical Engines' was never finished although much of it was printed during Babbage's life. Part was published by Major General Henry Babbage after his father had died and titled *Babbage's Calculating Engines*. The major section, which was not finished, was probably intended to give an outline of the technical history of the Engines. This is a serious loss as we might have had what is so conspicuously lacking: Babbage's mature views on the general powers of the Analytical Engines. It is the lack of such a statement that makes a definite comparison between Babbage's achievement and the modern stored-program computer so very difficult. A clear statement by Babbage on the powers of the Engines might have led to earlier recognition of his achievements, and even conceivably to earlier development of the modern computer.

There were many letters in these years stirring the embers of old interests. The mathematician J. J. Sylvester became interested in the mechanical notation and was thinking of teaching it.[16] He visited Dorset Street for a discussion with Babbage. In the autumn of 1862 Babbage met for the first time George Boole, the mathematician who developed the form of mathematical logic, Boolean Algebra, on which the logical structure of modern binary computers is based. Babbage explained the working of the Difference Engine and persuaded Boole to read Menabrea's paper which he had not previously seen.[17] It was unfortunate these two men did not meet earlier. There were personal links which might have effected the meeting, including Augustus de Morgan. But Babbage had long been a mandarin figure seeing little of young mathematicians and scientists.

Occasional glimpses of life in Dorset Street filter through. Babbage continued in close touch with his sister, and she and her children stayed with him when

---

[14] BL Add. Ms. 37,199, ff 132, 162–72.
[16] BL Add. Ms. 37,199, f 495.
[15] S. E. de Morgan, op. cit., 335.
[17] BL Add. Ms. 37,198, f 414.

they were in London. There is a draft of a letter, probably to his friend Jane Harley Teleki:

My lonely household has been relieved of some of its dreariness by the arrival of a fair young creature who gives me a joyous greeting every morning at my breakfast table. She sits quietly by my side whilst I am working in the drawing room and in the evening delicately reminds me that it is time to retire to rest by saying 'Polly wants to go to bed' on which I ring the bell and the servant covers up her cage with a curtain whilst I dream of another far away.[18]

Countess Harley Teleki, née Jane Bickersteth, daughter of Babbage's old friends, was a close friend of his last years, filling to some extent the gap left by Ada Lovelace's death. The Teleki family were Hungarian aristocrats who lost their estates during the 1848 revolution. Harley appears in her name because after the death of her father, Lord Langdale, her mother, eldest daughter of the Earl of Oxford, returned to her maiden name of Lady Jane Elizabeth Harley. Another close friend of Jane Harley Teleki was Panizzi, great librarian and creator of the British Museum reading room. Besides friendship with the Langdales, Babbage and Panizzi shared a close interest in the Italian liberal movement. Panizzi played an exciting and dangerous part in the secret societies in his youth and was associated with, if not a member of, the Carbonari. This was one thing Panizzi had in common with Louis Napoleon whom he also knew quite well. Panizzi had a flaming row with the Royal Society over cataloguing scientific books. The Royal behaved with a stupidity worthy of the Society Babbage had castigated in *The Decline of Science* and Panizzi won completely. No doubt Babbage sympathized. Jane Harley Teleki continued her parents' close association with the two men in their old age. Unable to stick the English winter she went to the Mediterranean, well provided with introductions by both Babbage and Panizzi. We find her in Turin in April, 1863, for only twenty four hours while returning to England. On receiving Babbage's note Menabrea promptly left his office to spend the evening with her. She died of cholera in Damascus in 1870 leaving Babbage and Panizzi greatly distressed.

Another good friend of Babbage's was Margaret, dowager Duchess of Somerset, also a friend of Jane Harley Teleki. Occasional letters from Jane to Babbage give us a glimpse of Babbage's social life.[19] On 5 April 1863 she wrote from Paris, 'I am quite shocked at the notion of your dividing your favours between the rival Duchesses in Park Lane [Somerset] and at the Admiralty but I should not think of mentioning the fact.' And again on 19 April from 10 Lownes Street: 'Are you aware that your cruel infidelity in "butterflying" at the Admiralty has made the Duchess in Park Lane ill? I called at her door yesterday, but found no admittance—her grace could not receive anyone owing to a bad cold?!'[20] However all continued happily as we find Margaret

18 Ibid., f 503.   19 Ibid., f 507.   20 Ibid., f 515.

Somerset writing on 2 October 1865: 'Pray come and meet the Turkish Ambassador—& the Spanish Ambassador just arrived & his charming lady—and be here at Dinner *next thursday* 5th of Oct at 8 o'clock and assist at a *magnifique haunch de venison* sent by our excellent friend Lord Dalhousie. Pray send a favourable answer to yours ever . . .'[21]

Harry Wilmot Buxton gives us a description of Babbage in old age.[22]

[He] sought and cultivated the society of educated women, in whose elegant accomplishments and lively conversation he endeavoured to temper the severe studies of his ordinary pursuits ... [He] could relish the *badinage* of the *salon* with all the enthusiasm of a man unused to the abstract theories of pholosophy, or the thorny paths of science.

He was, however, never flippant in his converse, though he delighted in refined banter, which in his hands was ever innocuous; and though his satire might be occasionally pungent, and always keen, still never intentionally wounded the most susceptible, or willingly gave pain to the most sensitive object of his raillery.

He, in fact, possessed an exquisite sensibility, and his affections, when once aroused were warm and generous. He was a steadfast friend, though tardy and cautious in the formation of his friendships.

He was little influenced by the opinion of others, and never hesitated to place himself in antagonism to vulgar clamour, when his own convictions pointed in that direction. If he believed in a principle, he would defend it in spite of opposition, even if the whole world were in array against him. ... in the instances in which he thought he had detected injustice, he was not so moderate in his indignation, nor did he think he was justified in withholding his severest censure, where anything like dishonesty or vain and empty pretension assumed the garb of philosophic teaching. ... he not infrequently drew upon himself some degree of enmity by the bold expression of his views, for he never abstained from the publication of his sentiments, when he thought that his silence might imply his approbation; nor did he ever take refuge in silence when he believed it might be interpreted as a sign of cowardice.

His colloquial instincts never failed him, and a desire to please was ever conspicuous in his intercourse; and if he occasionally approached severity in his criticism, or now and then gave his remarks a touch of causticity; it was only transient and superficial; for he nourished no malice in his heart; and what might appear to one who did not know him intimately, malignant, was at most only a playful impulse of caustic humour; for malignity found no place in his character and was altogether foreign to his nature. ...

He continued to reside in London except at rare intervals when he was induced to accept the invitations of friends residing in the country who were ever anxious to enjoy his society. He always appeared cheerful, and under the most varied aspects of fortune, retained the same outward show of gaiety to the world. He possessed an unlimited fund of humour and anecdote, and this caused his society to be much sought after and highly appreciated. His animal spirits were abundant and in his youth had been even boisterous, and never seemed to flag, nor did the small misfortunes of life appear to depress his temperament and those who had the rare good fortune to enjoy his intimate

21 BL Add. Ms. 37,199, f 260.
22 Buxton biography. Museum for the History of Science, Oxford.

friendship will not easily forget his ringing hearty laugh; and his honest appreciation of humour will long be remembered by his survivors.

Babbage's work on the Calculating Engines remained largely incomprehensible to all save a few. Many visitors went through the workshops in a complete fog, confusing hopelessly the Difference and Analytical Engines. However, Joseph Henry of the Smithsonian, who had visited Babbage in 1837, visited him again in 1870 and found his mental powers and activity little diminished. Joseph Henry also gave a picture of some of Babbage's minor contributions:

Hundreds of mechanical appliances in the factories and workshops of Europe and America, scores of ingenious expedients in mining and architecture, the construction of bridges and boring of tunnels, and a world of tools by which labor is benefited and the arts improved—all the overflowings of a mind so rich that its very waste became valuable to utilize—came from Charles Babbage. He more, perhaps, than any man who ever lived, narrowed the chasm [separating] science and practical mechanics.

Babbage's life had spanned a great era of British industrial development. Born as the industrial revolution was gathering momentum, he grew up while Britain was establishing herself as the unchallenged industrial leader of the world, and himself played a worthy role in developing Britain's industrial technology. But his great plan for the systematic application of science to industry had been ignored by successive governments. By the time he died the country's endemic weakness in the newer science-based industries—which was to have such serious consequences in the twentieth century—was beginning to be more widely recognized. To Babbage it seemed so unnecessary, the consequence of short-sighted policies.

In his technical work Babbage appears a very modern figure, pioneering operations research in the *Economy of Manufactures*, studying Brunel's wide gauge, and a host of other matters, such as postal services; above all in his work on digital computation. But, although his scientific genius leaps ahead to the middle of the twentieth century, it is misleading to think of Babbage as an early scientist, in the modern sense of the term: rather he was, and remained all his life, a natural philosopher. Babbage's intellectual roots lay deep in the Enlightenment, and it is the tension between his eighteenth century approach and the advanced nature of his technical developments that gives his work its peculiar flavour and fascination.

In March 1871, after playing a brave part in the Indian mutiny, Henry Babbage returned again on furlough. He was there to see his father through his last weeks. Babbage's old friend and brother-in-law, Edward Ryan, helped Babbage put his affairs in order. Money and property were equally divided between his three sons. Henry, who had helped his father over the Scheutz Engine, received the remains of the Calculating Engines and the associated Drawings, Work Books, and Notations.

Babbage passed away in the full consolation of his heterodox religious beliefs. He felt he was leaving for another world where, with the most acute senses, and purified by more exalted moral feelings, he would see once again the two Georgianas; and perhaps, as a reward for unremitting efforts on this earth, he might be permitted to glimpse the Celestial programs.

Babbage died on 18 October 1871, nearly eighty years old. The crowds which had thronged his salon had long since melted away. The old Duchess of Somerset's was the only carriage in the small procession which, on 24 October, followed the remains of Charles Babbage to Kensal Green cemetery.

EPILOGUE

—————————— o ● o ——————————

From the Analytical Engines to the Modern Computer

After Babbage died it fell to his youngest son to deal with the Engines. For a few months he continued construction of the mill of the minimal Engine on which his father was working when he died. This fragment, together with the drawings, notations, and most of the sketch-books, was placed in the Science Museum in London.

In 1879 a committee was appointed to report to the British Association on the feasibility of constructing an Analytical Engine.[1] The committee gave high praise to Babbage's ingenuity, endorsed the potential value of an Analytical Engine, but was unable to estimate the cost of construction. The main problem was that they were unsure how big a step it would be from Babbage's plans to the detailed workshop drawings necessary to fabricate a working engine. It appears that no record remained of all Babbage's feasibility studies. What mainly emerges from the report is that they were terrified of the whole project: without Babbage's knowledge and driving energy there was no possibility of constructing an Analytical Engine. However, when Henry Babbage retired permanently from the Indian service he designed a simple mill and had it constructed. In 1889 this mill made a table of multiples of π, but the machine kept sticking: Henry was not in the same class as his father as an engineer. This mill is also in the Science Museum.

In 1889 Henry published that part of his father's history of the Analytical Engines which had been printed during his life, together with additions of his own. This book, *Babbage's Calculating Engines*, has remained, with *Passages*, the principal source of knowledge of the engines. Actually the technical history was documented by Babbage in great detail in the Drawings, Sketch-Books, and Notations, but the unfamiliarity of the notation has made it accessible to few, and translation of Babbage's designs into modern terminology has only recently been undertaken.

Babbage's work continued to influence the development of calculating machinery in many ways. In Spain the work of Leonardo Torres y Quevedo at the beginning of this century on automata[2] derived directly from Babbage's engines, and several designers of calculators and tabulators knew of Babbage's work. Possibly Hollerith's original idea of using punched cards came in part

[1] Reprinted in *Babbage's Calculating Engines*, Spon, 1888.
[2] Brian Randell, *The Origins of Digital Computers—selected papers*, Springer, 1973.

from Babbage's use of a card input system. Such knowledge might, for example, have derived from the report of the Smithsonian Institution for 1873 or from an encyclopaedia. There are other threads to follow. Jevons, better known as the pioneer of marginal economic theory, in 1869 built his logical piano, a mechanical device for implementing boolean algebra, and Babbage's work was certainly one of his sources of inspiration. Allen Marquand, tutor in logic at Princeton, built an electromechanical version of Jevons's 'piano'.

Amongst the mathematical élite of the University of Cambridge Babbage's work was never entirely forgotten. It seems likely that it was one of the sources of inspiration for Alan Turing's 'Turing machine', a theoretical tool fundamental to modern theory of computation. Certainly Turing was familiar with Babbage's published work when he was working on the Colossus code-cracking machines at Bletchley Park during the second world war. Howard Aitken, who proposed the Automatic Sequence Controlled Calculator built by IBM during the war, was particularly enthusiastic about Babbage's engines. Since the development of the stored-program computer, Babbage's engines have been widely acknowledged and modern pioneers have paid generous tribute to his achievements.

However, in some respects his work has been curiously misunderstood. It has generally been thought that the Analytical Engine was a single machine, whereas it was really a class of machines, as is 'the computer' today. Even a list of the main types of Analytical Engine was not published until 1976.[3] And yet, in its versatility Babbage's achievement remains unrivalled. Where modern pioneers worked in teams and were part of a great movement, save for his assistants and occasional help from his sons, Babbage worked by himself, far ahead of contemporary thought. He had not only to elaborate the designs but to develop the concepts, the engineering, and even tools to make the parts. He had to develop the mathematics, and his notation through its successive stages, pioneer microprogramming and coding; even to conceive of the idea in the first place. Charles Babbage stands alone: the great ancestral figure of computing.

[3] Anthony Hyman, *Computing, a Dictionary of Terms, Concepts, and Ideas*, Arrow 1976, entry on 'calculating engines'.

PUBLISHED WORKS OF CHARLES BABBAGE

–––––––––––––––––––– ○ ● ○ ––––––––––––––––––––

Based on Babbage's own list printed in *Passages from the Life of a Philosopher*, corrected by A. W. Van Sinderen (*Annals of the History of Computing* (1980), 2, no. 2, 169–85). I have followed Babbage in including articles which he directly inspired, even though he was not the nominal author.

1813   *Memoirs of the Analytical Society*; 1 Preface, in collaboration with John Herschel, i–xxii; 2 On Continued Products, 1–31, Cambridge.

1815   'An Essay Towards the Calculus of Functions', *Phil. Trans.* **105**, 389–423.

1816   'An Essay Towards the Calculus of Functions', Part II, *Phil. Trans.* **106**, 179–256.
       'Demonstrations of Some of Dr. Matthew Stewart's General Theorems; To Which is Added, An Account of Some New Properties of the Circle', *Journal of Science*, 1, 6–24.
       S. F. LaCroix, *An Elementary Treatise on the Differential and Integral Calculus*, Tr. Charles Babbage, John Herschel, and George Peacock, Cambridge, J. Deighton.

1817   'Observations on the Analogy Which Subsists Between the Calculus of Functions and the Other Branches of Analysis', *Phil. Trans.* **107**, 197–216.
       'Solutions of Some Problems by Means of the Calculus of Functions', *Journal of Science*, 2, 371–79.
       'An Account of Euler's Method of Solving a Problem Relative to the Move of the Knight at the Game of Chess', *Journal of Science* 3, 72–7.
       'Note Respecting Elimination', *Journal of Science* 3, 355–7.

1819   'On Some New Methods of Investigating the Sums of Several Classes of Infinite Series', *Phil. Trans.* **109**, 249–82.
       'Demonstration of a Theorem Relating to Prime Numbers', *Edin. Phil. Jrl:.* 1, 46–9

1820   *Examples to the Differential and Integral Calculus*, in collaboration with John Herschel and George Peacock. Part III, 'Examples of the Solutions of Functional Equations', was by Babbage, Cambridge, J. Deighton.

1821   'An Examination of Some Questions Connected with Games of Chance', *Trans. Roy Soc. Edin.* **9**, 153–77.

1822   'Observations on the Notation Employed in the Calculus of Functions, *Trans. Camb. Phil. Soc.* 1, 63–76.
       'Barometrical Observations Made at the Fall of the Staubbach', in collaboration with John Herschel, *Edin. Phil. Jrl.* 6, 224–7.
       'A Note Respecting the Application of Machinery to the Calculation of Astronomical Tables', *Mem. Astron. Soc.* 1, 309.
       'A Letter to Sir Humphry Davy, Bart. PRS, on the Application of Machinery

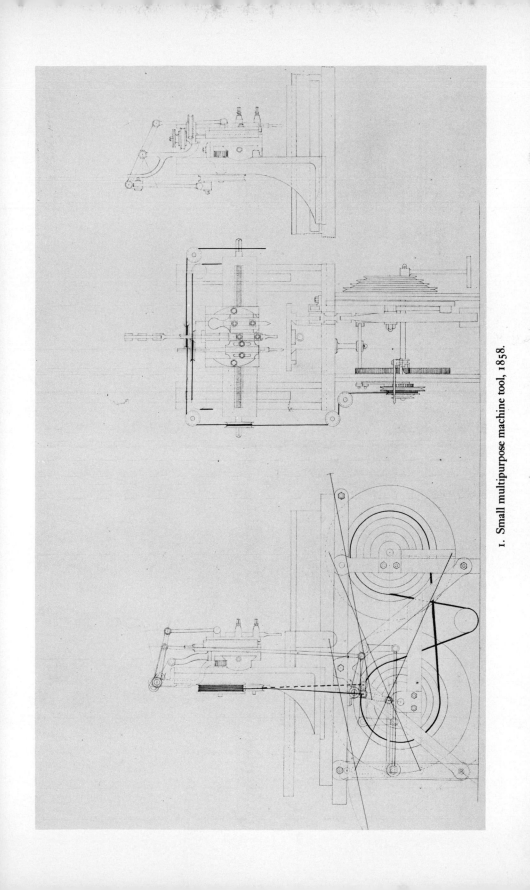

1. Small multipurpose machine tool, 1858.

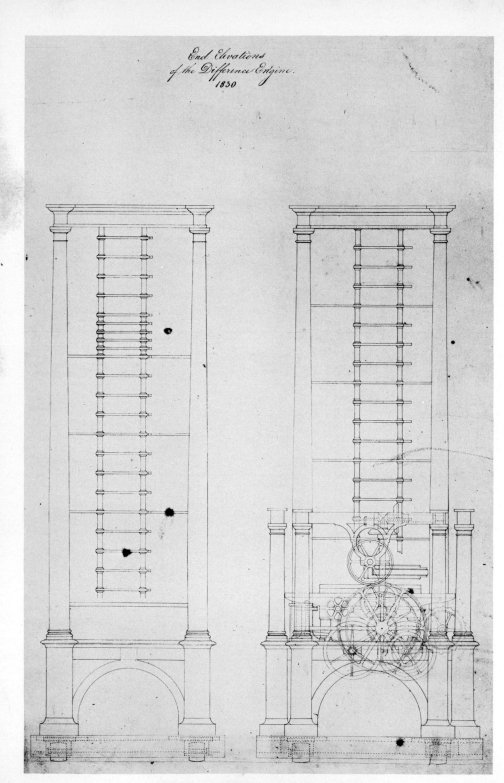

2. **Difference Engine No. 1. End elevation, 1830.**

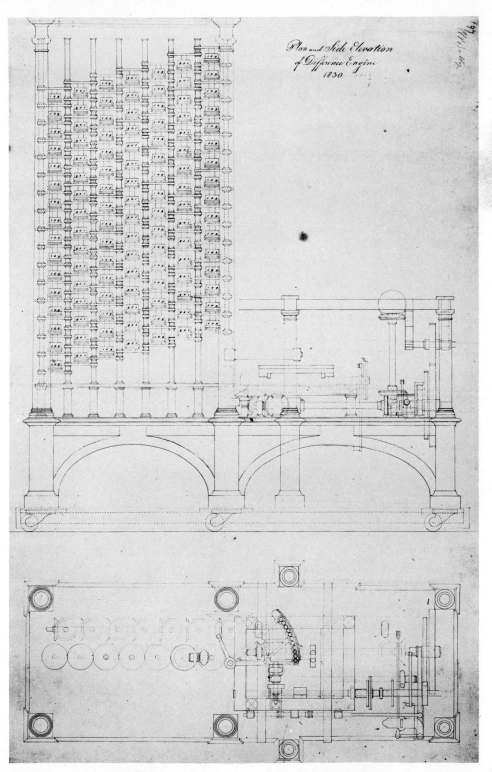

Plan and Side Elevation of Difference Engine 1830

3. Difference Engine No. 1. Plan and side elevation, 1830.

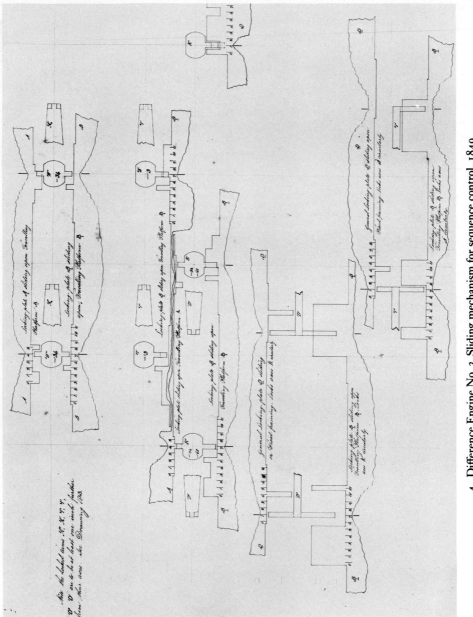

4. Difference Engine No. 3: Sliding mechanism for sequence control, 1849.

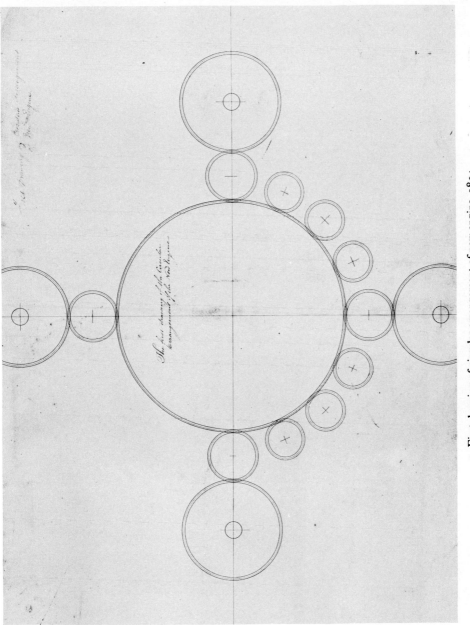

5. First drawing of circular arrangement of new engine, 1834.

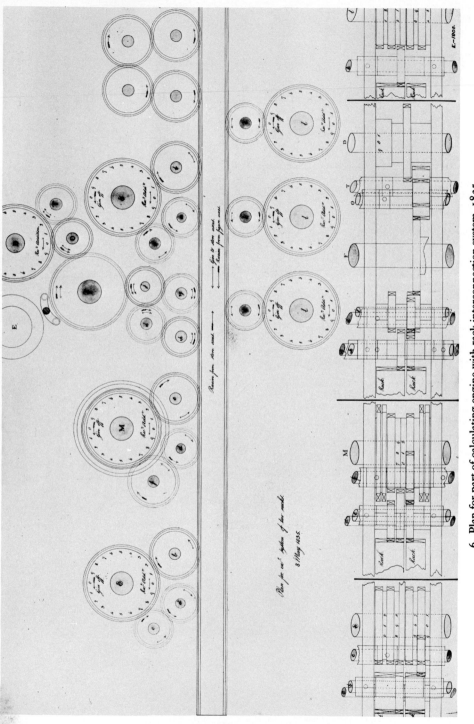

6. Plan for part of calculating engine with rack interconnection system, 1835.

7. Superposition of motion, 1836.

8. Two distinct engines with thirty figures each, 1836.

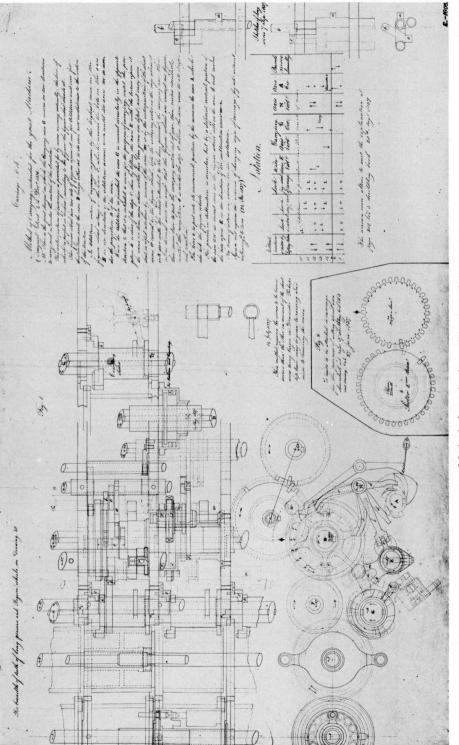

9. Method of carrying by anticipation, 1836.

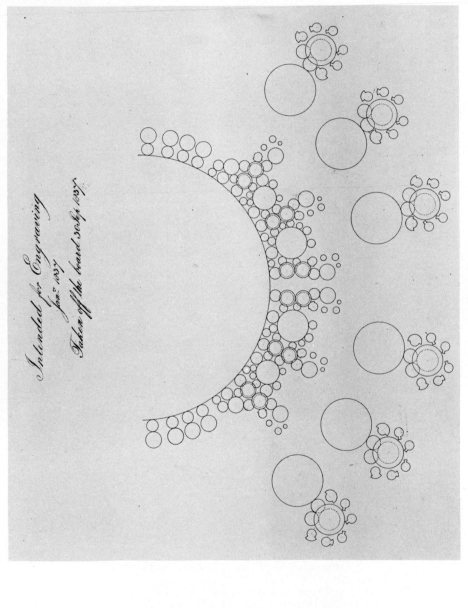

Intended for Engraving
Jan.ʸ 1837
Taken off the hard scale 1837.

10. General plan of mill, 1837.

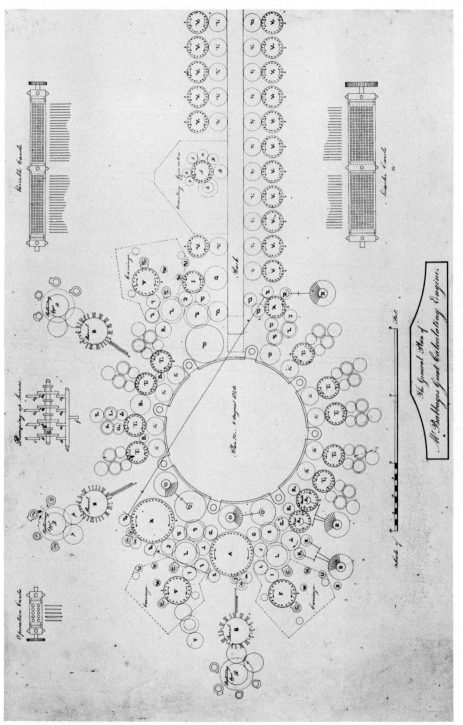

11. General plan of Mr. Babbage's Great Calculating Engine, 1840.

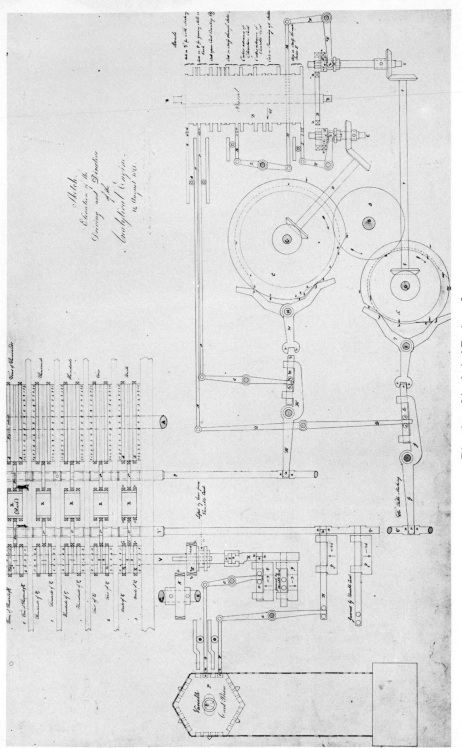

12. Direction of Analytical Engine, 1841.

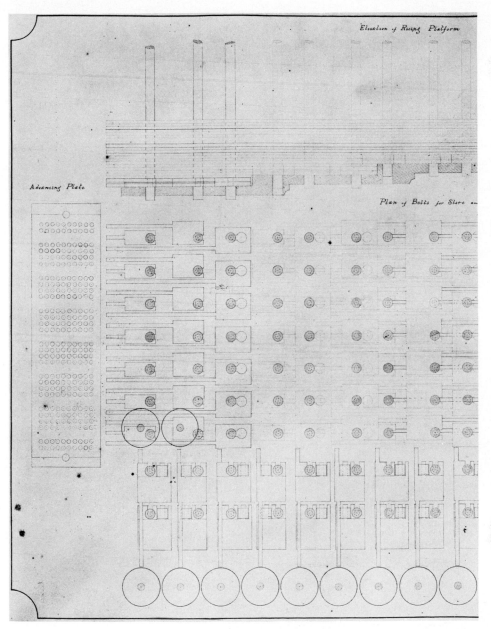

13. Plan of bolts for store and mill, *c.* 1858.

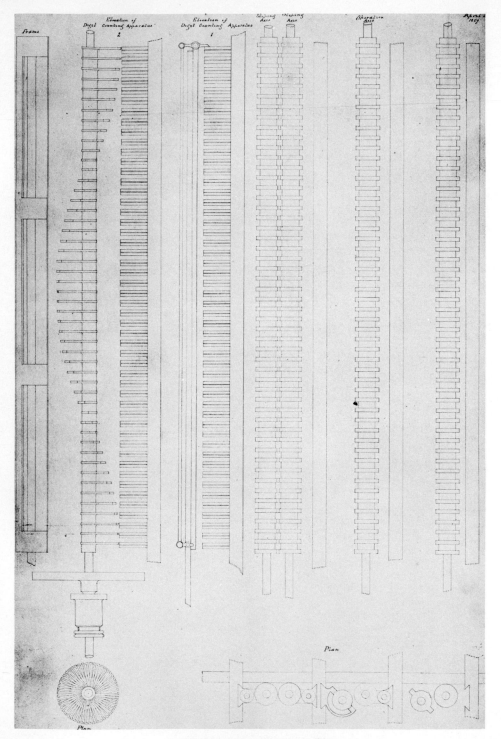

14. Digit counting apparatus, 1859.

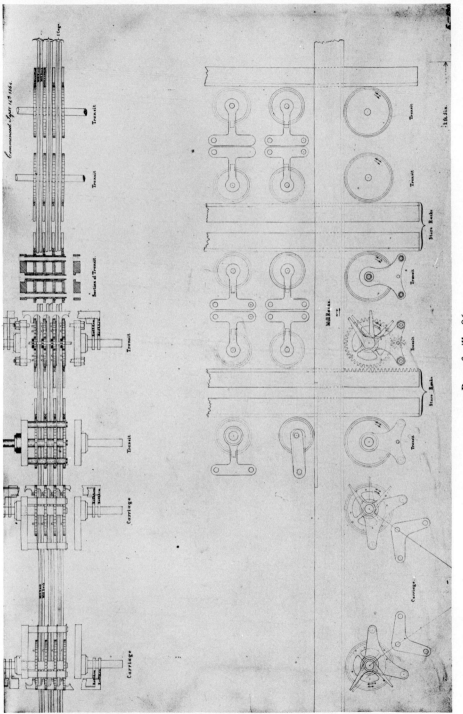

15. Part of mill, 1864.

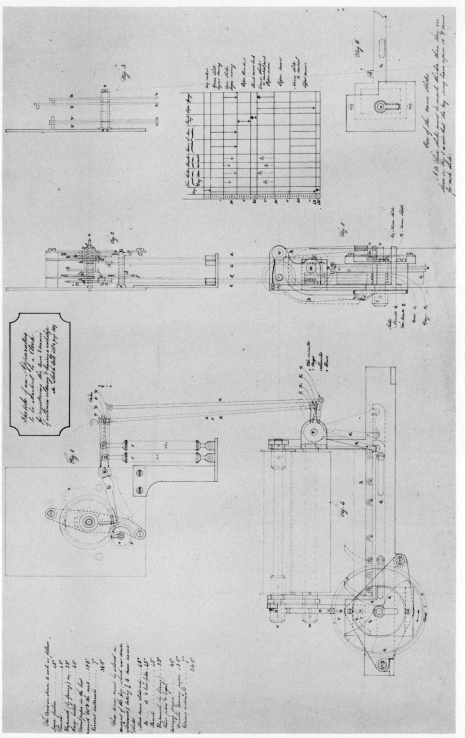

16. Apparatus for registering time of workmen.

to the Purpose of Calculating and Printing Mathematical Tables', J. Booth, London.

'Observations on the Application of Machinery to the Computation of Mathematical Tables', *Mem. Astron. Soc.* 1, 311–14.

1823   'On the Theoretical Principles of the Machinery for Calculating Tables', *Edin. Phil. Jrl.* 8, 122–8.

'On the Application of Analysis to the Discovery of Local Theorems and Porisms', *Trans. Roy. Soc. Edin.* 9, 337–52.

Scriptores Optici; or, A Collection of Tracts Relating to Optics, ed. in collaboration with Francis Masères—preface by Babbage—Baldwin, Craddock & Joy, London.

1824   'Observations on the Measurement of Heights by the Barometer', *Edin. Jrl. Sci.* 85–7.

1825   'Account of the Repetition of Mr. Arago's Experiments on the Magnetism Manifested by Various Substances During the Act of Rotation', in collaboration with John Herschel, *Phil. Trans.* 115, 467–96.

1826   'On a New Zenith Micrometer', *Mem. Astron. Soc.* 2, 101–3.

A Comparative View of the Various Institutions for the Assurance of Lives, J. Mawman, London.

'On a Method of Expressing by Signs the Action of Machinery', *Phil. Trans.* 116, 250–65.

'On Electrical and Magnetic Rotations', *Phil. Trans.* 116, 494–528.

'On the Determination of the General Term of a New Class of Infinite Series', *Trans. Camb. Phil. Soc.* 2, 217–25.

'Diving Bell', *Encyclopaedia Metropolitana*, 18, 157–67.

1827   'On the Influence of Signs in Mathematical Reasoning', *Trans. Camb. Phil. Soc.* 2, 325–77.

'Notice respecting some Errors Common to many Tables of Logarithms', *Mem. Astron. Soc.* 3, 65–7.

'Evidence on Savings Banks, before a Committee of the House of Commons', *Reports from the Committees of the House of Commons*, III, 1826–7, 869 558.

Table of Logarithms of the Natural Numbers from 1 to 108,000, J. Mawman, London.

1829   'Account of the Great Congess of Philosophers at Berlin on the 18th September 1828', *Edin. Jr. Sci.* 10, 225–34.

'A Letter to the Right Hon. T. P. Courtenay, on the Proportionate Number of Births of the Two Sexes under Different Circumstances', *Edin. Jrl. Sci. NS*, 1, 85–104.

'On the General Principles which Regulate the Application of Machinery to Manufacture and the Mechanical Arts', *Encyclopaedia Metropolitana*, 8, 1–84.

1830   'Notation', *Edinburgh Encyclopaedia*, 15, 394–9.

'Porisms', op. cit., 17, 106–14.

Reflections on the Decline of Science in England, and some of its Causes, (dedicated to 'a nobleman', probably Lord Ashley, later Earl of Shaftesbury) B. Fellowes, London.

Sketch of the Philosophical Characters of Dr. Wollaston and Sir Humphry Davy, B. Fellowes, London.

1831   'Sur l'emploi plus ou moins fréquent des mêmes lettres dans les différentes langues', (in a letter to L. A. J. Quetelet), *Corr. mathématique et physique*, 7, 135–7.

Specimen of Logarithm Tables (21 vols.): two pages from Babbage's edition of the logarithmic tables printed in many different coloured inks on many different colours and shades of paper—a single copy was printed (now in the Crawford Library), B. Fellowes, London.

Table of Logarithms of the Natural Numbers from 1–108,000 on Different Coloured Papers—single copy printed (28 volumes are now in the Crawford Library—Babbage said 35 were originally printed), B. Fellowes, London.

1832   'On the Advantage of a Collection of Numbers, to be Entitled the Constants of Nature and Art', *Edin. Jrl. Sci. NS6*, 334–40. (Expanded versions of this paper were later published in: *Compte Rendu des Travaux du Congrès Général de Statistique, 1835*, 222–30, Bruxelles; and *The Annual Report of the Board of Regents of the Smithsonian Institution, 1856*, 289–302).

On the Economy of Machinery and Manufactures, Charles Knight, London.
A second and third edition were published in 1832 and a fourth edition with index in 1835; translated into French, German, Italian, Spanish, Russian, and Swedish. (In 1833 three chapters were published as a pamphlet, *On Currency, on a New System of Manufacturing, and on the Effect of Machinery on Human Labour*, Charles Knight, London; a further extract was published in 1899 as *How to Invent Machinery*, ed. William H. Atherton, Manchester.)

1833   *A Word to the Wise*, John Murray, London. (reprinted in 1856 and subtitled: *Observations on Peerage for Life*.)

1834   Abstract of a Paper Entitled: 'Observations on the Temple of Serapis at Pozzuoli; with Remarks on Certain Causes which May Produce Geological Cycles of Great Extent' (full paper published in 1847), Richard Taylor, London.

1835   'Une lettre à M. Quetelet de M. Ch. Babbage relativement à la machine à calculer', *Acad. Roy. Bruxelles, Bulletins*, 2, 123–6.

'Letter from Mr. A. Sharp to Mr. J. Crosthwait, Hoxton, Feb. 2, 1721–2, Decyphered by C. Babbage, Esq. From the Original Letter in Shorthand'; in *An Account of the Revd John Flamsteed*, the First Astronomer-Royal, 348, by Francis Baily, Admiralty, London.

1837   *The Ninth Bridgewater Treatise; A Fragment*, John Murray, London. 2nd ed. 1838.

'On the Mathematical Powers of the Calculating Engine' (Ms. in Museum of History of Science, Oxford); published 1973, Brian Randell (ed), *The Origins of Digital Computers*, 17–52, Springer.

1838   'On Impressions in Sandstone Resembling Those of Horses' Hoofs', *Proc. Geol. Soc.* 2, 439.

1839   'Letter from Mr. Babbage to the Members of the British Association for the Promotion of Science', Richard Clay, London.

1842-3 'Notions sur la Machine Analytique de M. Charles Babbage' (by L. F. Menabrea), *Bibliothèque Universelle de Genève*, **41**, 352-76; tr. with additional notes by Ada Lovelace as 'Sketch of the Analytical Engine', (1843) *Scientific Memoirs*, iii, 666-731. (On page 373 of the Menabrea paper *le cas n* = ∞ was misprinted as *le cos n* = ∞. Ada translated this as 'when the cos of n = ∞', which is nonsense.)

1843 'Statement of the Circumstances Respecting Mr. Babbage's Calculating Engines' (by Sir Harris Nicolas) privately printed, London.
'Statement of the Circumstances attending the Invention and Construction of Mr. Babbage's Calculating Engines', *Phil. Mag.* **23**, 234-5.
'Description of the Boracic Acid Works of Tuscany'; *Handbook for Travellers in Central Italy*, 178-9, John Murray, London.

1847 'On the Principles of Tools for Turning and Planing Metals'; in *Turning and Mechanical Manipulation* (by Charles Holtzapffel), **2**, 984-7, Holtzapffel, London.
'Observations on the Temple of Serapis at Pozzuoli, Near Naples', Taylor, London.
'The Planet Neptune and the Royal Astronomical Society's Medal', *The Times*, 15 March 5d.

1848 *Thoughts on the Principles of Taxation with Reference to a Property Tax, and its Exceptions*, John Murray, London. 2nd ed. 1851; 3rd ed. 1852.

1851 'Laws of Mechanical Notation', privately printed, London.
The Exposition of 1851; Or Views of the Industry, the Science, and the Government of England, John Murray, London. 2nd ed. 1851.
'Notes Respecting Lighthouses', privately printed, London.

1852 'Note Respecting the Pink Projections from the Sun's Disc Observed during the Total Solar Eclipse in 1851', *Monthly Notices Astron. Soc.* **12**, 209-10.

1853 'On the Statistics of Lighthouses', *Compte Rendu des travaux du Congrès Générale de Statistique*, 230-7, Bruxelles.

1854 'Report on the Opthalmoscope', (by T. Wharton Jones), *British and Foreign Medical Rev.* **14**, 425-32.
'Mr. Thwaites's Cypher', *Jrl. Soc. Arts*, No. 93, 707-8.
Ibid., No. 98, 776-7.

1855 'Submarine Navigation', *Illustrated London News*, No. 749, 623-4.
'On the Possible Use of the Occulting Telegraph at Sebastapol', *The Times*, 16 July, **6** f.
'A Method of Laying the Guns of a Battery Without Exposing the Men to the Shot of the Enemy', *Illustrated London News*, No. 757, 210.
'Sur la machine suédoise de MM. Scheutz pour calculer les tables mathématiques', *Comptes rendus hebdomadaires*, **41**, 557-60, Académie des sciences, Paris.

1856 'Scheutz's Difference Engine and Babbage's Mechanical Notation' (by Henry Babbage), *Proc. Inst. Civil Eng.* **15**, 497-514. Also published as separate pamphlet.

'On the Action of Ocean Currents in the Formation of the Strata of the Earth', *Quart. Jrl. Geol. Soc.* **12**, 366–8.

'Analysis of the Statistics of the Clearing House During the Year 1839', *Jrl. Stat. Soc.* **19**, 28–48. Also published as a separate pamphlet.

Observations Addressed at the Last Anniversary to the President and Fellows of the Royal Society, John Murray, London.

1857   'Table of the Relative Frequency of Occurence of the Causes of Breaking of Plate Glass Windows', *Mechanics Mag.* **66**, 82.

1860   'Observations on the Discovery in Various Localities of the Remains of Human Art Mixed with the Bones of Extinct Races of Animals', *Proc. Roy. Soc.* **10**, 59–72.

'On Easily Recognizable Signs in Drawings', *Proc. Fourth International Statistical Congress*, 380, London.

'Letter to Dr. Farr, On the Origin of the International Statistical Congresses', *Proc. Fourth International Statistical Congress*, 505–7, London.

1864   *Passages from the Life of a Philosopher*, Longman, Green, London. ('A Chapter on Street Nuisances' was published by John Murray prior to publication of the book.)

1865   *Thoughts upon an Extension of the Franchise*, Longman, Green, London.

1868   'Observations on the Parallel Roads of Glen Roy', *Quart. Jrl. Geol. Soc.* **24**, 273–7.

(1889)   'History of the Analytical Engine' (incomplete), published posthumously with additions by Henry Prevost Babbage as *Babbage's Calculating Engines*, Spon, London.

APPENDIX A

○ ● ○

From Passages from the Life of a Philosopher

Explanation of the Difference Engine

Those who are only familiar with ordinary arithmetic may, by following out with the pen some of the examples which will be given, easily make themselves acquainted with the simple principles on which the Difference Engine acts.

It is necessary to state distinctly at the outset, that the Difference Engine is not intended to answer special questions. Its object is to calculate and print a *series* of results formed according to given laws. These are called Tables—many such are in use in various trades. For example—there are collections of Tables of the amount of any number of pounds from 1 to 100 lbs. of butchers' meat at various prices per lb. Let us examine one of these Tables: viz.—the price of meat 5*d.* per lb., we find

Number lbs	Table price s. d.	
1	0	5
2	0	10
3	1	3
4	1	8
5	2	1

There are two ways of computing this Table:

1st. We might have multiplied the number of lbs. in each line by 5, the price per lb., and have put down the result in *l. s. d.*, as in the 2nd column: or,

2nd. We might have put down the price of 1 lb., which is 5*d.*, and have added five pence for each succeeding lb.

Let us now examine the relative advantages of each plan. We shall find that if we had multiplied each number of lbs. in the Table by 5, and put down the resulting amount, then every number in the Table would have been computed independently. If, therefore, an error had been committed, it would not have affected any but the single tabular number at which it had been made. On the other hand, if a single error had occurred in the system of computing by adding five at each step, any such error would have rendered the whole of the rest of the Table untrue.

Thus the system of calculating by differences, which is the easiest, is much more liable to error. It has, on the other hand, this great advantage: viz., that when the Table has been so computed, if we calculate its last term directly, and if it agree with the last term found by the continual addition of 5, we shall then be quite certain that every term throughout is correct. In the system of computing each term directly, we possess no such check upon our accuracy.

Now the Table we have been considering is, in fact, merely a Table whose first difference is constant and equal to five. If we express it in pence it becomes—

	Table	1st Difference
1	5	5
2	10	5
3	15	5
4	20	5
5	25	

Any machine, therefore, which could add one number to another, and at the same time retain the original number called the first difference for the next operation, would be able to compute all such Tables.

Let us now consider another form of Table which might readily occur to a boy playing with his marbles, or to a young lady with the balls of her solitaire board.

The boy may place a row of his marbles on the sand, at equal distances from each other, thus—

●   ●   ●   ●   ●

He might then, beginning with the second, place two other marbles under each, thus—

He might then, beginning with the third, place three other marbles under each group, and so on; commencing always one group later, and making the addition one marble more each time. The several groups would stand thus arranged—

He will not fail to observe that he has thus formed a series of triangular groups, every group having an equal number of marbles in each of its three sides. Also that the side of each successive group contains one more marble than that of its preceding group.

Now an inquisitive boy would naturally count the numbers in each group and he would find them thus—

1   3   6   10   15   21

He might also want to know how many marbles the thirtieth or any other distant group might contain. Perhaps he might go to papa to obtain this information; but I much fear papa would snub him, and would tell him that it was nonsense—that it was useless—that nobody knew the number, and so forth. If the boy is told by papa, that he is not able to answer the question, then I recommend him to pay careful attention to whatever that father may at any time say, for he has overcome two of the greatest obstacles to the acquisition of knowledge—inasmuch as he possesses the consciousness that he does not know—and he has the moral courage to avow it.

If papa fail to inform him, let him go to mamma, who will not fail to find means to satisfy her darling's curiosity. In the meantime the author of this sketch will endeavour to lead his young friend to make use of his own common sense for the purpose of becoming better acquainted with the triangular figures he has formed with his marbles.

In the case of the Table of the price of butchers' meat, it was obvious that it could be formed by adding the same *constant* difference continually to the first term. Now suppose we place the numbers of our groups of marbles in a column, as we did our prices of various weights of meat. Instead of adding a certain difference, as we did in the former case, let us subtract the figures representing each group of marbles from the figures of the succeeding group in the Table. The process will stand thus:

	Table	1st Difference	2nd Difference
Number of the group	Number of marbles in each group	Difference between the number of marbles in each group and that in the next	
1	1	1	1
2	3	2	1
3	6	3	1
4	10	4	1
5	15	5	1
6	21	6	
7	28	7	

It is usual to call the third column thus formed *the column of first differences*. It is evident in the present instance that that column represents the natural numbers. But we already know that the first difference of the natural numbers is constant and equal to unity. It appears, therefore, that a Table of these numbers, representing the group of marbles, might be constructed to any extent by mere addition—using the number 1 as the first number of the Table, the number 1 as the first Difference, and also the number 1 as the second Difference, which last always remains constant.

Now as we could find the value of any given number of pounds of meat directly, without going through all the previous part of the Table, so by a somewhat different rule we can find at once the value of any group whose number is given.

Thus, if we require the number of marbles in the fifth group, proceed thus:

Take the number of the group . . . 5	
Add 1 to this number, it becomes . . . 6	
Multiply these numbers together . . . 2)30	
Divide the product by 2 . . . 15	

This gives 15, the number of marbles in the 5th group.

If the reader will take the trouble to calculate with his pencil the five groups given above, he will soon perceive the general truth of this rule.

We have now arrived at the fact that this Table—like that of the price of butchers' meat—can be calculated by two different methods. By the first, each number of the

Table is calculated independently: by the second, the truth of each number depends upon the truth of all the previous numbers . . .

I hope my young friend is acquainted with the fact—that the product of any number multiplied by itself is called the square of that number. Thus 36 is the product of 6 multiplied by 6, and 36 is called the square of 6. I would now recommend him to examine the series of square numbers

$$1, 4, 9, 16, 25, 36, 49, 64, \&c.,$$

and to make, for his own instruction, the series of their first and second differences, and then to apply to it the same reasoning which has been already applied to the Table of Triangular Numbers.

When he feels that he has mastered that Table, I shall be happy to accompany mamma's darling to Woolwich or to Portsmouth, where he will find some practical illustrations of the use of his newly-acquired numbers. He will find scattered about in the Arsenal various heaps of cannon balls, some of them triangular, others square or oblong pyramids.

Looking on the simplest form—the triangular pyramid—he will observe that it exactly represents his own heaps of marbles placed each successively above one another until the top of the pyramid contains only a single ball.

The new series thus formed by the addition of his own triangular numbers is—

Number	Table	1st Difference	2nd Difference	3rd Difference
1	1	3	3	1
2	4	6	4	1
3	10	10	5	1
4	20	15	6	
5	35	21		
6	56			

He will at once perceive that this Table of the number of cannon balls contained in a triangular pyramid can be carried to any extent by simply adding successive differences, the third of which is constant.

The next step will naturally be to inquire how any number in this Table can be calculated by itself. A little consideration will lead him to a fair guess; a little industry will enable him to confirm his conjecture.

It will be observed that in order to find independently any number of the Table of the price of butchers' meat, the following rule was observed:

Take the number whose tabular number is required.

Multiply it by the first difference.

This product is equal to the required tabular number.

Again, . . . the rule for finding any triangular number was:

Take the number of the group	.	.	.	5
Add 1 to this number, it becomes	.	.	.	6
Multiply these numbers together	.	.	.	2)30
Divide the product by 2	.	.	.	15

This is the number of marbles in the 5th group.

Now let us make a bold conjecture respecting the Table of cannon balls, and try this rule:

Take the number whose tabular number is
required, say 5
Add 1 to that number . . . 6
Add 1 more to that number . . 7

Multiply all three numbers together . . 2)210

Divide by 2 105

The real number in the 5th pyramid is 35. But the number 105 at which we have arrived is exactly three times as great. If, therefore, instead of dividing by 2 we had divided by 2 and also by 3, we should have arrived at a true result in this instance.

The amended rule is therefore—

Take the number whose tabular number is
required, say . . . . . n
Add 1 to it . . . . . . $n+1$
Add 1 to this . . . . . $n+2$
Multiply these three numbers
together . . . . $n \times (n+1) \times (n+2)$
Divide by $1 \times 2 \times 3$.

The result is . . . . $\dfrac{n(n+1)(n+2)}{6}$

This rule will, upon trial, be found to give correctly every tabular number.

By similar reasoning we might arrive at the knowledge of the number of cannon balls in square and rectangular pyramids. But it is presumed that enough has been stated to enable the reader to form some general notion of the method of calculating arithmetical Tables by differences which are constant.

It may now be stated that mathematicians have discovered that all the Tables most important for practical purposes, such as those relating to Astronomy and Navigation, can, although they may not possess any constant differences, still be calculated in detached portions by that method.

Hence the importance of having machinery to calculate by differences, which, if well made, cannot err; and which, if carelessly set, presents in the last term it calculates the power of verification of every antecedent term.

Of the Mechanical Arrangements necessary for computing Tables by the Method of Differences

From the preceding explanation it appears that all Tables may be calculated, to a greater or less extent, by the method of Differences. That method requires, for its successful execution, little beyond mechanical means of performing the arithmetical operation of Addition. Subtraction can, by the aid of a well-known artifice, be converted into Addition . . .

Description of the existing portion of Difference Engine No. 1

That portion of Difference Engine, No. 1, which during the last twenty years has been in the museum of King's College, at Somerset House, ... consists of three columns; each column contains six cages; each cage contains one figure-wheel.

The column on the right hand has its lowest figure-wheel covered by a shade which is never removed, and to which the reader's attention need not be directed.

The figure-wheel next above may be placed by hand at any one of the ten digits. In the woodcut it stands at zero.

The third, fourth, and fifth cages are exactly the same as the second.

The sixth cage contains exactly the same as the four just described. It also contains two other figure-wheels, which with a similar one above the frame, may also be dismissed from the reader's attention. Those wheels are entirely unconnected with the moving part of the engine, and are only used for memoranda.

It appears, therefore, that there are in the first column on the right hand five figure-wheels, each of which may be set by hand to any of the figures 0, 1, 2, 3, 4, 5, 6, 7, 8, 9.

The lowest of these figure-wheels represents the unit's figure of any number; the next above the ten's figure, and so on. The highest figure-wheel will therefore represent tens of thousands.

Now, as each of these figure-wheels may be set by hand to any digit, it is possible to place on the first column any number up to 99999. It is on these wheels that the Table to be calculated by the engine is expressed. This column is called the Table column, and the axis of the wheels the Table axis.

The second or middle column has also six cages, in each of which a figure-wheel is placed. It will be observed that in the lowest cage, the figure on the wheel is concealed by a shade. It may therefore be dismissed from the attention. The five other figure-wheels are exactly like the figure-wheels on the Table axis, and can also represent any number up to 99999.

This column is called the First Difference column, and the axis is called the First Difference axis.

The third column, which is that on the left hand, has also six cages, in each of which is a figure-wheel capable of being set by hand to any digit.

The mechanism is so contrived that whatever may be the numbers placed respectively on the figure-wheels of each of the three columns, the following succession of operations will take place as long as the handle is moved:

1st. Whatever number is found upon the column of first differences will be added to the number found upon the Table column.

2nd. The same first difference remaining upon its own column, the number found upon the column of second differences will be added to that first difference.

It appears, therefore, that with this small portion of the Engine any Table may be computed by the method of differences, provided neither the Table itself, nor its first and second differences, exceed five places of figures.

If the whole Engine had been completed it would have had six orders of differences, each of twenty places of figures, whilst the three first columns would each have had half a dozen additional figures.

This is the simplest explanation of that portion of the Difference Engine No. 1, at the Exhibition of 1862. There are, however, certain modifications in this fragment which render its exhibition more instructive, and which even give a mechanical insight into those higher powers with which I had endowed it in its complete state.

As a matter of convenience in exhibiting it, there is an arrangement by which the *three* upper figures of the second difference are transformed into a small engine which counts the natural numbers.

By this means it can be set to compute any Table whose second difference is constant and less than 1000,·whilst at the same time it thus shows the position in the Table of each tabular number.

In the existing portion there are three bells; they can be respectively ordered to ring when the Table, its first difference and its second difference, pass from positive to negative. Several weeks after the machine had been placed in my drawing-room, a friend came by appointment to test its power of calculating Tables. After the Engine had computed several Tables, I remarked that it was evidently finding the root of a quadratic equation; I therefore set the bells to watch it. After some time the proper bell sounded twice, indicating, and giving the two positive roots to be 28 and 30. The Table thus calculated related to the barometer and really involved a quadratic equation, although its maker had not previously observed it. I afterwards set the Engine to tabulate a formula containing impossible roots, and of course the other bell warned me when it had attained those roots. I had never before used these bells, simply because I did not think the power it thus possessed to be of any practical utility.

Again, the lowest cages of the Table, and of the first differences, have been made use of for the purpose of illustrating three important faculties of the finished engine.

1st. The portion exhibited can calculate any Table whose third difference is constant and less than 10.

2nd. It can be used to show how much more rapidly astronomical Tables can be calculated in an engine in which there is no constant difference.

3rd. It can be employed to illustrate those singular laws which might continue to be produced through the ages, and yet after an enormous interval of time change into other different laws; each again to exist for ages, and then to be superseded by new laws. These views were first proposed in the 'Ninth Bridgewater Treatise.'

APPENDIX B

○●○

On the Mathematical Powers of the Calculating Engine

26. Dec. 1837

The calculating part of the engine may be divided into two portions
1st The *Mill* in which all operations are performed
2nd The *Store* in which all the numbers are originally placed and to which the numbers computed by the engine are returned.

In a plan of the engine—those circles placed round the great central wheel constitute the Mill whilst that portion which adjoins the longitudinal part or rack represents the Store.

Of the Mill

Two Axes with figure wheels, I the *Ingress Axis* and ″A the *Egress Axis* connect the *Mill* with the *Store*.

The Mill itself consists of:

1. Three Figure Axes
2. Three Carriage Axes
3. Ten Table Figure Axes
4. Digit Counting Apparatus
5. Selecting Apparatus
6. Barrels
7. Reducing Apparatus for Barrels
8. Operation Cards
9. Repeating Apparatus
10. Combinatorial Counting Apparatus

1. Of the Figure Axes

The Figures A and ′A are connected with each other without the intervention of the central wheels so that a number on the figure wheels of one axis may be transferred to those of the other.

These figure wheels are considerably larger than any others in order to allow of sufficient space on their circumference for placing the pinions by which communications are made with other parts of the mill.

By means of some of these pinions a process called *Stepping down* and another called *Stepping up* may be performed. It consists in shifting each digit of a number one cage lower or one cage higher, which processes are equivalent to the arithmetical operations of dividing or multiplying the number by ten.

Other pinions are fixed on *register axes* R and R_1 and convey the two highest figures of the dividend to the *Selecting* apparatus. The third figure axis "A is placed near the Store and constitutes the egress axis. It is adjacent to the digit counting apparatus with which it communicates.

2. Carriage Axes

These Axes F, 'F, "F with their peculiar apparatus are employed to execute the carriage of the tens when numbers are added to or subtracted from each other. The carriages F and 'F can be both connected with the Figure Axis A or one of them with the Figure Axis A and the other with 'A or they may by means of the central wheels be connected with any other part of the Mill. The third carriage "F is connected both with the Mill and the Store and may be used with either.

Whenever the number subtracted is greater than that from which it is taken the resulting carriages would if effected, and if the mechanism admitted, produce a carriage in the forty first cage. This fact is taken advantage of for many purposes—it is one of very great importance and when it happens a *Running up* is said to occur. Connected with this part is a lever on which the *Running up* warning acts and this lever governs many parts of the engine according as the circumstances demand.

3. Table Figure Axes

These Axes are ten in number. Nine of them contain the table of the nine multiples of one factor in Multiplication and of the Divisor in Division. The tenth contains the complement of the Divisor in the latter operation. They are all connected with the central wheels and the number on each figure wheel can be *stepped up* or *down* upon the other figure wheel of the same cage. The figure which at each stepping goes off from the bottom wheel is transferred to the top wheel.

4. Of the Digit Counting Apparatus

This is a mechanism by which the digits of any number brought into the mill may be counted and certain calculations made as to the position of the decimal point in the result of multiplication and division. It is also used to limit the number of figures employed when the engine is making successive approximations either to the roots of equations or to the values of certain functions. It consists of three distinct systems nearly similar to each other.

5. Of the Selecting Apparatus

When a table of the nine multiples of a multiplicand has been made it becomes necessary in order to effect multiplication to select successively those multiples indicated by the successive digits of the multiplier. This mechanically is not difficult.

But when in the process of division it becomes requisite to select that multiple which is next less than the dividend from which it is to be subtracted the mechanical difficulty is of quite a different order and hitherto nothing but the most refined artifices have been found for accomplishing it. This refinement relates however entirely to the *nature* of those contrivances not to the certainty of their action nor to any delicacy of workmanship.

The apparatus consists of a portion of the carrying apparatus for three figure wheels which by the addition of another contrivance renders them available for the purpose of making the selection. This apparatus is placed immediately below the Table Axes.

6. *Of the Barrels*

The barrels are upright cylinders divided into about seventy rings the circumference of each ring being divided into about eighty parts. A stud may be fixed on any one or more of these portions of each ring. Thus each barrel presents about eighty vertical columns every one of which contains a different combination of fixed studs.

These barrels have two movements:

1st: They can advance horizontally by a parallel motion of their axis.

2nd: They can turn in either direction and to any extent on their axis.

When the barrels advance horizontally these studs act on levers which cause various movements in the mill, the stud belonging to each ring giving a different order.

Amongst these movements or rather these orders for movements the following may be more particularly noticed. The advance of a barrel may order:

a) A number with its sign to be received into the mill from the ingress axis.

b) A number with its sign to be given off from the mill. This number may thus be either altogether obliterated from the mill or it may at the same time be received on the egress wheel or the number may be given off from the mill to the egress wheel and at the same time be itself retained in the mill.

c) A Variable card to be turned.

d) An Operation card to be turned.

e) The circular movement of the Barrel itself or of any other barrel to another vertical. This always occurs at every step from the beginning to the end of what are called *operations*. The barrels when once ordered by the operation cards from their zero point to any given vertical always direct themselves to be turned to another vertical preparatory to their next advance. This circular motion is however occasionally changed by an action arising from another source.

7. *Of the Reducing Apparatus*

Behind each barrel is placed a reducing apparatus. It consists of six or eight sectors which can be made to act upon the barrel and give it a rotatory movement so as to make it pass over 1, 2, 3 or any required number of verticals previously to its next advance.

The levers which put these sectors into action are acted upon by:

a) The studs on their own barrel

b) The studs on any other barrel

c) The Operation cards
d) The Running up levers.
The first and third of these sources of action occur most frequently.

8. *Of the Operation Cards*

Those who are acquainted with the cards of a Jacards loom will readily understand the functions performed by these cards. To those who are unacquainted with that beautiful contrivance it may be necessary to state that the *Cards* consist of pieces of thick paste board, tin plate or sheet zinc pierced with a number of holes; these cards being strung together by wire or tape hinges pass over a square prism.

This prism is situated in front of a number of levers placed in rows which govern the Reducing Apparatus and consequently the barrels. The faces of the prism are perforated so as to present an opening opposite *every* lever.

If the prism alone is made to advance horizontally against these levers then the levers themselves will enter into the holes of the prism and be partly covered by it but they will not be moved out of their places.

Again if a card having as many holes as the prism has, or as there are levers opposite to it, is placed upon the advancing face of the prism no effect can be produced on the levers by this advance of the prism. But if a card having one hole less than the prism is placed on its face then when the prism advances the lever opposite that hole will be pushed and any order given for which that lever was appointed. Suppose after every order the levers to be replaced and let the prism be turned one quarter round then a new card will be presented to the levers and if one or more holes of this second card are stopped up a different order will be transmitted through the levers to the Reducing Apparatus and thence to the barrels.

Thus by arranging a string of cards with properly prepared holes any series of orders however arbitrary and however extensive may be given through the intervention of these levers.

The number of the levers acted upon by the operation card is small; they respectively direct the barrels to *commence* the following operations:

a) The *Addition* of two numbers
b) The *Subtraction* of one number from another
c) The *Multiplication* of two numbers
d) The same *Multiplication limited* to a given number of the first figures
e) The *Division* of one number by another
f) The same *Division limited* to a given number of figures in the quotient.

The levers numbered 4 and 6 are rarely used; the extraction of roots being the only case in which they are required.

These cards are called into action by orders from the barrels. What they shall order when acting depends on the nature of each individual card. What repetitions they shall be subject to depends on the orders communicated to them from the Combinatorial Cards and their Counting Apparatus. Many calculations are much simplified by having two sets of Variable Cards.

9. Of the Combinatorial Cards

One or more peculiar cards may be inserted among the operation cards of certain formulae. They are called *Combinatorial Cards*. The object of these Cards is:

To govern the *Repeating Apparatus* of the Operation and of the Variable Cards and thus to direct at certain intervals the return of those Cards to given places and to direct the number and nature of the repetitions which are to be made by those cards.

Whenever Combinatorial Cards are used other cards called index cards must occur amongst those of the formulae. The use of these cards is to compute the numbers which are to serve successively for the indices of the combinatorial cards. At what time the Combinatorial Cards shall act depends on the number of repetitions the last of those cards appointed. What orders each Combinatorial Card shall give depends on the nature of each individual Card.

Of the Store

The Store may be considered as the place of deposit in which the numbers and quantities given by the conditions of the question are originally placed, in which all the intermediate results are provisionally preserved and in which at the termination all the required results are found.

Various parts may be added to the store according to the purposes required. Some of them might perhaps with more convenience constitute distinct machines.

The store then may contain:

a) Figure Axes
b) Computing Apparatus
c) Number Cards
d) Card punching Apparatus
e) Printing Apparatus
f) Copper Punching App.s
g) Curve Drawing App.s
h) Variable Cards.

The Figure Axes and the Variable Cards alone are absolutely necessary for the mathematical enquiries in the present work and for the sake of simplicity the others will be only occasionally referred to.

1. Of the Figure Axes

A number of axes each having forty figure wheels placed in different cages one above another are connected with the rack of the Store.

These figure wheels are each numbered from 0 to 9; they may be turned by hand so that any digit may stand opposite a fixed index. Thus any number of not more than forty places of figures may be put upon the figure wheels of each axis.

Above the fortieth cage is another cage containing a wheel similar to a figure wheel and also having its circumference divided into ten parts. These parts have the signs (+) plus and (− 1) minus alternately engraved upon them. Above this wheel is a fixed character to distinguish each particular axis or rather the variable number which may be found upon its wheels. These fixed marks are $v_1, v_2, \ldots v_{32} \ldots$, as far as the number of quantities which can be contained in the Store.

Below the lowest or units figure wheel a small square frame appears in which may be inserted a card to be changed according to the nature of the calculation directed. On this card is written that particular variable or constant of the formula to be computed whose numerical coefficient and sign are expressed on the wheels above it.

The annexed representation will perhaps convey a clearer idea of this part of the engine.

The first line of quantities $v_1\ v_2, \ldots$ are never altered; they merely indicate the particular sets of figure wheels to which they are attached.

The next line contains the *signs* of the quantities which are themselves expressed by writing them upon pieces of card and placing them in the squares at the bottom.

The intermediate forty cages contain the numerical coefficients.

The first variable or v_1 is in the present figure equal to zero and previous for the setting of the engine to any problem, all the other variables $v_2, v_3, v_4, \ldots$ have the same symbols beneath them.

The second variable v_2 has beneath it the negative sign, the number 1758 and the algebraic quantity a.

Variables of the Engine		v_1	v_2	v_3	v_4	v_5	v_6
Signs	Cage 41	+	−	+	−	+	+
	Cage 40	0	0	0	0	0	0
	Cage 39	0	0	0	0	0	0
	. . . .						
	Cage 5 Tens of Thousands	0	0	0	2	0	0
	Cage 4 Thousands	0	1	4	3	0	0
	Cage 3 Hundreds	0	7	9	4	0	2
	Cage 2 Tens	0	5	7	1	3	4
	Cage 1 Units	0	8	1	0	6	7
Variables of the Question			a	x	x^3	$\sin \Theta$	$\sqrt{a^2 + x^2}$

The symbols placed on the engine are in the annexed figure

$$v_1 = 0 \qquad\qquad v_4 = -23410 \quad x^3$$
$$v_2 = -1758 \quad a \qquad v_5 = +36 \qquad \sin \Theta$$
$$v_3 = +4971 \quad x \qquad v_6 = +247 \qquad \sqrt{a^2 + x^2}$$

In this manner any function however complicated it may be if it is considered as a whole may be placed in the store with its proper sign and its numerical coefficient; the function itself being merely written on a card and placed in the square below its coefficient. The number of variables which can be contained within the store will depend on the length of the rack and number of figure axes which can be placed round it and although a large number of variables might with perfect safety be employed yet there is obviously a practical limit arising from the weight of the rack to be moved.

One hundred variables would not give an inconveniently large rack but still the calculations of such an engine would be limited. This limitation can be entirely removed by another set of cards called Number Cards which will presently be described.

If any of the coefficients contain decimals or if the result is required with any number of decimals then all the coefficients must be considered as having the same number of decimals. If an imaginary line is drawn between any two cages, the third and fourth for example, then all below it may be considered as decimals. In order to convey to the Engine this information there exists a wheel with the numbers from 1 to 40 engraved on its edge; this wheel being set at any number the Engine will treat all the numbers put into the Store as having that number of decimals.

2. *Of the Computing Apparatus*

One of the Figure Axes K has its figure wheel connected with the rack differently from the others. It is also connected with the Carrying Axis "F belonging to the mill from which however it may at times be disconnected.

The object of this is to enable the figure wheel on the rack to make certain simple calculations without the necessity of sending the numbers into the mill. During the process of one of the great operations in the Mill a series of numbers may be computed in the store by the method of differences and thus a considerable saving of time effected.

3. *Of the Number Cards*

The Number Cards have been introduced for the purpose of rendering the calculations absolutely unlimited by the too great number of variables and constants necessary for the solution of any problem.

The number cards are pierced with certain holes and stand opposite levers connected with a set of figure wheels placed on the Number Axis which can be made at intervals to communicate with the rack. When these Number Cards are advanced they push in those levers opposite to which there are no holes on the cards and thus transfer that number together with its sign which the holes on the card represent to the figure wheels of the axis opposite to which the Cards are placed. The number and its sign thus put upon these figure wheels may be immediately transferred to another part of the Store and the string of number cards being turned the next card conveys its number to the Store in the same way.

These number cards are for some purposes more convenient than figure wheels because the numbers upon a figure card are not obliterated by the act of giving them off. For by turning the string of figure cards back to any given one the number upon the card can be replaced in the Store as frequently as may be required and at any periods of time which the calculation may demand. On the other hand the numbers placed upon figure wheels are always obliterated in the act of giving them off. If it is necessary to retain on the Store any number which is to be given off to the Mill then it also must be given off through the rack to another store axis on whose figure wheels it must remain until at a second operation it is reconveyed by rack back to its original place.

4. Card Punching Apparatus

One mode of rendering permanent the results of any calculations made by the engine will obviously be by making it punch on cards certain holes similar to those just described as existing on the Number Cards. This plan will also enable us to use any intermediate computation which may be necessary in advancing towards the final results, for the cards so made may from time to time be removed from the punching apparatus and attached to the Number Cards. Other advantages will be observed when the subject of mathematical tables comes under consideration.

5. Printing Apparatus

It is desirable even when many copies of a calculation made by the engine are not wanted that the results should be themselves printed by the machine in order to ensure the absence of error from copying its answers.

A set of thin circular rings having metal types of the digits fixed at equal distances on their edges and themselves governed by the calculating wheels on which the result is placed are to be pressed down at intervals on a sheet of paper covered by another sheet of carbonised paper. This paper is fixed in a platform having proper motions for placing the printed results in right order.

Thus a single and correct copy may be produced although from the nature of the process the execution of the printing would not be of the highest order.

If however many printed copies are required then it is intended that the type so arranged shall be made to impress their characters on a soft substance from which mould a stereotype plate may be cast.

6. Copper Plate Punching App.[5]

If it should be deemed necessary to print tables or calculations upon copper-plate an apparatus has been contrived for that purpose. This process is necessarily slower in its operation than the former modes of rendering the calculated results permanent. It has however the advantage of possessing greater clearness; although the additional cost of taking off impression may in some instances be objectionable.

Since however the invention of the number cards these modes of printing or engraving have ceased to become essential parts of a Calculating Engine. The absolute certainty of every printed result can now be obtained although the printing mechanism be totally detached from the calculating portion of the Engine, an improvement which it was impossible to make until that point of enquiry was attained.

The cards on which the results are punched may themselves be placed in a distinct machine and from the holes formed in them the new machine may either engrave or print them as it may have been prepared to operate.

7. *Curve Drawing Apparatus*

The discovery of laws from the examination of a multitude of tabulated and reduced observations is greatly assisted by the representation of such tables in the form of curves.

As one of the employments of a calculating engine would be to reduce collections of facts by some common formula I thought that at the time it impressed the computed results it would be desirable that it should mark the point of a corresponding curve upon paper or copper if preferred. The three or four first figures of the table will be expressed by the curve. The contrivances for this purpose are not difficult and their employment does not lengthen the time of the calculation.

8. *Variable Cards*

The Variable Cards are appointed for the government of the various parts which constitute the Store. Like all the other cards they act by pushing forward certain levers placed in front of them. These levers cause the motions of the several parts of the Store which have been described.

It is necessary in the present work to consider only—the Figure Axes—the Number Cards—and the card punching App.ˢ, Variable Cards and Combinatorial Operation Cards.

With respect to these the principal functions of the Variable Cards will be to direct:

a) A Number and its sign to be given off from any Store Axis to the ingress Axis.
b) A number and its sign to be received upon any Store Axis.
c) Any number and its sign to be given off from the Number Cards to the wheels on the Number Axis.
d) Any number and its sign to be given off to the Card Punching Apparatus and a corresponding Card to be punched.

The number of levers necessary for these purposes is not so large as might at first appear, consequently the Cards need not approach an inconvenient magnitude. For example fourteen levers and their equivalent fourteen holes will be all that is required in the third of the above division for eight thousand variables.[1]

If the other appendages to the Store which have been already described should be thought necessary a small number of additional levers must be added.

---

[1] That is to say, Babbage initially planned binary coded store addressing. This he soon abandoned for *mechanical* simplicity, changing to one hole per variable. The binary coded addressing is of considerable interest in connection with the question of how close Babbage came to the modern concept of the computer.

INDEX

—o●o—

Aachen, 66, 67
Abel, Niels, Henrik, 36
Abercromby, George, 2nd Lord, 58
Académie Royale des Sciences, 3
Adams, John Crouch, 211, 212, 230
Aenaeid, The, 52
Airy, Sir George Biddell, 93, 149, 191, 211, 220, 221
Aitken, Howard, 255
Albert, Prince Consort, 217–219, 226
Alberti, Leon Battista, 227
Alembert, Jean le Rond d', 41
'Alethes and Iris', 206–208
Alexander I, Tsar, 144
Algebra Engine, 168, 210n., 244
Allen, John, 169
Alphington, 13–15
Althorp, John Charles Spencer, Viscount, 126
Amicable Society, 60
Amici, Giovan Battista, 99, 181
Analytical Engines, 1, 48, 120, 127, 135, 164–173, 181, 186, 193, 208–210, 221, 230–234, 241–246, 254, 255, 268–276; discussed in Italy, 181–186; Menabrea paper and Ada Lovelace's notes, 178, 210n., 242, 259; see also Menabrea
Analyticals, 24, 25, 36, 44, 46, 75, 88, 89, 99, 101, 102, 137n., 140n.
Analytical Society, 23–27, 49, 50, 88; formation of, 23, 24; *Memoirs of the,* 25, 26
Annales de Chimie, 91
Arago, François Jean Dominique, 2, 41, 42, 58, 73, 74, 74n., 181
Arcueil, 41; Society of, 30, 40–43, 74
Arnim, Bettina von, 74n.
Arnott, Dr Neil, 178
Array processor, 169, 242
Artillery, 67n., 232, 233
Ashburton, 6
Ashley, Lord; see Shaftesbury, Earl of
Ashley Combe, 178, 195, 197, 216, 234
Astronomical Society of London; see Royal Astronomical Society
Astronomy, Babbage's interest in, 34, 35

Athenaeum, The, 146, 151, 153, 154, 212, 221
Austen, Jane, 3
Authors, 118, 229, 248; proposals for society of, 203, 204
Automata, 233, 254

Babbage, Benjamin, Sr., 6, 10, 39
Babbage, Benjamin, 5, 8,11–13, 40; as bankers, 5, 9, 10–12; relations with Charles, 13, 28, 31–33, 39, 40, 47, 58, 63, 69; death of, 62–64
Babbage, Benjamin Herschel, 39, 40, 66, 75, 159, 162, 164, 198, 200, 201, 253; education, 64, 172, 173; marriage, 199–200; and Australia, 200
Babbage, Charles; see list of contents, and subject entries in index
Babbage, Charles, Jr., 39, 64, 65
Babbage, Dugald Bromhead, 39, 173, 174, 200, 253
Babbage, Elizabeth Plumleigh, 5, 10, 11, 62, 63, 65, 75; and her grandchildren, 66, 173, 200; death of, 200, 215
Babbage, Georgiana, (née Whitmore), 21, 25, 28–34, 39–40, 47, 61–63, 66, 69, 115, 173, 208, 253; death of, 65, 66
Babbage, Georgiana, (daughter of Charles), 39, 75, 173, 253
Babbage, Herschel; see Babbage, Benjamin Herschel
Babbage, Major General Henry Prevost, 39, 193, 200, 215, 227, 237–241, 252–254; youth of, 173, 174, 203; and Scheutz Engine, 239
Babbage, John, 7, 8, 8n.
Babbage, Dr John, 8, 11
Babbage, Laura, (née Jones), 199, 200
Babbage, Mary Anne; see Hollier, Mary Anne
Babbage Common, 192n.
Babbage's Bill, 247, 248
Babbage's Orchard, 7
Bach, Joseph von, 73
Bache, Alexander Dallas, 223

Bacon, Francis, Lord Verulam, 3, 4, 35
Bacon, Roger, 120, 227
Baily, Arthur, 45
Baily, Francis, 45, 45n., 59, 123, 258
Balard, Antoine Jerome, 91
Ballet, Babbage's design of, 206–208
Bank of England, 58, 59, 110
Banks, George, 78
Banks, Sir Joseph, P.C., 46, 89; relations with
    Babbage, 88, 89; views on formation of
    new scientific societies, 44, 45, 45n., 88,
    88n.
Baring, Francis, 9
Barlow, Nora, 179n.
Barrett, Elizabeth, 202
Barton, H. K., 246n.
Beaufort, Rear-Admiral Sir Francis, 95, 97,
    221n.
Belper, 106
Bentham, Jeremy, 38, 39, 64
Bentley, Richard, 22
Bentley's Miscellany, 153
Bernal, John Desmond, 159
Berthier, General Louis Alexandre, 30
Berthollet, Claude Louis, 30, 41, 42
Berzelius, Jöns Jacob, 74, 91
Betts, Miss, 148
Bezzi, 203
Bickersteth, Henry; see Langdale, Lord
Binary system, 167, 168, 245, 276n.
Biot, Edouard Constant, 41, 122
Biot, Jean Baptiste, 41–43
Birmingham, 62, 81, 140n., 144, 146, 152,
    153
Blackett, Prof. Patrick Maynard Stuart, 159
Blake, Mr, 203
Blake, William, 110
Bleeding Heart Yard, 195, 224n.
Board of Longitude, 89, 96
Board of Woods and Forests, 193
Boissy, Marquis de, 203
Bologna, 67, 69
Bonaparte, Charles Lucien, Prince of
    Musignano, 69, 186, 186n.
Bonaparte, Charlotte, 72
Bonaparte, Joseph, formerly King of Spain,
    72
Bonaparte, Louis, Comte St Leu, (formerly
    King of Holland), 69, 72
Bonaparte, Louis Napoleon, Napoleon III,
    Emperor of France, 69, 225, 226, 250
Bonaparte, Lucien, Prince of Canino, 28–30,
    69, 72n, 80

Bonaparte, Napoleon, Emperor, 4, 6, 28–30,
    41–43, 212, 226
Bonaparte, Napoleon Louis, 69, 72
Bonham-Carter, Victor, 229n.
Booksellers, 117, 118, 229
Boole, George, 88, 244, 249
Boolean Algebra, 58, 249
Boughton, 40, 65
Box tunnel, the, 147–149
Boyle, Robert, 88
Bradford, 107
Breguet, Abraham Louis, 145
Brewster, Sir David, 99, 127, 127n., 150,
    150n., 152, 176, 181, 188, 192, 137
Bridgewater, Earl of, 99n., 136
Bridgewater Treatises, 136, 137
Bristol, 6, 39, 81, 146–148, 152, 156n., 157,
    162, 176, 199
British and Foreign Bible Society, 23
British Association for the Advancement of
    Science, 73, 119, 151–156, 163, 174, 179,
    181, 199, 239, 254; foundation of, 102,
    102n., 127, 127n., 150, 151
British Museum, 127
Brixham, 6
Brock, Michael, 75n.
Brockenden, William, 178
Bromhead, Sir Edward Thomas Ffrench, 21,
    24–27, 36, 39n., 40, 46, 99, 102
Bromhead, Sir Gonville, 21, 39
Brougham, Henry Peter, Baron Brougham
    and Vaux, 52, 129, 179
Brown, Samuel, 107
Browne, Isaac Hawkins, 36
Browning, Robert, 202
Bruce Castle School, 64, 65, 173
Brunel, Isambard Kingdom, 99, 106, 143–
    149, 200–202, 216, 220, 227, 232, 238;
    and Thames Tunnel, 57, 66; battle of the
    gauges, 143, 155–163
Brunel, Marc Isambard, 55, 86, 106, 110,
    113, 117, 143–145, 169, 232; and
    Difference Engine, 53; Thames Tunnel,
    56, 57, 145
Brussels, 111, 226
Buckfastleigh, 6
Buckingham and Chandos, Richard
    Grenville, first Duke of, 99
Buckland, Revd. William, 148, 149, 149n.,
    152, 190, 191
Buddle, John, 155
Bulferetti, Luigi, 182n.
Burdett, Sir Francis, 39, 177

Burdett-Coutts, Angela Georgina, 177
Burthogge, Dr Richard, 7, 208
Buxton, Harry Wilmot, 245, 251, 252
Byron, Anne Isabella, Lady Noel, 23, 177,
178, 197; and Ada's death, 178, 234–237
Byron, Augusta Ada; see Lovelace, Augusta,
Ada, Countess of
Byron, George Gordon, 6th Lord, 36, 177,
178, 203

Calais, siege of, 5
Calculating Engines, 26, 37, 47, 48, 51, 122,
138, 191, 242–245, 248, 253–255; see
also: Difference Engine; Analytical
Engine; Algebra Engine; Array Processor
Caldwell, John Bernard, 238n.
Calèche, Babbage's design of, 72, 74
Caledonian Canal, the, 190
Cambridge, 17, 18, 151, 174
Cambridge, University of, 1, 2, 4, 18–28, 32,
40, 92, 119, 120, 137, 140, 140n, 149, 255
Cambridge Philosophical Society, 44, 45,
88n.
Cambridge Union, 23, 38
Campbell, Lord John, 229
Campbell, Thomas, 203
Carbonari, 250
Card-copying machine, 168n.
Carew, Bamfylde Moore, 9
Carlyle, Thomas, 179
Carnot, Lazare Nicolas Marguerite, 30, 40
Carpenter, Revd. Dr. L., 158
Carry systems, 48, 50, 170–172, 209;
hoarded carry, 48; anticipatory carry, 171,
172, 183n., 245
Casting, 172n., 245–6, 246n.
Castle, Mr R., 156
Catholics, Roman, 84, 248, 249
Cartography, Babbage's interest in, 73, 229
Cauchy, Augustin Louis, 36, 36n.
Cavendish, William, later Duke of
Devonshire; Babbage organizes election
campaigns for, 2, 76–80
Cavour, Camillo Benso, Conte di, 175
Chain, principle of the, 245
Champernowne, Arthur, 33, 34
Chantrey, Sir Francis Legatt, 179
Charles Albert, King of Sardinia, 182–184,
187, 188
Chartism, 79, 186, 211
Chaucer, Geoffrey, 140, 227
Cheetham Hill Library, 121

Chemistry, Babbage's interest in, 21, 28, 31,
33
Chézy, Antoine de, 43
Circumlocution Office, 193–195
Cirencester Agricultural College, 238
Clark, Sir George, 190
Clement, Joseph, 53, 57, 66, 123–128, 130–
132, 164, 231
Clements, Paul, 56n.
Coach building, Babbage's interest in, 67, 72,
74
Cobbett, William, 38
Cobden, Richard, 229, 229n.
Cockerell, Charles Robert, 203
Codes, 225, 227, 228, 237, 238
Colby, Thomas Frederick, 45, 60, 62, 66,
115
Colden, David, 178
Colebrooke, Henry Thomas, 45
Collège de France, 42
Collingwood, 212
Committee on the Export of Tools and
Machinery, 119
Committee on the Fluctuation of
Manufacturers Employment, 116
Computer, The stored-program; see stored-
program computer
Computers, 43, 44, 49
Condorcet, Jean Antoine Nicolas de Caritat,
Marquis de, 2, 30, 83
Constants of nature and art, Babbage's
proposals for tables of, 4, 74
Conybeare, Revd. William Daniel, 129, 130
Cooper, John, 148
Co-operatives, 106, 107
Copernicus, Nicolaus, 35
Copley Medal, 89
Cotes, Robert, 22
Cournot, Augustin, 120
Courtenay, Thomas Peregrine, 13, 257
Cousins of Charles Babbage, 5, 63
Coutts, Thomas, 177
Cow-catcher, Babbage's suggestion of, 143,
144
Crediton, 7
Creevey, Thomas, 169
Crimean War, 177, 226, 231
Croker, John William, 52n.
Crosby Row, 10
Crosland, Maurice, 41n.
Crosse, Andrew, 234
Crosse, John, 234–236
Crystal Palace, 219–221, 221n., 242

Cullen, Michael J., 151n.
Curve plotter, 209, 276

Daguerre, Louis Jacques Mandé, 176
Dainton, 13, 64, 192, 201; see also Ippelpen
Dalhousie, Fox Maule, 11th Earl Dalhousie, 251
Dalton, John, 90n., 179, 180
Dancing lady, 175
Dangan, Mr, 221n.
Daniell, Prof. John Frederic, 200
D'Arblay, Alexander Charles Louis, 25
D'Arblay, Mme, (Fanny Burney), 25
Dart, river, 6, 19, 21, 55
Dartington Hall, 6, 33, 34
Dartington Hall School, 64
Dartmouth, 5, 6
Darwin, Charles, 71, 136, 155
Da Vinci, Leonardo, 209, 222
Davy, Sir Humphry, 46, 51, 89, 143, 206, 249; views on the decline of science, 88, 90, 91, 93
Decimal coinage, 103, 113
Decline of Science campaign, 2, 24, 55, 74, 88–102, 105, 147
De la Beche, Sir Henry Thomas, 71, 200, 223
De Morgan, Augustus, 23, 88, 202, 221, 237, 247–249
De Morgan, Sophia Elizabeth, (née Frend), 23, 237, 249n.
Dent, Edward John, 160
Dent, Sir W. Walson, 40
Derby, 2, 106
Derby, Edward Geoffrey Stanley, 14th Earl of, 230
Derwent, river, 2
Deutsche Naturforscher Versammlung, 73, 74, 150, 150n.
Dexter, Walter, 204n.
Dickens, Charles, 140, 152, 153, 177–179, 193–195, 202, 204, 214, 216, 222, 223, 229n.; see also: *Little Dorrit*; *Pickwick Papers*; Podsnap
Difference Engine, 1, 4, 14, 48–55, 65, 75, 90, 110, 134, 164–167, 175, 177, 178, 180, 192, 193, 210, 221, 221n., 239, 261–267; origin of, idea for, 44, 49, 50; start on, 50–55, 103; government support for, 52, 53, 123–128; construction of, 50, 51, 53, 54, 62, 66, 84, 99, 105, 123–128, 150, 241; engineering spin-off from 48, 54, 55, 133, 134, 145, 170, 230–232; cessation of

work on, 57, 107, 126, 130–135, 193; Babbage seeks decision on, 132–135, 169, 170, 181, 190, 191, 195; and Circumlocution Office, 193–195
Difference Engine, Second, 197, 208, 209, 230, 231
Diffference Engine, Third, 208, 209
Difference Engine of Scheutz, 239, 240
Diffraction gratings, 66, 67
Directive; see Operations section of Analytical Engines
Dirichlet, Gustav Peter Lejeune, 73
Disraeli, Benjamin, Lord Beaconsfield, 230, 231
Diving bell, 56
Division of labour, 43, 112, 121; and mental work, 44
Doctor Thorne, 76n.
Dodd, Revd. Philip, 21n., 23n.
Don Giovanni, Babbage attends performance of, 205
Donkin, Bryan, 124, 125, 155, 240
Don Quixote, 15
Drinkwater, John Elliot, (later Bethune), 78, 119
Ds and dots, controversy of the, 23–27
Drummond, Henry, 248
Drummond, Thomas, 206
Dublin, 66, 154, 174, 201, 221n.
Dudley Observatory, 239
Dulau, B., and Co., 20
Duncome, Thomas Slingsby, 87

Earth, internal movements of the, 69–71, 71n.
East Horsley, 195, 196
East India Company, directors of the, 58, 59, 84
Eastlake, Sir Charles Lock, 203
École des Ponts et Chaussées, 43
École Polytechnique, 21, 43
Economist, The, 119
Economy of Machinery and Manufactures, On the, 18, 38, 41, 55, 65, 67, 75, 82, 90, 90n., 103–122, 137
Edgeworth, William, 174
Edinburgh, 39, 55, 153, 174
Edinburgh, University of, 39
Edinburgh Review, 239
Education, Babbage's views on, 83, 91, 92, 224, 233
Education of Charles Babbage, 3, 13–24

Elections: Babbage's campaigns in Finsbury, 82–87, 110, 181, 204, 205; urged to stand in Cambridge, 180, 181
Electromagnetic induction, 58
Electromechanical systems, Babbage considers use of, 172, 172n., 227, 227n.
Elephant and Castle, 5, 10, 126
Encyclopaedia Metropolitana, 56, 105, 105n.
Enfield, 15–17, 64
Engels, Friedrich, 122
Enlightenment, the, 3, 74, 252
Equitable Society, 59, 60
Escher Fils, 186
Essex, Catherine Stephens, Countess of, 203
Eternal punishment, Babbage's views on, 141, 142, 208
Etna, Mt, 71
Eton, 45, 148
Euston, Henry Fitzroy, Earl of, (later 5th Duke of Grafton), 76
Everest, George, 72n., 98
Examples to the Differential and Integral Calculus, 27
Exeter, 6, 7
Exeter Bank, 9
Exhibition of 1862, 245
Export of capital, 213, 213n.

Fairbairn, Sir William, 2, 232, 239
Fairman, Col. 45
Falcon, 166
Falk, William, 192
Faraday, Michael, 10, 58, 71, 191, 206
Fellowes of Ludgate St, 98, 99n., 118
Field, Joshua, 131, 132, 145, 146
Fitton, William Henry, 29, 71, 101, 129, 130, 150n., 174
Fitzclarances, 82
Florence, 69, 71, 72, 184, 186, 189
Florin, Origin of the, 113
Forbes, Edward, 71
Forbes, James, 127, 150n.
Follett, Sir William Webb, 190
Forgery; see Royal Society of London
Forster, John, 194n.
Foster, James, 81
Fox, Charles Richard, 169
Fox, William Johnson, 229
Fox Talbot, William Henry, 176
France, 4, 23, 24, 29, 30, 40–43, 92, 216, 217; Babbage's visits to, 40, 41, 61, 105, 181, 182, 186, 216, 226, 239
Frankfurt, 67

Franklin, Sir John, 95, 123
Freeman, Stephen, 16, 17, 64, 64n.
French revolution, 2, 23, 28–30
French science, 34, 22, 23, 24, 29, 30, 41–44, 91, 179
Frend, William, 23
Friedenburg, Dr. G., 122
Friends, Society of, 180
Fulford, Roger Thomas Baldwin, 190n.

Galignani's Newspaper, 27, 69
Galileo Galilei, 181, 183
Galle, Johann Gottfried, 211
Games, Babbage's interest in the theory of, 233
Garibaldi, Giuseppe, 187
Gauss, Carl Friedrich, 74
Gay-Lussac, Joseph Louis, 58
Geihe, Sir Archibald, 55n.
Geological Society, 34, 71, 71n., 93
George St., Adelphi, 12
Geothermal sources of power, Babbage's interest in, 70, 187
Gilbert, Davies, (formerly Giddy), 52, 53, 88, 89, 94, 95, 97–100, 147
Gilbert, Keith Reginald, 144n.
Gilbert, Mary Ann, 89
Gioja, Melchiorre, 112n.
Glasgow, 39, 108, 110
Gloucester, Bishop of; see Gray, Robert
Goderich, John Frederick Robinson, Lord, (later Earl of Ripon), 52, 53, 123
Goldsmiths, 8, 9, 68
Gombart, M., 61
Gompertz, Benjamin, 45
Goodall, Joseph, 45
Goulburn, Henry, 79, 80, 126, 191
Gowing, Prof. Margaret Mary, 224n.
Grant, Sir Robert, 86, 87
Gravatt, William, 246
Gray, Robert, Bishop of Bristol and Gloucester, 180
Great Exhibition, 211, 216–223, 228, 231; Industrial Commission for, 216, 223
Great Western Railway, 147–149, 157–163
Gregory, Olinthus Gilbert, 45
Green, George, 25
Greville, Charles, 190
Grey, Charles, Earl, 78, 79, 81, 82, 124
Groombridge, Stephen, 45
Grote, George, 39, 64, 174, 229
Grote, Mrs Harriet, 37, 174
Guest, Sir Josiah John, 61, 215

Guiccioli, Contessa Teresa, 148, 202, 203
Guizot, François, 180

Habakkuk, Prof. Hrothgar John, 224n.
Haileybury, 39
Hallam, Henry, 172, 180, 203
Halloran, Mr, 13
Hannaford, John, 7n.
Hansen, Christian, 36n.
Harcourt, William Vernon, 127, 156
Harley, Lady Jane Elizabeth, later
  Bickersteth, and then Lady Langdale, 38,
  250
Harness, William, 178
Harris, Sir William Snow, 223
Hartlib, Samuel, 1, 3
Harvey, George, 100
Hawes, Benjamin, 191, 232
Hawks, Francis Lister, 210n.
Hawkshaw, John, 157, 158
Haytor quarry, 55
Hazlewood School, 64, 65
Health of Charles Babbage, 18, 21, 50, 65,
  126
Henry, Joseph, 4, 49, 252
Hero of Alexandria, 47
Herschel, Caroline, 34
Herschel, Sir John Frederick William, 25–27,
  28n., 30–34, 40, 40n., 41, 45, 46, 49, 60–
  63, 65, 66, 71, 71n., 86, 88, 119, 123, 130,
  139, 140n., 146, 149, 161, 163, 176, 180,
  188, 192, 200, 212, 230, 247; and the
  Analytical Society, 25, 26; and the Royal
  Society, 58, 89–91, 95, 97–102, 150; and
  the British Ass., 102, 155, 156
Herschel, Sir William, 25, 34, 40, 42
Hill, Rowland, 64, 64n., 65, 115
Hill, Thomas Wright, 64, 65
Hobson, John Atkinson, 213n.
Hodgskin, Thomas, 84
Hodgson, Mr, 143
Holes in glass, Babbage's method for making,
  68, 69
Holland, Edward, 238
Holland, Elizabeth Vassall Fox, Lady, 169
Hollerith, Herman, 254, 255
Hollier, Henry, 61
Hollier, Mary Anne, (née Babbage), 11, 200,
  215, 249; marriage, 61
Hollis, Patricia, 83n.
Holtzapffel and Co., 159, 231
Hooker, Joseph, 71
Hoskins, W. G., 5n.

Huguenots, 5
Hudson, John, 20
Hull, 108
Humboldt, Friedrich Wilhelm Heinrich
  Alexander von, 2, 41, 72–74, 74n., 186,
  204n.
Humboldt, Karl Wilhelm, Freiherr von, 73
Hume, David, 139
Hume, Joseph, 64
Huskisson, William, 99, 144
Hutchinson, Capt. (George?), 67n.
Huxley, Thomas Henry, 155
Hyman, Robert Anthony, 164n., 242n.

IBM, 242, 255
Industrial revolution, 1, 2, 3, 10
Input/output system, 164, 166, 182, 209
Interchangeable parts, manufacture of
  equipment with, 4, 57, 144, 145, 231
Ipplepen, 13n., 192n.; see also Dainton
Ireland, 66, 108, 130, 154, 174, 201, 221
Isaac, Elias, 40, 80, 173
Isaac, Harriet, (née Whitmore), 65, 173
Ischia, 70
Italian scientific meetings, 72, 181–189
Italy, 2, 60, 67–72, 91, 181–190, 222, 250
Ivory, James, 36, 90n., 100

Jacquard, Joseph Marie, 166, 182, 184
Jarvis, C. G., 127, 131–133, 167, 168, 171,
  173, 208, 210, 215, 240
Jearrad, Charles, 126
Jevons, William Stanley, 120, 255
Johnstone, Miss, 203
Jones, Richard, 119
Jones, Thomas Wharton, 259
Jouberthon, Mme, 30

Kahn, David, 227n.
Kater, Edward, 179
Kater, Capt. Henry, 95–97
Katz, Prof. Bernhard, 95
Kepler, Johannes, 47
King, William Lord; see Lovelace, Earl of
King's College, London, 192, 193, 202n.,
  239, 246
Kintner, Elvan Erwin, 202n.
Knight, Charles, 118, 203, 229, 229n.

La Croix, Silvestre François, 20, 24, 27, 40n.
Lagrange, Joseph Louis, 22, 22n.
Lambton, Hedworth, 169
Landlord, Babbage's reputation as, 192n.

Landseer, Charles, 203
Landseer, Sir Edwin Henry, 203
Lane, Mr, 33
Langdale, Henry Bickersteth, Lord, 38, 39, 176, 250
Lansdowne, Henry, Marquess of, 58, 129, 151, 174
LaPlace, Pierre Simon, Marquis de, 29, 41. 42, 61, 137n.
Lardner, Revd. Dionysus, 98, 120, 147, 148, 155, 161–163, 178, 239
Leeds, 106, 108
Leibniz, Gottfried Wilhelm, 22, 50; mathematical notation of, 22–27, 137
Leigh, Augusta, (née Byron), 178
Lenin, Vladimir Il'ich, 213n.
Leslie, Prof. Sir John, 39, 41
Lethbridge, Sir Thomas Buckler, 52
Leudon, Revd., 130
Le Verrier, Urbain Jean Joseph, 211, 212
Levin, Rahel, 74n.
Liberal party, 117
Liebig, Justus von, 43
Liége, 66
Life assurance, Babbage's interest in, 58–60
Life peerage, 86, 87
Lighthouses, 56, 225, 226; see also Occulting lights
Lind, Johanna Maria, known as Jenny, 203
Linnaen Society, 71, 93
Little Dorrit, 193–195
Liverpool, 6, 99, 108, 143, 152, 153, 163
Locke, Capt., 237
Locke, John, 7, 208
Lock-picking, Babbage's interest in, 228
Lombe, John and Thomas, 2
London Bridge, 5, 10, 145
Longford, Elizabeth Pakenham, Countess of, 144n.
Losano, Mario Giuseppe, 183n.
Louvain, 66
Lovelace, Augusta Ada, Countess of, (née Byron), (also Lady King), 36, 177, 178, 185, 186, 195–198, 202, 203, 221, 259; death of, 233–237, 241, 250
Lovelace, William King, Earl of, (Also Lord King), 178, 185, 195, 196, 212, 235–7
Lowry, Wilson, 45
Loyd, Samuel Jones, Lord Overstone, 119
Lubbock, John William, 86, 150n.
Lucasian Professorship, 27, 28, 69, 85, 93, 105
Luddites, 79, 116

Lumley, Bernard, 206, 208
Lunar caustic, 80
Lunar society, 4
Lunn, John Richard, 175n.
Lunn, Revd. Francis, 69, 99n.
Lyell, Sir Charles, 71, 129, 130, 149n., 174, 181
Lyons, 181, 182

Maastrich, 66
Macaulay, Thomas Babington, Lord, 76, 190
Macclesfield, George Parker, Earl of, 45
MacCullagh, Prof. James, 182, 203
Machine tools, Babbage's designs of, 219, 232
Machinery and Production Engineering 246n.
Machinery, 106
Maclaren, James, 199
Maclise, Daniel, 203
Macready, William Charles, 148n., 178, 179, 203
Maddicott, Fisher, 192n.
Maddicott, Robert, 192n.
Magnus, Heinrich Gustav, 73
Malthus, Revd. Thomas Robert, 119
Manchester, 2, 99, 104, 106, 110, 143, 152, 179, 231
Manchester City News, 231
Manchester Literary and Philosophical Society, 179
Somerset, Margaret, Duchess of, (née Shaw-Stewart), 192, 250, 251, 253
Marginal value theory, Babbage's interest in, 119, 120, 122
Marquand, Allen, 255
Marriage of Charles Babbage to Georgiana Whitmore, 31–33
Marryat, Frederick, 16, 17, 64
Martineau, Harriet, 129; *Illustrations of Political Economy* 104, 114
Marx, Karl, 103, 109, 121, 122
Masères, Francis, 257
Mathematics, Babbage's work on, 1, 23–25, 31, 34, 36, 37, 51, 88, 120, 227, 255
Maudslay, Henry, 53, 57, 124, 125, 131, 145, 231
Maule, Frederick, 25
Maupertuis, Pierre Louis Moreau de, 22
Maxwell, Herbert, 169n.
Mayne, Sir Richard, 247n.
Mechanical notation, 58, 62, 89n., 165, 167, 173, 241, 254
Mechanics Institute, 178
Mechanics Magazine, 82n.

Medhurst, George, 201
Melbourne, William Lamb, Viscount, 87,
    132, 134, 169, 180, 221
Mellon, Harriot, 177
Menabrea, Luigi Federico, 2, 182, 187, 198,
    213, 250; paper on Analytical Engine,
    183, 185, 190, 196, 221, 242, 249, 259
Mendelssohn, Felix, 204
Mendicants, Babbage's attitude to, 214
Mensdorf, Count, 218
Merlin, 175
Metric system, 4, 43, 229
Microprograms, 48, 168, 209, 242, 255
Mill, John Stuart, 103, 104, 117, 121
Mill of Analytical Engine, 164, 166, 209,
    242–245, 254
Milman, Henry Hart, 178, 190, 203
Minchin, James, 77, 78
Mining, 6, 7, 108
Miracles, 138
Moll, Gerard, 88n.
Monge, Gaspard, 30
Monthly Chronicle, 163
Moore, David, 45
Morishima, Micio, 122
Morland, Samuel, 48
Morning Chronicle, 76n., 83n.
Morning Post, 80
Mosotti, Ottavino Fabrizio, 182, 183
Mudfog Papers, 152–154, 153n.
Mulready, William, 203
Munich, 67
Murchison, Sir Roderick Impey, 155, 156
Murray, John, 248
Museum of the History of Science, Oxford,
    246, 251n.
Music, Babbage's interest in, 21, 130, 204,
    205
Musical composition by Analytical Engines,
    198

Naples, 69–71
Napoleon, see Bonaparte, Napoleon
Napoleon, Louis; see Bonaparte, Louis
    Napoleon
Nasmyth, James, 170n., 230, 230n., 232
Navarino, Battle of, 71
Nelson, Horatio, 199
Neptune, discovery of, 211, 212
Newcastle, Henry Pelham, 4th Duke of
    Newcastle-under-Lyme, 81
Newcastle-upon-Tyne, 144, '52–156, 161
Newcomen, Thomas, 6

Newington, 5, 11
New River Company, 173
New System of Manufacturing, Babbage's;
    see Profit sharing
Newton, Isaac, 3, 4, 22–27, 35, 85, 88, 104;
    mathematical notation of, 22–27
Newtonian science, 3, 22, 22n., 23, 137, 140
Nicolas, Sir Nicholas Harris, 82, 86
Nightingale, Florence, 177
North Star, the, 162
Norton, Mrs Caroline Elizabeth Sarah, (née
    Sheridan), 148
Number bases, 167, 168; see also Binary
    system

O'Brien, Denis Patrick, 93n.
Occulting lights, 217, 217n., 219, 225, 226
Ockham Park, 178, 195
Oerstead, Hans Christian, 74, 91
Oken, Lorenz, 73
Oliver Twist, 153
Operations research, 37, 65, 115, 120, 157–
    163, 252
Operations section of Analytical Engines,
    164, 209, 244, 245
Opthalmoscope, Babbage's invention of, 233
Order of Merit, Babbage's proposal for, 90,
    99
Oxford, 174
Oxford, University of, 22, 63, 179, 246
Oxford Movement, 136

Palmerston, Henry John Temple, 3rd
    Viscount, 79
Panizzi, Sir Anthony, 250
Parma, 67, 110
Pascal, Blaise, 47, 50
Patriotic Fund, 81
Paxton, Joseph, 219
Peacock, Revd. George, 25–27, 36, 45, 102,
    140n., 149
Pearson, Dr. George, 45
Pearson, Revd. William, 45
Peel, Sir Robert, 52, 80, 134, 149, 149n.,
    190–192, 221
Peel, William Yates, 79, 80
Peterhouse, Cambridge, 25
Petty, William, 4, 97, 151
Pickwick Papers, 76n., 152
Pipelining, 209, 274
Pitt, William, 4
Plana, Giovanni Antonio Amadeo, 181–183,
    185, 186, 188

Plantamour, Émile, 182
Platt, Prof. Desmond Christopher St. Martin, 213n.
Playfair, Prof. John, 39
Playfair, Lyon, 216, 217, 217n., 220, 224, 232, 239
Plymouth, 6, 37, 56, 100
Podsnap, 224
Poisson, Siméon-Denis, 42
Political economy, Babbage's interest in, 9, 38, 47, 48, 92, 103–122, 188
Political Economy Club, 103
Political interests of Charles Babbage, 2, 24, 35, 38, 39, 81, 187; see also: Elections: Babbage's campaigns in Finsbury; and Cavendish, William
Politics and Poetry on The Decline of Science, 204, 205
Pont Charles Albert, 184, 185
Portsmouth block-making machinery, 4, 57, 144, 145
Postal services, Babbage's interest in, 47, 65, 115, 252
Powell, Mrs, 173
Powell, Revd. Baden, 63, 136
Praed, William Mackworth, 9, 12
Praeds Bank, 9, 12
Praeds Bank of Truro, 9
Prandi, Fortunato, 182, 188
Pressnell, Leslie Sedden, 9n.
Priestley, Joseph, 4, 74n.
Printing, Babbage's interest in, 47, 51, 54, 55, 110–112, 275
Privy Council, Babbage proposed for membership of, 192, 217
Profit sharing, 116, 117, 188
Programs, 48, 198, 198n.; Celestial, 138, 139, 253
Prony, Baron Gaspard Clair François Marie Riche de, 4, 43, 44, 50, 61
Protector Life Assurance Company, 58, 59
Prussia, 2, 73, 74, 92, 106, 211
Pugsley, Sir Alfred, 147n.
Punched cards, 166, 168, 182, 183, 245, 254, 255, 272, 274–276

Quetelet, Lambert Adolphe Jacques, 61, 151, 258

Rachel, Elisa Félix, known as Mlle, 203
Racogni, estate of, 184
Railways, 55, 99, 99n., 115, 143–149, 155–163, 181, 187, 200–202, 220, 252; atmospheric, 200–202

Ramsgate, 56
Randell, Prof. Brian, 49n., 258
Reed, Prof. Henry, 226, 227
Reed, Wemyss, 217n.
Reggio, 67
Reichenbach, Mr, 67n.
Religious beliefs of Babbage, 13, 14, 19; of Babbage's forebears, 10, 14; his plans to enter the Church, 28, 31
Reichenbach, Mr, 67n.
Rennie, George, 2, 125, 145, 180, 186
Rennie, John, 2, 56
Rennie, John, (Jr.), 2, 57, 125, 145
Ricardo, David, 103
Roberts, Capt., 45
Robespierre, Maximilien Marie Isidore de, 42
Robinson Crusoe, 15
Rocket, the, 144
Roebuck, John Arthur, 87
Rogers, Samuel, 172, 177, 179, 203
Roget, Peter Mark, 36, 98
Rolt, Lionel Thomas Caswell, 147n., 149
Rome, 69, 71, 72
Romilly, Sir John, 140n.
Rosse, William Parsons, Earl of, 230, 237
Rowdens, The, 12, 40
Royal Astronomical Society, 51, 52, 176; foundation of, 45, 46, 93
Royal Institution, The, 34–36
Royal Medal, 89, 89n., 96
Royal Society Dining Club, 36, 55
Royal Society of London, The, 3, 36, 45, 46, 58, 123, 136, 137n., 150, 150n., 151, 181, 221, 250; reform of, 36, 44–46, 88–102; early proposal for reform, 97; forgery of Council minutes, 89, 95; see also: Copley Medal; Royal Medal; Banks, Sir Joseph; Davy, Sir Humphry
Russell, John Scott, 238
Russell, Lord John, 79, 81, 221n.
Russia, 67, 144, 145, 222, 226
Ryan, Sir Edward, 25, 29, 40, 62, 80, 102, 237, 253

Sabine, Capt. Edward, 89, 95, 96
St Albans, William Aubrey de Vere. Beauclerk, 9th Duke of, 177
St John's College, Cambridge, 36, 36n., 40n.
St Mary Newington, 11
Sardinia, Kingdom of, 181–189, 260
Savery, Thomas, 6, 187
Schikard of Tübingen, Wilhelm, 47

Scheutz, George, 239–241
Schneider and Co., 186
Schumpeter, Joseph Alois, 121
Science Museum, 48n., 128, 193, 239, 240, 256
Scientific Advisers to The Admiralty, 89, 95
Scientific methods applied to industry and commerce, Babbage's interest in, 1, 2, 3, 24, 30, 55, 91, 92, 102–105, 111, 184, 224, 252
Scotland, 39, 55, 153, 156, 174, 239
Scott, Sir Walter, 100
Seaward, John, 147
Sedgwick, Revd. Adam, 40, 149
Senior, Nassau, 119
Serapis, Temple of, 71
Serullas, Georges Simon, 91
Servants of Babbage, 215
Shaftesbury, Anthony Ashley Cooper, Baron Ashley and Earl of, 90n., 106; supports Difference Engine, 90n., 123, 126; Babbage's influence on, 90n.; *Decline of Science* dedicated to, 90n.
Shea, David, 100
Sheepshanks, Revd. Richard, 149, 221
Sheil, Richard Lalor, 148
Shinners Bridge, 6
Silk weaving, 2, 166, 198
Simeon, Revd. Charles, 17, 18, 23
Simms, Mr W. V., 157
Sismonda, Angelo, 185
Slawinski, Peter, 45
Slegg, Michael, 23, 24
Smeaton, John, 56, 105
Smiles, Samuel, 104, 143n., 145n.
Smith, Adam, 43, 103, 112
Smith, John, 26
Smith, Robert, 22
Smith, Sydney, 81, 178, 190
Smithsonian Institution, 4, 49, 49n., 252, 255
Socialism, Babbage's views on, 187, 213
Soirées, Babbage's, 75, 81, 129, 130, 174–177, 203, 204
Somerset, Edward Adolphus, Duke of, 29, 45, 45n., 55, 57, 98, 123, 129, 134, 169, 181, 192, 197
Somerset, Margaret, Duchess of, (née Shaw-Stuart), 192, 250, 251, 253
Somerville, Mary, (née Fairfax), 41, 41n., 137n., 178
South, Sir James, 45, 90n., 97, 146, 177, 221, 249
South Devon Railway, 200–202

Spankie, Robert, 86
Spencer, Herbert, 229, 229n.
Spring-Rice, Thomas, (later Lord Monteagle), 169
Staffordshire, potteries of
Stage lighting, 205–208
Stanfield, Clarkson, 203
Stanhope, Mrs Elizabeth, (née Green), 148
Stanley, Arthur Penrhyn, 178
Statistical methods applied to commerce and industry, Babbage's interest in, 2, 48, 111, 114, 119, 120, 220
Statistical Society, The Manchester, 152, 152n.
Statistical Society of London, The, 103, 114, 120, 151, 152n., 230
Staverton, 6
Steam engines, 2, 6, 7, 49, 50, 55, 107–109, 187, 201
Steamships, 147, 148, 155, 161, 238
Stephenson, George, 143–145, 155, 163
Stephenson, Robert, 143–145, 157, 215, 220
Stepney, Catherine, Lady, 178, 179
Stewart, Dugald, 39, 55
Stomach pump, 66, 67
Store, 164, 166, 209, 242, 245, 272
Stored-program computer, 138, 183; Analytical Engines as precursors of, 1, 164, 210n., 242, 255; Babbage's conception of, 244, 245
Storeycombe Quarry, 192n.
Stover Canal, 55
Strachey, Giles Lytton, 190n.
Street nuisances, 246–248
Strutt, Anthony R., 106, 106n.
Stuart, Hon James, 132
Submarine, Babbage's proposal for a, 56
Sussex, H.R.H. Augustus Frederick, Duke of, 100, 101, 150, 180
Swift, Dean Jonathan, 104
Swing, Captain, 78, 79
Sylvester, James Joseph, 241

Tables, Mathematical, 1, 43, 44, 49, 60
Talfourd, Sir Thomas Noon, 178, 179
Taxation, Babbage's views on, 83, 85, 212, 213
Teape family, 10; see also Babbage, Elizabeth Plumleigh
Teignmouth, 5, 11, 12, 18, 19, 20, 21, 33, 40, 62
Telegraph, 61; electric, 172, 184, 227

Teleki, Jane Frances, (later Harley-), (née Bickersteth), Countess, 38, 248n., 250
Telford, Thomas, 100
Tell-tale, Babbage's design for, 110
Temple, Christopher, 86
Tennant, Smithson, 28, 33
Tennyson, Alfred Lord, 202
Thackeray, Col. Frederick Rennell, 45
Thames, 2; Brunel's tunnel under, 56, 57, 66
Thenard, Louis Jacques, 91
Thermo-electricity, 187
Thiers, Louis Adolphe, 212
Thompson, Beilby, Lord Wenlock, 69
Thorburn, Robert, 203
Thwaites, Mr, 237, 259
Times, The, 82n., 83, 83n., 88n., 152, 156n.; printing of, 112
Tiverton, 7
Tocqueville, Charles Alexis Henri Maurice Clérel de, 175, 216
Tooke, Thomas, 119
Torbay, 5, 6
Torquay, 6, 39; see also Torbay
Torrens, Col. Robert, 87
Torres y Quevedo, Leonardo, 254
Totnes, 6, 7, 10, 11, 18, 33, 40, 55, 64; and Babbage's forebears, 5, 6, 7, 9; his continuing interest in, 7, 55, 192
Totnes Grammar School, 18, 20, 33
Tozer, Mrs, (née Maddicott), 192n.
Trade unions, 106, 117, 174
Travers, Benjamin, 178
Trevelyan, Sir Charles Edward, 228, 229
Trevelyan, George Otto, 39
Trinity College, Cambridge, 19, 22, 23, 137n., 149, 176, 178
Trinity College, Dublin, 66, 174, 182, 203
Trinity House, 226
Turing, Alan Matheson, 255
Tuscany, Leopold II, Grand Duke of, 69, 72, 181, 182, 186
Tyndall, John, 239

Union of theory and practice, Babbage's doctriné of, 1, 102, 120, 122
University College, 173, 237
University College School, 173
Ure, Andrew, 121
Utilitarianism, 38, 39, 64, 82

Van Sinderen, Alfred W., 21n., 90n., 223n., 233n., 256

Vaucanson, Jacques de, 166
Venice, 67, 72
Vesuvius, Mt, 69, 70, 208
Victor Emmanuel II, King of Italy, 182, 183
Victoria, Queen, 180, 181, 222, 223
Vidocq, François Eugène, 228
Vienna, 67n., 72
Viète, François, 227
Voltaire, François-Marie Arouet de, 22
Volterra, region of, 187

Wakley, Thomas, 86
Walpole, Edmund, 126n.
Walworth, 5, 10, 11
Weisbrod, Dr, 67
Weld, Charles Richard, 221
Wellington, Arthur Wellesley, 1st Duke of, 78, 81, 82, 87, 106, 129, 132–134, 199, 202, 218, 228; supports Difference Engine, 123, 124, 169, 193; supports engineers, 56, 57, 169
Werner, Abraham Gottlob, 33
Weyer, Jean Sylvain Van de, 169
Wheatstone, Sir Charles, 172, 178, 200, 227, 227n.
Whewell, Revd. William, 23, 88n., 93, 137n., 149, 150n., 152, 152n., 216; *Open Letter to Charles Babbage*, 137, 137n., 139, 140
Whitmore, Thomas, 80
Whitmore, Wolryche, 29, 40, 99, 103, 115, 123, 156, 215, 238; political opinions of, 80; election campaigns of, 61, 80
Whittaker, John William, 36, 36n., 40n.
Whitworth, Sir Joseph, 2, 57, 197, 210n., 223, 231–233, 246n.
Wilkinson, Henry, 40n.
William IV, King, 179, 180
Wilson, Mary, 235, 236
Winstanley, Denys Arthur, 23n., 140n., 149n.
Wollaston, William Hyde, 41n., 45, 46, 51, 89, 90, 97, 101
Wood, Nicholas, 157
Woodhouse, Robert, 22, 26, 44
Wren, Sir Christopher, 97
Wright, Richard, 66, 107, 245
Wyon, Mr, 203

York, 127, 127n., 150, 151, 153, 174
Young, Dr Thomas, 97, 101
Yule, John David, 136n.